Agricultural Economics and Agribusiness Management

Agricultural Economics and Agribusiness Management

Edited by **Adler Bryant**

SYRAWOOD
PUBLISHING HOUSE
New York

Published by Syrawood Publishing House,
750 Third Avenue, 9th Floor,
New York, NY 10017, USA
www.syrawoodpublishinghouse.com

Agricultural Economics and Agribusiness Management
Edited by Adler Bryant

International Standard Book Number: 978-1-68286-339-8 (Hardback)

Printed in China.

Contents

Preface

This book has been a concerted effort by a group of academicians, researchers and scientists, who have contributed their research works for the realization of the book. This book has materialized in the wake of emerging advancements and innovations in this field. Therefore, the need of the hour was to compile all the required researches and disseminate the knowledge to a broad spectrum of people comprising of students, researchers and specialists of the field.

Agriculture is a primary source of occupation in many countries. Agricultural economics is a rapidly expanding field of research. It studies the methods for managing the production of food and fiber, regulating the natural resources, marketing agricultural products and managing large agribusiness corporations. This book aims to shed light on the vital aspects of this discipline such as crop yield, food policies, soil ecosystems, etc. The data included in this book has been contributed by industry experts. It will help students and researchers in better understanding this field.

At the end of the preface, I would like to thank the authors for their brilliant chapters and the publisher for guiding us all-through the making of the book till its final stage. Also, I would like to thank my family for providing the support and encouragement throughout my academic career and research projects.

Editor

Evaluation of chloroplast relationship between some apple genotype from Azerbaijan of Iran and their comparison with other local genotypes, cultivars and rootstocks

Shahin Jahangirzadeh Khiavi[1], Zabihollah Zamani[1], Mohsen Mardi[2] and Mohamadreza Fatahi Moghdam[1]

[1]Department of Horticulture Science, University of Tehran, Karja 31587, Iran.
[2]Agricultural Biotechnology Research Institute (ABRII), Karaj, Iran.

Cleavage Amplification Polymorphism (CAP) technique was used to detect chloroplast DNA (cpDNA) diversity in 26 Iranian apple genotypes (19 from Azerbaijan) as well as two commercial cultivars and two dwarf rootstocks. Study of approximately 4.13% of the total chloroplast genome by using three chloroplast universal primer pairs and two restriction enzymes (*Eco*R I and *Mse* I) detected 8 haplotypes. The most-abundant and widely distributed haplotype was H4 (frequency = 26.66% approximately). Three haplotypes (H1, H3 and H6) were exclusive haplotypes of Azerbaijan population. All the mutations detected were insertion-deletions in the range of 20 to 50 bp. The total cpDNA diversity in populations was h_T=0.467, and a major portion of it was within populations (h_S=0.445). The genetic differentiation among populations was low (G_{STC}=0.048) and no genetic structure was observed. This result proposes the possible utilization of the CAP technique for the characterization of apple genotypes and cultivars. The polymorphism in cpDNA of *Malus* has to be considered carefully when planning phylogenetic studies involving this species.

Key words: *Malus,* chloroplast DNA (cpDNA), haplotype, cleavage amplification polymorphism (CAP), chloroplast universal primerc, restriction enzymes, phylogenetic.

INTRODUCTION

Domesticated apple is one of the most important fruit crops of the colder and temperate parts of the world (Zohary and Hopf, 2001). The family Rosaceae includes the genus *Malus* (apples) with about 55 species, which are divided into infrageneric groups (sections); the cultivated species are divided into intra-species groups or cultivars (Phipps et al., 1990). *Malus* taxonomy is sometimes problematic because the distinction between wild species or cultivars could come with uncertainty thereby, making the recognition of distinct categories difficult (Mabberley et al., 2001); and this problem of understanding the maternal relationship between cultivars can be better solved.

The genetic information present in the chloroplast DNA (cpDNA) is of great interest in phylogeny and population genetics, largely because of non Mendelian mode of inheritance of this genome (Sugiura, 2003). In many plants (Mariac et al., 2000) and most angiosperms, it has been identified that this organelle was inherited from the mother with approximately one-third having some degree of biparental inheritance (Whatley, 1982; Ishikawa et al., 1992). This DNA molecule (cpDNA) has been shown to be highly conserved in structure between plants (Palmer and Stein, 1986), thus, universal primer pairs are designed to amplify some regions of this molecule(Taberlet

*Corresponding author. E-mail: shjahangirzadeh@gmail.com.

Table 1. Studied genotypes, cultivars and rootstocks, their origin, place of collection and determined haplotype.

S/N	Sample name	Origin	Place of collection	Haplotype
1	Boomi Meshkin Shahr	Azerbaijan	Seed and Plant Improvement Institute of Iran	H1
2	Azerbaijan1	Azerbaijan	"	H2
3	Azerbaijan5	Azerbaijan	"	H3
4	Azerbaijan6	Azerbaijan	"	H3
5	Azerbaijan7	Azerbaijan	"	H3
6	Azerbaijan 8	Azerbaijan	"	H2
7	Azerbaijan14	Azerbaijan	"	H4
8	ME2	Azerbaijan	"	H3
9	ME3	Azerbaijan	"	H5
10	ME4	Azerbaijan	"	H2
11	ME5	Azerbaijan	"	H4
12	ME6	Azerbaijan	"	H4
13	Ghermez Ahar	Azerbaijan	"	H6
14	Mahali Beyghi Urmia	Azerbaijan	"	H2
15	Ghara Yarpagh	Azerbaijan	"	H4
16	Khan Almasi	Azerbaijan	"	H4
17	Israeli	Azerbaijan	"	H7
18	Torsh Mesri	Azerbaijan	"	H4
19	Gami Almasi	Azerbaijan	The University of Urmia apple collection	H8
20	T3	Taleghan	Seed and Plant Improvement Institute of Iran	H4
21	Boshghabi Taleghan	Taleghan	"	H4
22	Torshe Damavand	Damavand	"	H7
23	Torsh Chalus	Chalus	"	H7
24	Torsh Meygon	Shemiranat	"	H2
25	Golab Damavand	Damavand	"	H2
26	Shemirani Tabestaneh	Shemiranat	"	H8
27	Red Delicious	USA	"	H8
28	Fuji	Japan	"	H5
29	M9	UK	The University of Urmia apple collection	H8
30	M4	UK	"	H5

et al., 1991; Demesure et al., 1995; Dumolin-Lapegue et al., 1997; Fofona et al., 1997; Heinze, 1998; Weising and Gardner, 1999; Grivet et al., 2001). Digesting of these amplified genome regions in many investigation on chloroplast DNA variation has been performed for understanding the phylogeny and population relationship in Rosaceae (Mohanty et al., 2001, 2003; Panda et al., 2003; Coart et al., 2006; Turkec et al., 2006; Bouhadida et al., 2007).

In the present study for the first time in Iran, Cleavage Amplification Polymorphism (CAP) method was used to identify the maternal relationships between some Iranian apple genotypes (cytoplasmic diversity) and the results obtained were compared with some cultivars and rootstocks to make this relationship more transparent on the basis of chloroplast DNA analysis.

MATERIALS AND METHODS

All the *Malus* genotypes used in this investigation were obtained from the collection of Seed and Plant Improvement Institute of Iran and the University of Urmia apple collection. These samples mostly belonged to two endemic populations of Iran including Azerbaijan (19 accessions) and Central Alborz (7 accessions) from Taleghan, Damavand, Shemirant and Chalus. Two commercial cultivars as well as two Malling rootstocks were present in the study (Table 1). Young and well expanded leaves were selected and stored at -80°C until they were used for the study.

DNA isolation

Total DNA (nuclear and organelle) was isolated by using the procedure described by the Diversity Arrays Technology Pty Ltd (DArT P/L) company (http://www.diversityarrays.com). Required amount (0.2 g approximately) of plant materials was grinded by mortar and pestle under liquid nitrogen to fine powder. They were transferred to a 2 ml tube and 1 ml of fresh buffer (Tables 2 and 3) added and incubated at 65°C for 1 h. It was then allowed to cool down for 5 min on ice and 1 ml of chloroform: isoamyl alcohol (24: 1) mixture was added. This suspension was well mixed for 30 min. After centrifugation at 10000 g for 20 min, water phase was transferred to the fresh tube and same volume of ice cold isopropanol was added. And tubes were inverted until nucleic acids became

Table 2. Chloroplast universal primers, their sequence, PCR conditions, size of amplified fragment and quality of amplification, applied on apple accession in this study.

Name of cpDNA primers*	Sequence 5' ⟷ 3'	PCR conditions: Annealing temperature (°C)	PCR conditions: Extension time (min)	Reported amplified fragment size in tobacco** (bp)	Obtained amplified fragments size (bp)	Degree of amplification***
K1K2	F: GGGTTGCCCGGGACTCGAAC R:CAACGGTAGAGTACTCGGCTTTTA	54.5	3	2585	2470	++
CS	F: CCAGTTCAAATCTGGGTGTC R: GGGATTGTAGTTCAATTGGT	56.5	2	1611	1800	++
HK	F: ACGGGAATTGAACCCGCGCA R: CCGACTAGTTCCGGGTTCGA	56	2:30	1831	1650	++
TF	F: CATTACAAATGCGATGCTCT R: ATTTGAACTGGTGACACGAG	58.5	2	1754	-	NA

*Abbreviations are the same as in Dumolin-Lapegue et al. (1997) and Grivet et al. (2001); ** the amplified fragments size reported in tobacco were from Grivet et al. (2001); *** NA, no amplification; ++, good amplification.

Table 3. Major patterns and variants (in bp) of fragments revealed in each polymorphic site detected with different primer pair– restriction enzyme combinations in 30 accessions of studied apple accessions.

Polymorphic sites	Length or point mutations: Major pattern	Length or point mutations: Variant
K1K2- *Eco*RI	KE_A (610)	KE_B (650)
HK- *Eco*RI	HE_A (260)	HE_B (300)
HK- *Mse*I (1)	$HM1_A$ (300)	$HM1_B$ (280)
HK- *Mse*I (2)	$HM2_A$ (280)	$HM2_B$ (260), $HM2_C$ (230)

visible. Following centrifugation at 10000 g for 30 min, the supernatant was discarded and the pellet of DNA washed with 2 ml 70% ethanol.

Ethanol was discarded and pellet dried and dissolved in 250 µl of 1 X TE buffer (10 mM TrisHCl pH 8.0, 1 mM EDTA pH 8.0).

PCR amplification and digestion

Extracted DNAs were PCR- amplified using four universal primer pairs for chloroplast DNA (K1K2, CS, HK and TF and Table 4) (Dumolin- Lapegue et al., 1997; Grivet et al., 2001). The amplification of cpDNA regions was performed in 30 µl of reaction mixture containing 30 ng of total DNA, 10 ng of each primer pair, 200 µM of each four dNTPs, one unit Taq DNA polymerase (recombinant, Fermentas, Canada), 2 mM of $MgCl_2$ and 1X PCR buffer with KCl. The thermocycle system used was Gene Amp PCR system 9700.

Table 4. Nei's analysis of gene diversity in subdivided populations.

Combination	Digestion pattern	Sample size	Ht	Hs	Gst	na	ne	h	I
K1K2-*Eco*RI	KE_A	30	0.4992	0.4851	0.0283	2.0000	1.9651	0.4911	0.6842
K1K2- *Eco*RI	KE_B	30	0.4992	0.4851	0.0283	2.0000	1.9651	0.4911	0.6842
HK- *Eco*RI	HE_A	30	0.3948	0.3023	0.2344	2.0000	1.9231	0.4800	0.6730
HK- *Eco*RI	HE_B	30	0.3948	0.3023	0.2344	2.0000	1.9231	0.4800	0.6730
HK-*Mse*I	$HM1_A$	30	0.4640	0.3295	0.2900	2.0000	1.9912	0.4978	0.6909
HK-*Mse*I	$HM1_B$	30	0.4640	0.3295	0.2900	2.0000	1.9912	0.4978	0.6909
HK-*Mse*I	$HM2_A$	30	0.4908	0.4851	0.0118	2.0000	1.9231	0.4800	0.6730
HK-*Mse*I	$HM2_B$	30	0.3410	0.2912	0.1461	2.0000	1.7241	0.4200	0.6109
HK-*Mse*I	$HM2_C$	30	0.4548	0.4320	0.0501	2.0000	1.7241	0.4200	0.6109
Mean	-	30	0.4447	0.3824	0.1401	2.0000	1.9033	0.4731	0.6657
St. Dev	-	-	0.0031	0.0076	-	0.0000	0.1051	0.0309	0.0319

Table 5. Nei's (1972) genetic identity (above diagonal) and genetic distance (below diagonal) in studied apple accessions.

Population	Azerbaijan	Central Alborz	Breeding population
Azerbaijan5	-	0.9561	0.7844
Central Alborz	0.0449	-	0.8521
Breeding population	0.2428	0.1600	-

PCR was carried out as described by Mohanty et al. (2001) (Table 4), using an initial cycle of 4 min at 94°C, followed by 30 cycles of 45 s at 94°C, 45 s at 54.5°C to 58.5°C and 2 or 3 min at 72°C (annealing temperature and extension time depending on the primer and the length of the fragment to be amplified) (Table 4), and finally a 10 min extension at 72°C. Tow restriction enzymes (*Eco*R I and *Mse* I; Fermentas, Canada) were used for digesting 5 µl PCR products in volume of 30 µl containing 10 unit of restriction enzyme following the manufacturer's instructions (each reaction mixture was incubated for 10 h).

cpDNA restriction fragments were separated in 3% agarose gels cast in a Tris-borate-EDTA buffer (1×), by running at 70 V for 4 h, and after that, fragments were visualized in UV light. Each mixture loaded in agarose gel contained 10 µl of restriction mixture, 3 µl of loading buffer and 2 µl Gel Red (Biotium, USA). The size markers used for analyzing the size of polymorphic bands were 1 kb and 50 bp ladders (Fermentas, Canada).

Data analysis

NTSYSpc (Rohlf, 1992) and POPGENE (Weir, 1996) were used and a dendrogram was constructed by UPGMA clustering analysis using the DICE's coefficient (Sneath and Sokal, 1973) for the statistical analysis of the results.

RESULTS AND DISCUSSION

From four universal chloroplast primer pairs tested on *Malus* spp. (K1K2, CS, HK and TF) only primer pair TF that amplified fragment *trn*T- *trn*F in tobacco did not amplify any band in this work. Other primer pairs amplified the fragments. Chloroplast DNA size in apple has been reported by Ishikawa et al. (1992) to digest this molecule by restriction enzyme (*Bam*HI) in Golden Delicious, McIntosh and Red Delicious cultivars which are respectively 143.4, 143.35 and 143.7 kb. We considered this molecule (143.48 kb) by averaging the reported sizes. Therefore, the fragment that primers amplified (5920 bp) by approximately 4.13% of the chloroplast genome was analyzed (Table 4).

From six combinations of primer- restriction enzyme used in this study for CAP analysis of cpDNA, two combinations, K1K2-*Mse* I and CS-*Mse* I resolved monomorphic fragments and the region amplified by CS primer did not have any site for *Eco*R I. The remaining combinations showed polymorphic pattern. Combinations of K1K2-*Eco*R I and HK-*Mse* I had two sites for enzymes.

HK-*Mse* I showed two polymorphic patterns but in combination with K1K2-*Eco*RI having two sites, one of the fragments was monomorph and the other one was polymorph (Table 5). All of these polymorphic patterns resulted because of insertion-deletion (indel) mutation between 20 to 50 bp ranges. Indel mutations in the combinations of K1K2-*Eco*R I and HK- *Eco*R I are shown in Figure 1.

From the four universal chloroplast primer pairs tested on *Malus* spp. (K1K2, CS, HK and TF) only primer pair TF that amplified fragment *trn*T- *trn*F in tobacco did not work in *Malus* genus. Similar result was reported by Mohanty et al. (2001) that used TF primer pairs on *Prunus avium* having the same family with genus *Malus* and did not obtain any amplification. According to polymorphism pattern, all genotypes were grouped into eight haplotypes (Table 1). Haplotype H4 was the larger

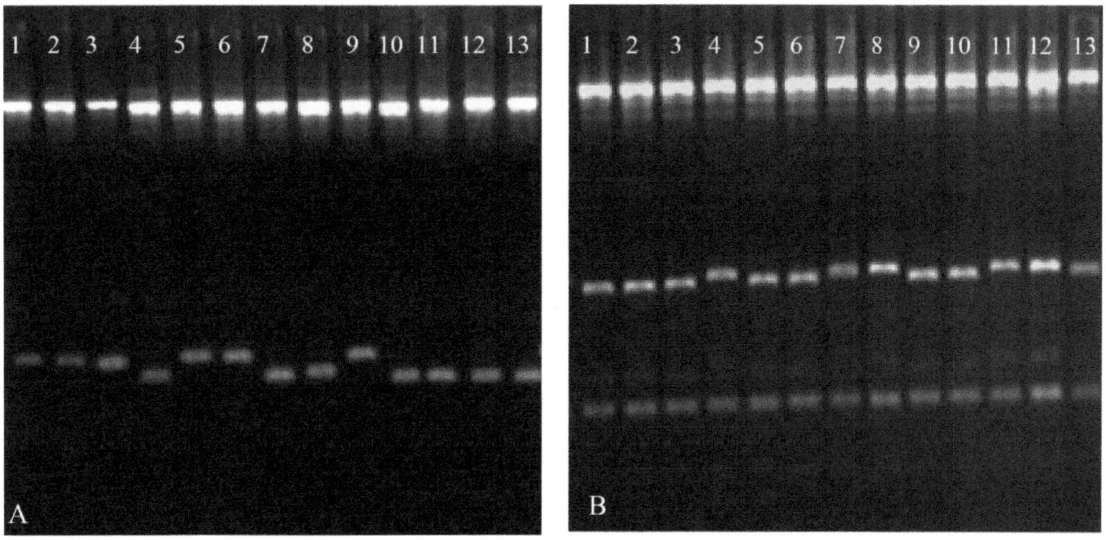

Figure 1. Restriction patterns in two primer-enzyme combinations HK-EcoRI (A) and K1K2-EcoRI (B) resolved on agarose gel. Lines refer to Azerbaijan5, Azerbaijan7, Azerbaijan14, ME3, ME5, Ghara Yarpagh, Israeli, Torsh Damavand, Boshghabi Taleghan, Shemirani Tabestaneh, Fuji, M4 and Gami Almasi respectively.

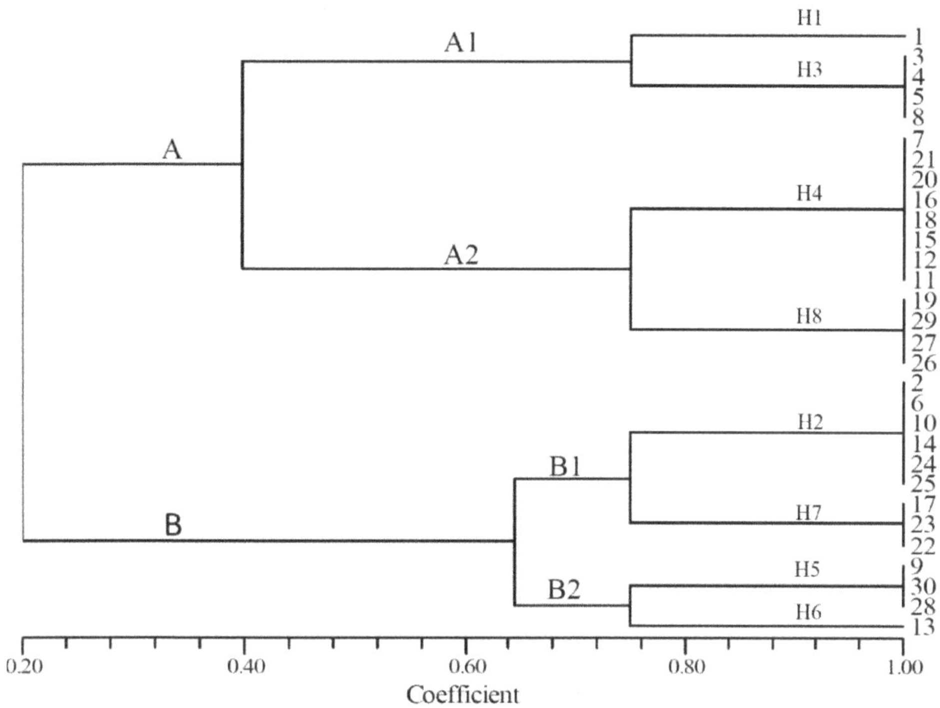

Figure 2. Dendrogram of studied Iranian apple genotypes, commercial cultivars and rootstocks generated by UPGMA clustering analysis using the DICE coefficient based on cpDNA polymorphism. (Numbers refer to samples according to Table 1).

and common haplotype with eight genotypes, two haplotypes (H1 and H6) each covering one genotype and were the smaller haplotypes. The members of the remaining haplotype were from one to eight. From the eight haplotypes that were identified, only three haplotypes (H1, H3 and H6) were specific for Azerbaijan population. Other haplotypes had members from populations, cultivars or rootstocks.

In the dendrogram (Figure 2) at a similarity coefficient value of 0.2, samples were divided into two main groups: A and B. The main group, A, comprises mostly Azerbaijan population samples and forms haplotypes H1,

H3, H4 and H8. This main group itself is divided into two groups at the similarity coefficient value of 0.40. In this division, two haplotypes, H1 and H3 were located together and all of their members were from Azerbaijan population samples. Group B versus the first main group (A), at similarity coefficient value of 0.64 discriminated the haplotypes, H2 and H7 from H5 and H6. The important point that should be mentioned is the coefficient of similarity value of 0.75 where all eight haplotypes were in individual groups.

It is clear from the dendrogram that two accessions Boshghabi Taleghan and T3 that originated from Taleghan area in Central of Alborz Region, which is the nearest region to Azerbaijan, were beside six samples of Azerbaijan in one haplotype (H4). This means that these genotypes probably have been transferred from Azerbaijan that has a vast area of culture and history of apple compared to Taleghan. Similar condition was present in haplotype H2, where two accessions from Alborz population (Torsh Meygon and Golab Damavand) were with four samples of the Azerbaijan population. However, occurrence of these four samples (Boshghabi Taleghan, T3, Torsh Meygon and Golab Damavand) from mutations should also be considered. Meanwhile, the same maternal parent cannot be completely rejected because introduction from other areas and selection for better genotypes has been a factor for transmission of genotypes between regions from the past.

Two haplotypes (H1 and H6) each had only one member, both from Azerbaijan. The commercial cultivars and rootstocks located in haplotypes, H5 and H8 and some Iranian accessions were grouped with them. Two dwarf rootstocks, M9 and Gamy Almasi (Iranian native dwarf rootstock) were in haplotype H5 and this means that they could have common maternal parent.

One of the main reasons for the absence of genetic structure among different populations appears to be cytoplasmic gene flow among them (that is, seeds migration). The migration rate of seeds may be higher than expected (Mohanty et al., 2001).

Total diversity (h_T), average intrapopulation diversity (h_S) and level of population subdivision (G_{STC}) calculated by POPGENE program were 0.467, 0.4451 and 0.0481 respectively. According to the genetic differentiation among populations in *Malus* spp (G_{STC} = 0.0481) and comparison of it by similar reports such as in *Prunus avium* (G_{STC} = 0.08; Mohanty et al. 2001) and *Prunus spinosa* (G_{STC} =0.19; Mohanty et al. 2000) which are from the same Family with the genus *Malus* (Rosaceae), it could be determined that genetic differentiation among populations in present study (G_{STC}) is low and is much lower compared to forest species such as *Quercus petraea* (G_{STC} =0.90; Petit et al. 1993a), *Fagus sylvatica* (G_{STC} =0.83; Demesure et al., 1996), *Argania spinosa* (G_{STC} =0.60; El Mousadik and Petit, 1996) and *Alnus glutinosa* (G_{STC} =0.87; King and Ferris, 1998).

Petit et al. (1993b) reported that in some species the G_{STN} (genetic differentiation among populations using nuclear markers) is much lower than the G_{STC}. It could be deduced that cytoplasmic genomes which have inheritance from maternal parents are relatively much more structured than the nuclear genome because cytoplasmic gene flow is limited just by seed dispersal.

More comparison with other similar studies was impossible, because of small number of common samples in similar studies, different primers for amplification of different regions of chloroplast DNA and the multiple numbers of enzymes that cause many primers- enzymes combination. The Azerbaijan population samples did not completely separate from other samples, and this could be due to migration of genotype by mostly human activity.

Generally, identification of haplotypes by defining evolutionary paths, geographical distribution and division of large populations into smaller groups makes the genetic studies and research with marker that take more time (such as SSR and AFLP) easier. By developing new molecular techniques, especially DNA sequencing, DNA based molecular study like phylogenic study shifts to the use of new methods.

REFERENCES

Bouhadida M, Martín JP, Eremin G, Pinochet J, Moreno MÁ, Gogorcena Y (2007). Chloroplast DNA diversity in *Prunus* and its implication on genetic relationships. J. Am. Soc. Horticult. Sci. 132(5):670–679.

Coart E, Van Glabeke S, De Loose M, Larsen AS, Roldan-Ruiz I (2006). Chloroplast diversity in the genus *Malus*: New insights into the relationship between the European wild apple (*Malus sylvestris* (L.) Mill.) and the domesticated apple (*Malus domestica* Borkh.). Molec. Ecol. 15:2171-2182

Demesure B, Sodzi N, Petit RJ (1995). A set of universal primers for amplification of polymorphic non-coding regions of mitochondrial and chloroplast DNA in plants. Mol. Ecol. 4:129–131.

Demesure B, Comps B, Petit RJ (1996). Chloroplast DNA phylogeography of the common beech (*Fagus sylvatica* L.) in Europe. Evolution 50:2515–2520.

Dumolin-Lapegue S, Pemonge MH, Petit RJ (1997). An enlarged set of consensus primers for the study of organelle DNA in plants. Molec. Ecol. 6:393–397.

El Mousadik A, Petit RJ (1996). Chloroplast DNA phylogeography of the argan tree of Morocco. Mol. Ecol. 5:547–555.

Fofona B, Harvengt L, Baudoin JP, Du Jardin P (1997). New primers for the polymerase chain reaction of cpDNA intergenic spacers in *Phaseolus* phylogeny. Belgian J. Bot. 129:118-122.

Grivet D, Heinze B, Vendraminà GG, Petit RJ (2001). Genome walking with consensus primers: Application to the large single copy region of chloroplast DNA. Mol. Ecol. Notes 1:345-349.

Heinze B (1998). PCR based chloroplast DNA assay for the identification of native *Populus nigra* and introduced poplar hybrids in Europe. For. Genet. 5:33-40.

http://www.diversityarrays.com/sites/default/files/pub/DArT_DNA_isolation.pdfhttp://www.diversityarrays.com/sites/default/files/pub/DArT_DNA_isolation.pdf.

Ishikawa S, Kato S, Imakawa S, Mikami T, Shimamoto Y (1992). Organel DNA polymorphism in apple cultivars and rootstocks. Theoret. Appl. Genet. 83:963-967.

King RA, Ferris C (1998). Chloroplast DNA phylogeography of *Alnus glutinosa* (L.) Gaertn. Molec. Ecol. 7:1151-1161.

Mabberley DJ, Jarvis CE, Juniper BE (2001). The name of the apple. Telopea J. 9:421-430.

Mariac C, Trouslot, P, Poteaux C, Bezancon G, Renno JF (2000). Chloroplast DNA extraction from herbaceous and woody plants for direct restriction fragment length polymorphism analysis.

Biotechniques 28:110-113.

Mohanty A, Martín JP, Aguinagalde I (2001). Chloroplast DNA study in wild populations and some cultivars of *Prunus avium* L. Theoret. Appl. Genet. 103:112–117

Mohanty A, Martin JP, Gonzalez LM, Aguinagalde I (2003). Association between chloroplast DNA and mitochondrial DNA haplotypes in *Prunus spinosa* L. (Rosaceae) populations across Europe. Ann. Bot. 92:749-755.

Mohanty A, Martín JP, Aguinagalde I (2000). Chloroplast DNA diversity within and among populations of the allotetraploid *Prunus spinosa* L. Theoret. Appl. Genet. 100:1304–1310.

Palmer JD, Stein DB (1986). Conservation of chloroplast genome structure among vascular plant. Curr. Genet. 10:823-833.

Panda S, Pedro MJ, Aguinaglde I (2003). Chloroplast DNA study in sweet cherry cultivars (*Prunus avium* L.) using PCR-RFLP method. Genet. Resourc. Crop Evolution. 50:489-495.

Petit RJ, Kremer A, Wagner DB (1993a). Geographic structure of chloroplast DNA polymorphisms in European oaks. Theoret. Appl. Genet. 87:122–128.

Petit RJ, Kremer A, Wagner DB (1993b). Finite island model for organelle and nuclear genes in plants. Heredity 71:630-641.

Phipps JB, Robertson KR, Smith PG, Rohrer JR (1990). A checklist of the subfamily Maloideae (Rosaceae). Can. J. Bot. 68(10): 2209-2269.

Rohlf FJ (1992). Ntsys.pc, Numerical Taxonomy and Multivariate Analyses System, Exeter Software, New York. Sears, B.B. (1980). Elimination of plastids during spermatogenesis and fertilization in the plant kingdom. Plasmid. 4:233-255.

Sneath PHA, Sokal RR (1973). Numerical Taxonomy. Freeman, SanFrancisco p. 573.

Sugiura M (2003). History of chloroplast genomics. Photosyn. Res. 76:371-377.

Taberlet P, Gielly L, Pautou G, Bouvet J (1991). Universal primers for amplification of three non-coding regions of chloroplast DNA. Plant Molec. Biol. 17:1105–1109.

Turkec A, Sayar M, Heinze B (2006). Identification of sweet cherry cultivars (*Prunus avium* L.) and analysis of their genetic relationships by chloroplast sequence-characterised amplified regions (cpSCAR). Genet. Resourc. Crop Evol. 53:1635-1641.

Weir BS (1996). Interspecific differentiation. In: D.M. Hillis et al. (Eds). Molecular systematics, 2nd edition. Sinauer Associates, Sunderland pp. 385-403.

Weising K, Gardner RC (1999). A set of conserved PCR primers for the analysis of simple sequence repeat polymorphisms in chloroplast genomes of dicotyledonous Angiosperms. Genome 42:9-19.

Whatley JM (1982). Ultrastructure of plastid inheritance: Green algae to Angiosperms. Biol. Rev. 57:527-569.

Zohary D, Hopf M (2001). Domestication of Plants in the Old World: The Origin and Spread of Cultivated Plants in West Asia, Europe, and the Nile Valley. Oxford University Press p. 328.

2

Analysis of histone deacetylase families of *Arabidopsis thaliana* and *Oryza sativa*

Joseena Mariam Iype, Rashmi Mishra, S. Karthikeyan, S. Babu and K. M. Gothandam*

School of Bio Sciences and Technology, VIT University, Vellore-632014, TamilNadu, India.

Histone deacetylases (HDACs) plays critical roles in the regulation of chromatin structure and gene expression. HDACs are highly conserved in yeast, animals and plants, suggesting a fundamental mechanism in transcriptional regulation. Sequence characterization and phylogenetic relationships reveals the existence of several distinct families. In plants, the genome-wide identification of HDACs has been performed in several species. These genes are emerging as crucial players in all aspects of plant development, including embryogenesis, abaxial/adaxial polarity determination, flowering and senescence as well as responses to day length and environmental stresses. In this study, we analysed the HDACs from *Oryza sativa* and *Arabidopsis thaliana*. This analysis revealed the presence of 21 and 82 genes coding for putative HDACs in rice and arabidopsis, respectively and found out Poly amphipathic helix domain (PAH) and HDAC interact domains are present in these genes. In this work, we analysed the HDAC families and classified based on their structure.

Key words: Histone deacetylases, *Oryza sativa, Arabidopsis thaliana,* poly amphipathic helix domain (PAH) domain, Histone deacetylases (HDACs) domain.

INTRODUCTION

Gene expression in eukaryotes involves a complex interplay among transcription factors and chromatin proteins that pack chromosomal DNA into the confined space of the nucleus while poising genes for activation or repression (Kadonaga, 1998). The basic unit of chromatin is the nucleosome core particle, a structure in which ~146 bp of DNA is wrapped around a protein octamer made up of two subunits each of the core histones H2A, H2B, H3 and H4 (Kornberg and Lorch, 1999). Core histones can exist in multiple alternative states of acetylation, methylation, phosphorylation, ubiquitination or ADP - ribosylation (Strahl and Allis, 2000). The regulatory significance of these modifications for processes including gene repression, gene activation and replication is increasingly clear (Grunstein, 1997; Ng and Bird, 2000; Struhl et al., 1998).

Acetylation and deacetylation of histones are known to play important roles in the regulation of gene expression in eukaryotes (Meyer, 2000; Alinsug et al., 2009; Aquea et al., 2010). Histone acetyltransferases (Grant and Berger, 1999; Strahl and Allis, 2000) and histone deacetylases (Johnson and Turner, 1999; Knoepfler and Eisenman, 1999; Ng and Bird, 2000) have been identified in large, multisubunit complexes that target these enzymes to specific sites in nuclear DNA. In the nucleosomes of tightly coiled, condensed chromatin, positively charged lysine residues in the N-terminal tails of core histones interact with DNA and other chromosomal proteins. Acetylation of the lysine residues by histone acetyltransferases disrupts these interactions, resulting in relaxation of the chromatin structure, which allows gene expression to occur (Meyer, 2000). Conversely, the deacetylation of lysine residues in histones by HDACs results in the condensation of chromatin structure and the repression of gene expression (Meyer, 2000). HDAC genes and proteins have been isolated and characterized from a variety of animals, fungi, and plants. There are two subgroups of HDACs found in all eukaryotes. One subgroup is most closely related to the yeast RPD3 HDAC,

---

*Corresponding author. E-mail: gothandam@yahoo.com.

Abbreviations: HDAC, Histone deacetylase; **PAH,** Poly amphipathic helix domain.

and the other is more closely related to yeast HDA1 (Johnson and Turner, 1999). In addition to these two subgroups of HDACs, plants contain a family of HDACs (HD2) that appear to be unique and unrelated to yeast RPD3 (Lusser et al., 1997 ; Kölle et al., 1999; Lechner et al., 2000; Wu et al., 2000). HDACs generally are found as part of multiprotein complexes that contain transcriptional repressors, corepressors, and a variety of other proteins (Ng and Bird, 2000). In some cases, HDAC complexes contain ATP-dependent chromatin-remodeling proteins and proteins that bind to methylated DNA (Ahringer, 2000).

MATERIALS AND METHODS

Protein identification and domain prediction

Thorough NCBI (www.ncbi.nlm.nih.gov) searches with several divergent HDACs proteins of rice and arabidopsis were performed to retrieve HDACs genes of Arabidopsis through and of rice. Using the *Arabidopsis thaliana* and *Oryza sativa* sequences of HDACs, we performed a blast search for each sequence using BLASTp algorithm in the uniprot browser (http://www.uniprot.org/). The similar sequences obtained of all HDAC proteins were analysed for recognizable domains in PFAM (http://pfam.sanger.ac.uk/).

Phylogenetic analysis

Phylogenetic analyses were performed using *Arabidopsis thaliana* and *O. sativa* HDAC protein sequences. The evolutionary history was inferred using the neighbour joining method; made by using ClustalW (align.genome.jp/). Conserved domains were obtained from the multiple sequence alignment.

RESULTS AND DISCUSSION

Identification and domain prediction of HDACs from *A. thaliana* and *O. sativa*

Histone acetylation is a reversible process that plays a vital role in epigenetic regulation. Therefore, histone acetylation and deacetylation are of particular importance to plant growth, development, defense and adaptation. HDAC proteins sequences retrieved from different databases as described in the Materials and Methods. Totally, 21 and 82 genes coding for putative HDACs in rice and Arabidopsis were identified. The individual domain structures and number of identical proteins from other plants are presented in Table 1. The retrieved sequences were subjected to blastclust (http://toolkit.tuebingen.mpg.de /blastclust) to make them into clusters. The different clusters for rice and Arabidopsis are given Tables 2 and 3 respectively. PAH: poly amphipathic helix repeat, this family contains the paired amphipathic helix repeat. The family contains the yeast SIN3 gene (also known as SDI1) that is a negative regulator of the yeast HO gene. The family contains the eukaryotic Sin3 proteins, which have at least three PAH domains (PAH1, PAH2, and PAH3). Sin3 proteins are components of a co-repressor complex that silences transcription, playing important roles in the transition between proliferation and differentiation. Sin3 proteins are recruited to the DNA by various DNA-binding transcription factors such as the Mad family of repressors, MeCP2, p53, and Ume6. Sin3 acts as a scaffold protein that in turn recruits histone-binding proteins RbAp46/RbAp48 and histone deacetylases HDAC1/-HDAC2, which deacetylate the core histones resulting in a repressed state of the chromatin. The PAH domains are protein-protein interaction domains through which Sin3 fulfils its role as a scaffold. The PAH2 domain of Sin3 can interact with a wide range of unrelated and structurally diverse transcription factors that bind using different interaction motifs; HDAC INTERACT: HDAC_interact super family, this domain is found on transcriptional regulators. It forms interactions with histone deacetylases.

Classification of HDAC in rice and arabidopsis

To determine the evolutionary relationship of HDAC proteins of *A. thaliana* and *O. sativa* sequences, all HDAC protein sequences from these plants were used to construct a neighbour-joining phylogenetic tree. The tree is shown in Figures 1 and 2. The HDACs in *O.sativa* were divided into 2 classes:

Class I with 4 domains- 3 PAH and 1 HDAC interact
Class II with 2 domains- 1 PAH and 1 HDAC interact

The HDACs in *Arabidopsis thaliana* were divided into 4 classes:

Class I with 3 domains- 2 PAH and 1 HDAC
Class II with 2 domains- 1 PAH and 1 HDAC
Class III with 4 domains- 3 PAH and 1 HDAC
Class IV with 1 domains - 1 HDAC

Multiple sequence alignment

Amino acid sequence of histone deacetylases were used for multiple sequence alignment using clustalW. Highly conserved regions were obtained from the alignment. As shown in Figure 3. The HDAC sequence is conserved in all the 21 sequences. This domain represents the HDAC_interact super family as shown in Figure 4.

Function of histone deacetylases genes in plants

The homeostatic balance of nucleosomal histone acetylation is maintained by actions of histone

Table 1. PFAM structure and domains of *O. sativa* 's HDAC gene.

S/N	Accession No.	PFAM structure	Domains
1	CBD19699		3 PAH; 1 HDAC
2	CBD25483	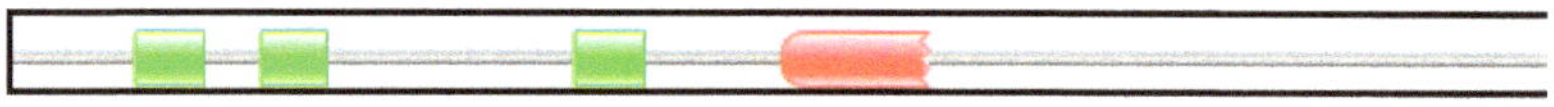	3PAH; 1 HDAC
3	CBD23005		3 PAH; 1 HDAC
4	CBD20685		3 PAH; 1 HDAC
5	EEE64875	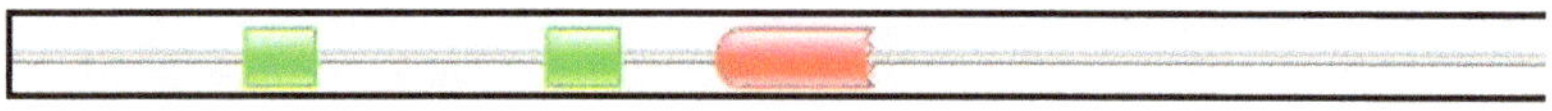	2 PAH; 1 HDAC
6	EEE61986		3 PAH; 1 HDAC
7	EEE53729		3 PAH; 1 HDAC
8	EEC78333		3 PAH; 1 HDAC
9	EEC69795		3 PAH; 1 HDAC
10	EAY99233		2 PAH; 1 HDAC
11	NP_001056476		1 PAH; 1 HDAC
12	NP_001041795	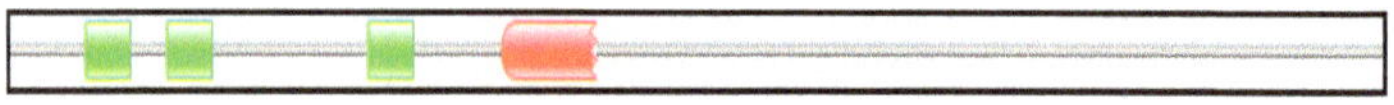	3 PAH; 1 HDAC
13	AAO72669	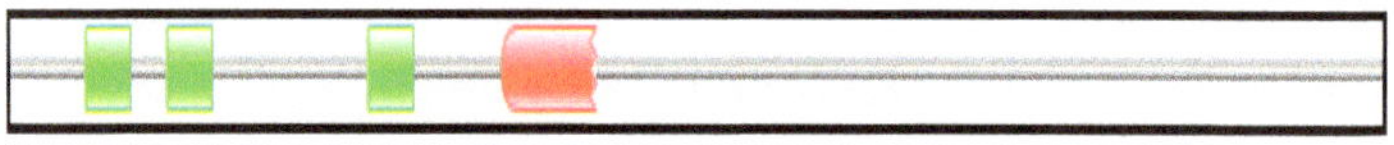	3 PAH; 1 HDAC
14	BAF18390	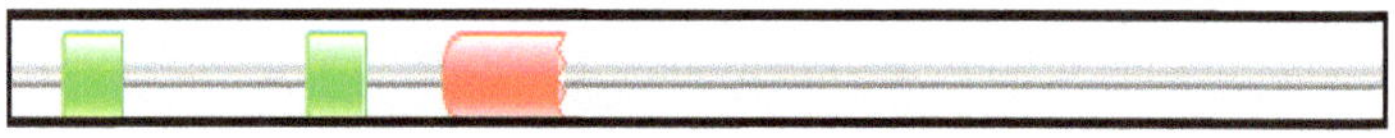	2 PAH; 1 HDAC
15.	BAF03709	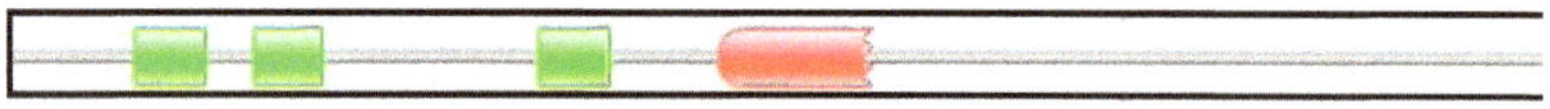	3 PAH; 1 HDAC
16	CBD19688		3 PAH; 1 HDAC
17	CBD25474		3 PAH; 1 HDAC
18	CBD22994		3 PAH; 1 HDAC
19	CBD20674		3 PAH; 1 HDAC
20	BAD44865		3 PAH; 1 HDAC
21	BAH00519		3 PAH; 1 HDAC

Table 2. Rice HDAC clusters and their pfam structure.

S/N	Number of similar structures	Domain structures and Accession no:	Number and name of domains
1	17	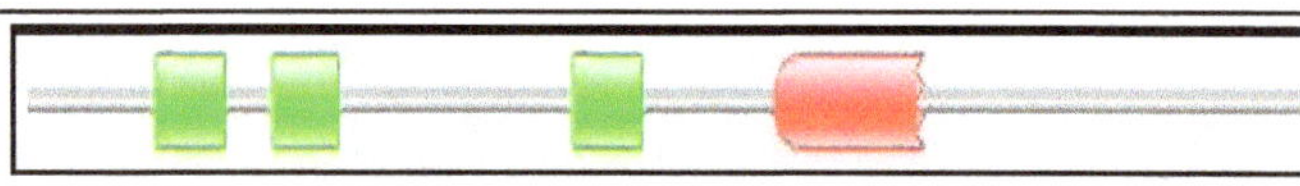 EEE53729, EEE61986, EEC69795, BAF03709, NP_001041795, BAH00519, CBD20685, CBD23005, CBD25483, CBD19699, CBD20674, CBD22994, CBD25474, CBD19688, BAD44865 EEC78333, AAO72669	2 PAH 1 HDAC
2	4	 BAF18390, NP_001056476, EAY99233, EEE64875	2 PAH 1 HDAC

Table 3. *A. thaliana* HDAC clusters and their pfam structure.

S/N	Number of similar structures	Domain structures and accession no.	Number and name of domains
1	45	 CBD20673, CBD22993, CBD25471, CBD19687, CAC01821, CBD20672, CBD22992, CBD25470, CBD19686, Q9SRH9, NP_186781, NP_197006, Q9LFQ3, CBD20681, CBD23001, CBD25479, CBD19695, AAF03494, O48686, CBD31870, CBD20669, CBD22989, CBD25466, CBD19683.1,NP_173829, CBD20676, CBD22996, CBD25476, CBD19690, AAC00578, ABR13973, AAB61107, CBD20675, CBD22995, CBD25475, CBD19689, CBD20671, CBD22991, CBD25469, CBD19685, O04539, NP_177163, BAF03709, NP_001041795, BAC43533	3 PAH 1 HDAC INTERACT
2	16	 CBD20683, CBD23003, CBD25481, CBD19697, AAD39565, CBD20670, CBD22990, CBD25468, CBD19684, Q9XIE1, NP_176197, CBD20682, CBD23002, CBD25480, CBD19696, AAD39330	2 PAH I HDAC INTERACT
3	13	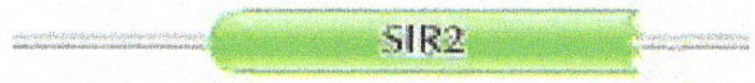 NP_001078550, NP_568207, NP_850795, NP_974753, NP_850796, NP_850797, NP_974754, BAB02925, NP_189069, NP_189529, BAB02130, NP_200387, AAX92703	1 SIR2
4	8	 BAF18390, NP_001056476, NP_172515, CBD20668, CBD22988, CBD25467, CBD19682, Q9XIK6	1 PAH 1 HDAC INTERACT

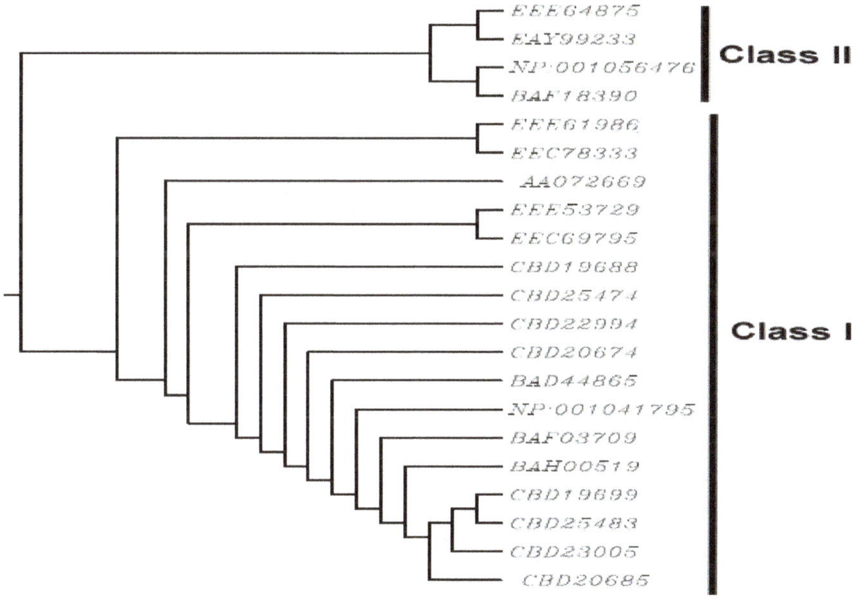

Figure 1. Classification of HDAC families **of** *O. sativa.*

Acetyltransferases (HAT) and histone deacetylases (HDAC). Most of the HDAC genes are responsive to drought or salt stresses. Down-regulation of most of HDAC genes may be important for stress-inducible gene expression in rice. These HDACs may repress stress responsive gene expression in the absence of stress signals. Therefore, histone acetylation / deacetylation switch is suggested to be an important mechanism of short-term gene regulation in plants (Zhou, 2009), may be largely involved in stress responses in rice. It has been

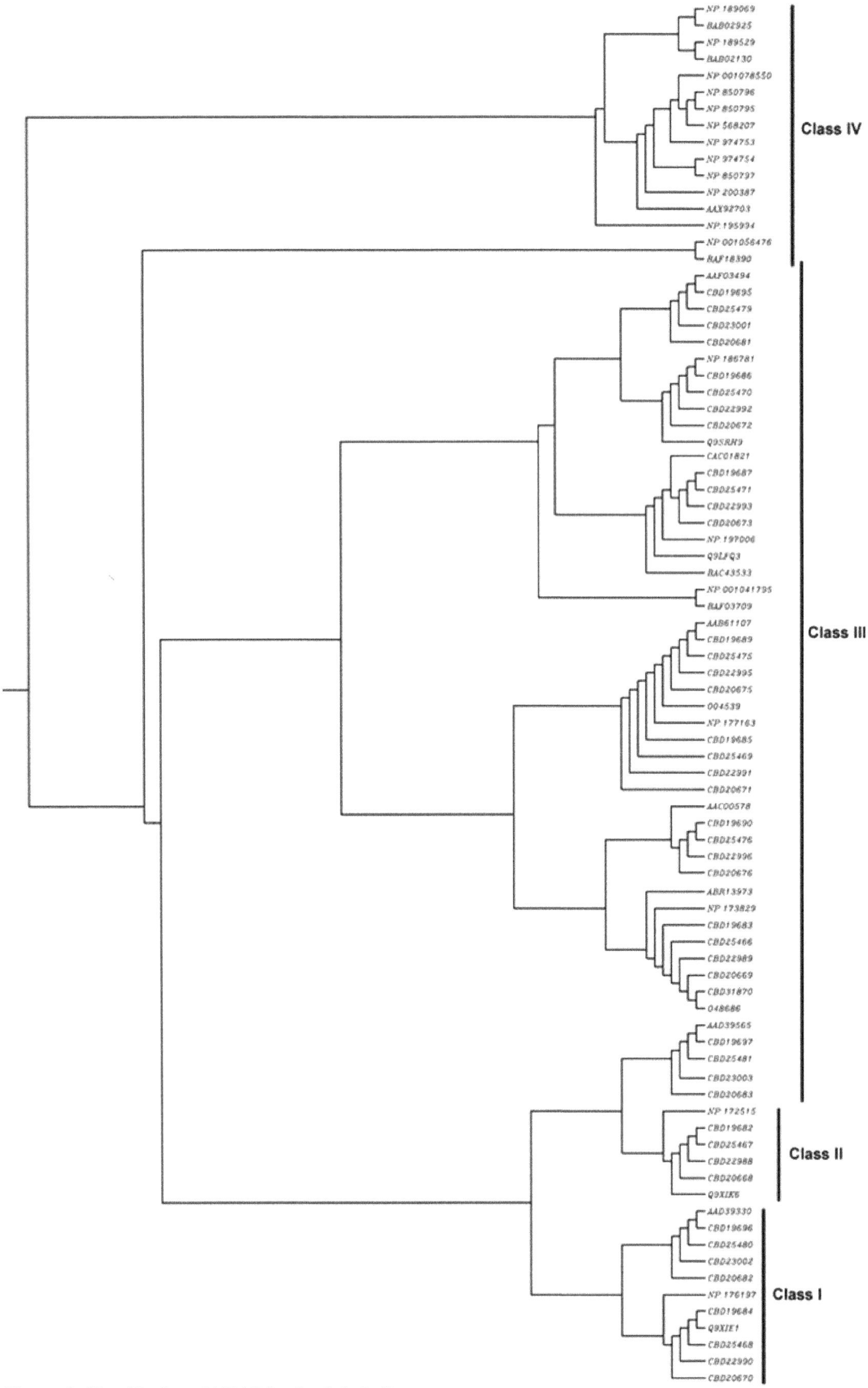

Figure 2. Classification of HDAC family of *A. thaliana.*

```
CBD19699        CTPSYRLLPKHYPMPPAGNKTELGASVLNDHWVSVTSGSEDYSFKHMRKN 437
CBD25483.1      CTPSYRLLPKHYPMPPAGNKTELGASVLNDHWVSVTSGSEDYSFKHMRKN 437
CBD23005.1      CTPSYRLLPKHYPMPPAGNKTELGASVLNDHWVSVTSGSEDYSFKHMRKN 437
CBD20685.1      CTPSYRLLPKHYPMPPAGNKTELGASVLNDHWVSVTSGSEDYSFKHMRKN 437
BAH00519.1      CTPSYRLLPKHYPMPPAGNKTELGASVLNDHWVSVTSGSEDYSFKHMRKN 437
BAF03709.1      CTPSYRLLPKHYPMPPAGNKTELGASVLNDHWVSVTSGSEDYSFKHMRKN 438
NP_001041795    CTPSYRLLPKHYPMPPAGNKTELGASVLNDHWVSVTSGSEDYSFKHMRKN 438
BAD44865.1      CTPSYRLLPKHYPMPPAGNKTELGASVLNDHWVSVTSGSEDYSFKHMRKN 355
CBD20674.1      CTPSYRLLPKHYPMPPAGNKTELGASVLNDHWVSVTSGSEDYSFKHMRKN 355
CBD22994.1      CTPSYRLLPKHYPMPPAGNKTELGASVLNDHWVSVTSGSEDYSFKHMRKN 355
CBD25474.1      CTPSYRLLPKHYPMPPAGNKTELGASVLNDHWVSVTSGSEDYSFKHMRKN 355
CBD19688.1      CTPSYRLLPKHYPMPPAGNKTELGASVLNDHWVSVTSGSEDYSFKHMRKN 355
EEE53729.1      CTPSYRLLPKHYPMPPAGNKTELGASVLNDHWVSVTSGSEDYSFKHMRKN 530
EEC69795.1      CTPSYRLLPKHYPMPPAGNKTELGASVLNDHWVSVTSGSEDYSFKHMRKN 472
AAO72669.1      CTPSYRLLPKHYPMPPAGNKTELGASVLNDHWVSVTSGSEDYSFKHMRKN 136
EEE61986.1      CTPSYRLLPKNYPMPPASCRTDLGASVLNDLWVSVTSGSEDYSFKHMRKN 513
EEC78333.1      CTPSYRLLPKNYPMPPASCRTDLGASVLNDLWVSVTSGSEDYSFKHMRKN 482
EEE64875.1      CTPSYCLLPKNCVTLQSSYQTELGRSILNDSLVSVTSGRED-CYKFRTKN 313
EAY99233.1      CTPSYCLLPKNCVTLQSSYQTELGRSILNDSLASVTSGRED-CYKFRTKN 313
NP_001056476    CTPSYCLLPKNCVTLQSSYQTELGRSILNDSLVSVTSGRED-CYKFRTKN 341
BAF18390.1      CTPSYCLLPKNCVTLQSSYQTELGRSILNDSLVSVTSGRED-CYKFRTKN 341
                ***** ****:      :. :*:** *:***  .***** ** .:*.  **
```

Figure 3. Multiple sequence alignment of HDACs in *O. sativa*. Multiple sequence alignment of Histone deacetylases was generated using ClustalW. HDAC interact Domains CTPSYCLLPK and SVTSGRED were found (shown in red).

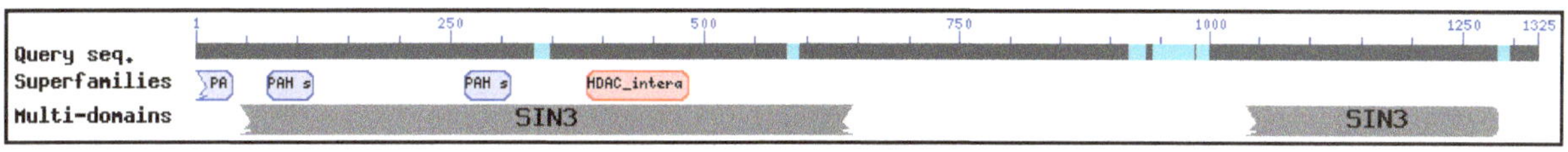

Figure 4. Shows conserved domain of a HDAC protein sequence of *O. sativa*.

shown that over-expression of OsHDAC1 (HDA702) leads to increased growth rate and altered architecture in transgenic rice (Jang et al., 2003). HDAC induces DNA fragmentation, cell death, and lesions mimicking plant hypersensitive responses induced by pathogen attacks. In this work, the HDACs of *A. thaliana* and *O. sativa* have been identified and their conserved domains are analyzed. Phylogenetic analysis of histone deacetylases of *A. thaliana* and *O. sativa* is done and classified.

REFERENCES

Ahringer J (2000). NuRD and SIN3. Trends Genet. 16:351-356.

Alinsug MV, Yu CW, Wu K (2009). Phylogenetic analysis, subcellular localization, and expression patterns of RPD3/HDA1 family histone deacetylases in plants. BMC Plant Biol. 9:37 doi:10.1186/1471-2229-9-37.

Aquea F, Timmermann T, Arce-Johnson P (2010). Analysis of histone acetyltransferase and deacetylase families of *Vitis vinifera*. Plant Physiol. Biochem. 48:194-199.

Grant PA, Berger SL (1999). Histone acetyltransferase complexes. Semin. Cell Dev. Biol. 10:169-177.

Grunstein M (1997) Histone acetylation in chromatin structure and transcription. Nature 389:349-352.

Jang IC, Pahk YM, Song SI, Kwon HJ, Nahm BH, Kim JK. (2003). Structure and expression of the rice class-I type histone deacetylase genes OsHDAC1-3: OsHDAC1 over expression in transgenic plants leads to increased growth rate and altered architecture. Plant J. 33:531-41.

Johnson CA, Turner BM (1999). Histone deacetylases: Complex transducers of nuclear signals. Semin. Cell Dev. Biol. 10:179-188.

Kadonaga JT (1998) Eukaryotic transcription: an interlaced network of transcription factors and chromatin - modifying machines. Cell 92:307-313.

Knoepfler PS, Eisenman RN (1999). Sin meets NuRD and other tails of repression. Cell 99:447-450.

Kölle D, Brosch G, Lechner T, Pipal A, Helliger W, Taplick J, Loidl, P (1999). Different types of maize histone deacetylases are distinguished by a highly complex substrate and site specificity. Biochemistry 38:6769-6773.

Kornberg RD, Lorch Y (1999) Twenty - five years of the nucleosome, fundamental particle of the eukaryote chromosome. Cell 98:285-294.

Lechner T, Lusser A, Pipal A, Brosch G, Loidl A, Goralik- Schramel M, Sendra R, Wegener S, Walton, JD, Loidl P (2000). RPD3-type histone deacetylases in maize embryos. Biochemistry 39:1683-1692.

Lusser A, Brosch G, Loidl, A, Haas H, Loidl P (1997). Identification of maize histone deacetylase HD2 as an acidic nucleolar phosphoprotein. Science 277:88-91.

Meyer P (2000). Transcriptional transgene silencing and chromatin components. Plant Mol. Biol. 43:221-234.

Ng HH, Bird A (2000). Histone deacetylases: Silencers for hire. Trends Biochem. Sci. 25:121-126.

Strahl BD, Allis CD (2000). The language of covalent histone modifications. Nature 403:41-45.

Struhl K, Kadosh D, Keaveney M, Kuras L, Moqtaderi Z (1998).Activation and repression mechanisms in yeast. Cold Spring Harb. Symp. Quant. Biol. 63:413-421.

Wu K, Tian L, Malik K, Brown D, Miki B (2000). Functional analysis of HD2 histone deacetylase homologues in *Arabidopsis thaliana*. Plant J. 22:19-27.
Zhou DX (2009). Regulatory mechanism of histone epigenetic modifications in plants. Epigenetics 4:15-18.

3

Analysis of factors affecting technical efficiency of smallholder farmers: Comparing time-varying and time invariant inefficiency models

Kisu Simwaka[1], Stuart Ferrer[2] and Geoff Harris[1]

[1]School of Economics and Finance, University of KwaZulu-Natal, J Block, 3rd Floor, Westville Campus, Durban, South Africa.
[2]South African Cane Growers Association, KwaZulu-Natal, South Africa.

This paper estimates technical efficiency and identifies factors that explain differentials in technical efficiency for affected and non-affected households. We use time-varying and time-invariant inefficiency models of production. The results show that fertilizer, labour, seeds, and age are found to have contributed significantly to technical efficiency for affected farm households under both time varying and time-invariant models. On the other hand, only fertilizer contributed significantly to technical efficiency of non-affected farm households. In general, results show that Malawian farmers are generally technically efficient. Nevertheless, there is still scope for improvement of the productivity of small-scale farmers, as some farm households, particularly female headed households with mortality, are still operating at low levels of efficiency. Two main policy issues emerge from the results of this study. Firstly, all types of obstacles that could limit the use of farm inputs should be removed. This should include complete liberalisation of purchase and distribution of such inputs and the development of some low-cost technology to reduce labour constraints on the farm. Secondly, there is need to develop social capital in smallholder farming through the recommencement of farmers' clubs or by setting up agricultural cooperatives.

Key words: Technical efficiency, smallholder agriculture, integrated household surveys.

INTRODUCTION

The prevention and mitigation of human immunodeficiency virus (HIV)/acquired immune deficiency syndrome (AIDS) require a multisectoral approach. Research into the socio-economic impact of HIV/AIDS on households and communities is essential for formulation of current and future policies and intervention strategies intended to mitigate this impact. From theory, HIV/AIDS has large effects on agriculture and rural welfare. The role of researchers is to assess micro-level and macro-level consequences of HIV/AIDS. In doing this, there is need to distinguish between mortality and morbidity that is HIV/AIDS related and that which is not. Empirical knowledge on how rural households respond to HIV/AIDS remains weak.

A review of literature shows that a rising number of studies in Africa are beginning to offer micro-level insights on the impacts of HIV/AIDS on farm households and their responses. However, most of these studies are faced with at least three major problems. Firstly, the few available micro-level studies use data from specific geographical sites, deliberately chosen because they were associated with high infection rates. Although they suggest valuable insights into how affected households respond to the disease, such studies are limited in their ability to be generalised in order to understand national level impacts (Chapoto and Jayne, 2005).

Secondly, only a few studies on this topic are based on panel data. Unfortunately, cross-sectional surveys cannot adequately measure the dynamic effects of mortality and morbidity let alone control for unobserved heterogeneity. Cross-sectional studies cannot provide accurate insights on impacts. In addition, almost all quantitative micro-level studies to-date have estimated the effects of mortality in affected households compared to non-affected households. However, there are no studies that have distinguished morbidity and mortality that is HIV/AIDS related from one that is not. Furthermore, relatively few studies provide adequate focus and empirical evidence on gender aspects of the HIV/AIDS impact on households (Chapoto and Jayne, 2005).

In Malawi, research into the impact of HIV/AIDS is still at an early stage. The only notable contributions on HIV/AIDS impact on agriculture in Malawi is Arrehag et al. (2006). However, there is absence of discussion on the impact of HIV/AIDS on technical efficiency of farmers. This study is of both practical and theoretical relevance. At the practical level, measuring the technical efficiency of small-scale farmers, and identifying the factors that affect it, may provide useful information for the formulation of economic policies likely to improve producer technical efficiency. Moreover, from the microeconomic standpoint, identifying the factors that may improve farm profitability is of major significance since, by using information derived from such studies, farms may become more efficient and hence more profitable. At the theoretical level, the study aims to bring some contribution to the understanding of producer technical performance in developing countries.

Research objectives and questions

The broad goal of this study is to estimate technical efficiency and identify factors that explain differentials in technical efficiency of affected [HIV/AIDS affected households refers to households where at least one prime-age adult family member has died of or suffered from HIV/AIDS-related illnesses of chronic TB, chronic pneumonia and chronic diarrhoea over the last three to five years. Epidemiological studies demonstrate that in Eastern and Southern Africa, HIV is the leading cause of disease-related death among adults between 15 and 49 years of age (UNAIDS/WHO, 1998; Ngom and Clark, 2003)] and non-affected farm households (Non-affected households are those where at least one prime-age adult family member has died of or suffered from chronic non-HIV/AIDS related illnesses over at least three to five years) using time varying and time-invariant inefficiency models. Specifically, the study seeks to address three objectives. First, the study estimates mean and plot specific technical efficiency levels in smallholder farms producing maize, the main staple crop. Second, it establishes the differentials in technical efficiency between affected and non-affected small-scale farm households. Third, it examines the impact of gender, education, fertilizer applications, seed, labour, and land on the technical efficiency of the farmers.

HIV/AIDS and smallholder agriculture in Malawi

Malawi is one of the countries hardest hit by HIV/AIDS (Arrehag et al, 2006). It is ranked eighth in terms of highest global prevalence. The national adult HIV/AIDS prevalence in the productive age group of 15 to 49 years declined from 14.4% in 2003 to 11.9% in 2007. About 840,000 individuals are currently living with HIV/AIDS. Women are disproportionately affected by the epidemic. In 2007, about 490,000 women above the age of 14 were living with HIV/AIDS. The most common means of transmission are through multi-partner heterosexual sex and mother-to-child transmission. Prevalence rates are significantly higher in urban areas (20.4%) than in semi-urban (17.0%) and rural areas (13.0%). However, there is evidence that infection rates are growing in rural areas and going down in urban areas. At regional level, the Southern region of Malawi is the most densely populated and has the highest prevalence rate among pregnant women (21.7%). The prevalence rates for pregnant women in Northern and Central regions are at 14.0 and 14.3%, respectively (Arrehag et al., 2006).

Malawian smallholder agriculture [This section relies on 2008 final report of the evaluation of the 2006/07 agricultural input subsidy program by Ministry of Agriculture and Food Security (MoAFS) report] is mostly dominated by large numbers of poor farmers. These farmers are usually engaged in low input maize production on small land holdings, about around 0.35 ha, without the application of chemical fertilizer. Maize production by these farmers is usually not adequate enough to meet annual consumption needs, and they depend upon casual labour and other income generating activities to meet their needs.

From 1998, first universal "starter packs" and then "targeted inputs" of free packs of fertilizer and matching maize seed for 0.1 ha of land were distributed. Maize production and prices fluctuated markedly. During the 2003/2004 season, when targeted input programs (TIPs) were implemented, about 40% of smallholder households bought chemical fertilizer at commercial prices, with mean purchases of around 65 kg per household. Food insecurity problems for such farmers have worsened in recent years. This resulted in national food shortages and expensive food imports by government and consequently rising maize prices (MoAFS, 2008).

National maize production during 2004/2005 season

sales by subsidized sales, and assumptions about the was low at 1.2 million tonnes. This was due to poor rainfall, late distribution and limited scale of the targeted inputs programme for the 2004/2005 season. With slow official importation and emergency response measures, this low production resulted in very serious food shortages and high maize prices in 2005/2006.

In 2005/2006 season, government put in place a large-scale input subsidy with the objectives of promoting access to and use of fertilizers in maize production in order to increase agricultural productivity and food security. The main objective of the input subsidy program was to raise smallholder productivity, food and cash crop production, and reducing vulnerability to food insecurity and hunger. Other objectives were to support food self-sufficiency, growth of the private sector input markets, and wider economic growth and development. About 2 million seeds and 3 million fertilizer coupons were initially planned for distribution to targeted households within districts and areas. This was followed by delivery of two sets of nitrogen (N), phosphorus (P), potassium (K) and urea coupons. Village Development Committees were charged with the responsibility of assigning coupons to the targeted households at the rate of one NPK (23:21:0) and one urea coupon per household (MoAFS, 2008).

Altogether, about 75,000 tonnes of fertilizer and 4,500 tonnes of improved maize seed were distributed. However, distribution of inputs in the southern region was delayed for a number of reasons. These included late purchase of fertilizer, late issue of coupons, and late opening of markets. This together with inadequate stocks in some markets resulted in the majority of farmers spending long periods queuing up for their inputs. This resulted in late planting and/or fertilizer applications. The total expenditure amounted to MK10.3 billion (about US$91 million) was spent, of which 87% came from the Malawi government. As a result of supplementary coupons, fertilizer sales were 17% above budget while the government expenditures were 25% over the budgeted cost (MoAFS, 2008).

The analysis of available evidence shows that the programme has a lot of potential to contribute positively to government's objectives of increasing crop production, food security and pro-poor growth. For instance, the 2005/06 and 2006/07 TIPs made a positive contribution toward the achievement of those objectives. However, there are also several improvements that could increase the programme's effectiveness and efficiency. The Ministry of Agriculture and Food Security (MoAFS) estimated total maize production of 2.7 and 3.4 million tonnes in 2005/6 and 2006/7, respectively, both record harvests and markedly higher than the 1.2 million tonne estimate for 2004/5. The increase in maize production that can be attributed to the 2006/7 subsidy is estimated at 670,000 tonnes, with a low estimate of just over 5000,000 tonnes and a high estimate of just below 900,000 tonnes. The wide range is because the estimates are sensitive to estimated rates of displacement of commercial fertilizer effectiveness of additional fertilizer and seed in improving yields

LITERATURE REVIEW AND STUDY FRAMEWORK

Empirical literature

A couple of studies have analyzed determinants of technical efficiency of smallholder farmers. Chirwa (2007) uses a small sample of smallholder maize farmers in one district in Southern Malawi. He estimates technical efficiency among smallholder maize farmers in Malawi and identifies sources of inefficiency using plot-level data. The results show that all the coefficients of the inputs in the production function are positive, but only labour was statistically significant at the 10% level. The econometric results based on the stochastic production function also show that maize production is done under constant returns to scale. Many smallholder maize farmers are technically inefficient, with mean technical efficiency scores of 46.23% and technical scores as low as 8.12%. Obwona (2006) uses cross-section data from a sample of 65 small- and medium-scale farmers. A stochastic production frontier approach is used to estimate the farmer-specific technical efficiencies. The estimated efficiencies are then explained by socioeconomic and demographic factors. The results show that education, credit accessibility and extension services contribute positively towards the improvement of efficiency.

Ogundele (2006) establishes the differentials in technical efficiency between farmers planting improved rice varieties and those planting traditional varieties in Nigeria. The results of frontier analysis indicate that farm size was the most significant determinant of technical efficiency. Other variables that contributed to technical efficiency included hired labour, herbicides and seeds. Education and farming experience were found to influence technical efficiency in traditional technology rice farms.

Theoretical model

In this study, we will follow Battese and Coelli (1995) stochastic production frontier for panel data. We consider data on small holder maize farmers in Malawi for the two agricultural seasons from which production data was obtained in the Integrated Household panel survey. The general functional form for the stochastic frontier for maize farmers in Malawi is the trans-logarithmic function:

$$\ell n y_{it} = \beta_0 + \beta_0^* D_{it} + \beta_1 \sum_{j=1}^{5} \beta_j x_{jit} + \sum_{j\leq}^{5} \sum_{k=1}^{5} \beta_{jk} x_{jit} x_{kit} + (v_{it} - u_{it}) \quad (1)$$

Where, the subscripts *i* and *t* represents the *i*-th farm household and the *t*-th year of observation, respectively;

y stands for the logarithm of quantity of maize harvested (in kilograms) (*ly*), *D* is the dummy variable for use of hybrid maize, which takes value 1 if hybrid maize was used, zero otherwise; x_1 is the logarithm of the total amount of land (in hectares) on which maize was grown (*lh*); x_2 is the logarithm of the total amount of labour (in man days) from both family and hired labour (*lb*); x_3 is the logarithm of the amount of fertilizer (kgs) applied to the maize field (*lf*); x_4 is the logarithm of the quantity of maize seed sown (in kgs) (*ls*); x_5 is the year of observation, accounts for Hicksian neutral technological change; where x_5= 1, 2 for years, 2004/05 and 2006/07, respectively. v_{it} and u_{it} are the random variables defined previously.

The disturbance term in a stochastic frontier model is assumed to have two components. The model makes the following assumptions about the error terms: One component is assumed to have a strictly non-negative distribution while the other component is assumed to have a symmetric distribution. In the econometrics literature, the nonnegative component is usually referred to as the inefficiency term. The component with the symmetric distribution is referred to as the idiosyncratic error. The Stata estimation of panel stochastic production functions permits two different parameterizations of the inefficiency term: A time-invariant model and the Battese-Coelli parameterization of time-effects. In the time-invariant model, the inefficiency term is assumed to have a truncated-normal distribution. On the other hand, in the Battese-Coelli parameterization of time effects, the inefficiency term is estimated as a truncated-normal random variable multiplied by a specific function of time. In both models, the idiosyncratic error term is assumed to have a normal distribution. The only panel-specific effect is the random inefficiency term (Battese and Coelli, 1995).

This stochastic frontier model (1) includes year of observation in such a way that non-neutral technical change is specified. However, neutral technical change is present if the coefficients of the interactions between year of observation and the input variables, land, labour, fertilizer and seed, are zero, that is, $\beta_{i2} = 0$, *i* = 1,2. There would be no technical change among the maize farmers if the coefficients of all variables involving year of observation were zero, that is, $\beta_2 = \beta_{i2} = 0,\ i = 1,\ 2$. Further, the Cobb-Douglas production frontier is a special case of the translog frontier in which the coefficients of the second-order terms are zero, that is, $\beta_{jk} = 0,\ j \le k$ =1, 2.

The first-order terms in the translog frontier, defined by Equation 1, are included in the modified Cobb-Douglas frontier considered in Battese et al. (1996), but the latter paper includes additional variables involving the ratio of hired labour to total labour, the logarithm of land preparation, the logarithm of the number of ploughings and dummy variables associated with the use of mechanical traction and the tenancy status of farmers. The model for the technical inefficiency effects in the stochastic frontier of Equation 1 is defined by:

$$u_{it} = \delta_0 + \delta_1(age_{it}) + \delta_2(edu_{it}) + \delta_3(t_i) + w_{it} \qquad (2)$$

Where, age_{it} and edu_{it} represents the age and years of formal education of the primary decision maker at the *t*-th year of observation.

MATERIALS AND METHODS

This study used a two-year panel data (panel data was used because cross sectional studies have distinct limitations. Such studies cannot capture the impacts other than those immediately preceding the interview with the respondent. So, for example, a study of the impact of adult deaths may miss much of the impact of the illness preceding the death) from Integrated Household Surveys that were conducted by the Malawi National Statistical Office in collaboration with the World Bank in 2004/05 and 2006/07.

The 2004/05 survey collected information from a nationally representative sample of 11,280 households. It was designed to cover a broad range of issues, with primary objective of providing a complete and integrated data set to better understand the socio-economic status of the population in Malawi [Malawi Economic Planning and Development (MEPD), National Statistical Office (NSO) and World Bank, 2005]. The questionnaire covered the socio- economic characteristics of the household on the following aspects: Demographic, education, health, agriculture, and anthropometric information, among other attributes (Integrated Household Survey 2004/05 report). The demographic characteristics observed include; age, sex, relationship to household head, marital status and place of residence, household size, and deaths in the households (NSO, 2005).

Each of the twenty-seven districts in Malawi was treated as a separate sub-section of the main rural stratum (except for Likoma district). The household survey used a two-stage stratified sample selection process. The primary sampling units (PSU) were the enumeration areas (EAs). These were chosen for each stratum based on probability proportional to size (PPS). The second stage involved randomly selecting 20 households in each EA. Every listed household in an EA had the same chance of being selected to be enumerated (NSO, 2005).

Out of the total 11,280 households selected, 10,777 households were occupied and successfully interviewed, resulting in a response rate of 98%. Of the selected households, 507 replacements were made. The main reason for replacement was that the dwelling could be located but no household member was available after repeated attempts or the dwelling was not occupied. There were only 41 refusals from respondents (NSO, 2005).

A follow-up national survey was carried out in 2007. About 3,298 households were re-interviewed in 175 enumeration areas in 28 districts. Of these, 3100 were previously sampled and interviewed in the 2004 Integrated Household Survey. Households and enumeration areas within each district were chosen randomly. After excluding households with missing information, obvious data errors, those who stated that they farmed over 20 hectares of land, and those that could not be properly matched between the two surveys, the sample was reduced to 2,431 households (NSO, 2008). Thus,

Table 1. All affected households – time varying inefficiency model.

ly	Coef.	Std. Err.	z	P>|z|	[95% Conf. Interval]	
lh |	0.1059402	0.1031257	1.03	0.304	-0.0961825	0.3080629
lb |	0.1173416	0.0896894	1.31	0.191	-0.0584464	0.2931296
lf |	0.4318217	0.0441411	9.78	0.000	0.3453068	0.5183367
ls |	0.1530484	0.0914311	1.67	0.094	-0.0261534	0.3322501
t |	0.2525243	0.1411019	1.79	0.074	-0.0240303	0.5290789
sex |	0.0596876	0.114341	0.52	0.602	-0.1644166	0.2837919
age |	0.0012207	0.0027493	0.44	0.657	-0.0041679	0.0066093
edu |	-0.0653894	0.0735749	-0.89	0.374	-0.2095936	0.0788147
cons |	4.229109	0.4277983	9.89	0.000	3.390639	5.067578
/mu |	-2.513486	8.383145	-0.30	0.764	-18.94415	13.91718
/eta |	-0.4898009	0.3190093	-1.54	0.125	-1.115048	0.1354459
/lnsigma2 |	0.8361765	1.670105	0.50	0.617	-2.43717	4.109523
/ilgtgamma |	1.100469	2.203094	0.50	0.617	-3.217515	5.418454
sigma2 |	2.307527	3.853814			0.0874079	60.91765
gamma |	0.750348	0.4126965			0.0385119	0.9955856
sigma_u2 |	1.731449	3.84247			-5.799655	9.262552
sigma_v2 |	0.5760787	0.0635129			0.4515958	0.7005617

Time-varying decay inefficiency model; number of obs, 410; group variable, id; number of groups, 263; time variable, t; Obs per group min, 1; average, 1.6; maximum, 2; Wald chi^2 (8), 196.51; Log likelihood, -515.64965; Prob > chi^2, 0.0000.

final analysis is based on the balanced panel of 2,431 households in the smallholder sector that were both interviewed in 2006/07 and either 2002/03 or 2003/04. For 1,101 of these households, information on crop production and input use relates to the 2002/03 and 2006/07 years. Information on the remaining 1,330 households pertains to the 2003/04 and 2006/07 years (NSO, 2008).Data used in the study will be inputs and output of production and social economic variables. Input variables include cultivated land area (in hectares, ha) or farm size. In Malawi, farm sizes have been described as small. Chirwa (2003) found that the average farm size was 0.35 ha. Labour (person-hours) constitutes the most important input in small-scale agriculture. Hence any constraint on availability of labour is detrimental to farm productivity. Labour input can be sourced from within the family (family labour) or from the commercial pool (hired labour). The amount of person-days of family labour that can be engaged by farmers depend on household size, the age structure of the household and the primary occupation of the household members if family labour is in short supply, farmers resort to hired labour. The amounts of persons-days of hired labour that can be committed to production depend on the availability of hired labour, farm wage rate among other factors. The quantity and type of seed applied depend on size of the farm, availability of seed, seed variety, and price per kg. Fertilizer is measured in kg/ha. Maize output is measured in kg/ha. Social-economic variables include age, gender and education. Education plays an important role in skill acquisition and technological transfer. Farmers with higher levels of education are likely to be more efficient in the use of inputs than their counterparts.

RESULTS AND DISCUSSION

Determinants of technical efficiency

Tables 1 to 4 show estimation results for determinants of technical efficiency for affected and non-affected households using time varying and time-invariant inefficiency models. The results show that fertilizer, labour, seeds and age are found to have contributed significantly to technical efficiency for affected households under both time varying and time-invariant models. These results are largely attributed to male headed households, as fertilizer and age are the only significant variables for affected female headed households (Appendix Tables 1 to 7). On the other hand, only fertilizer contributed significantly to technical efficiency of non-affected smallholder farmers under time varying model. This is mainly attributable to non-affected male headed households (Appendix Tables 1 to 7). These findings, especially for non-affected smallholder farmers, differ with Chirwa (2007) whose small sample of small-scale farmers in one of the districts in Southern Malawi find only labour as the statistically significant variable. They also differ with Obwona (2006) whose finding showed that education has a significant impact on technical efficiency. The estimated return to the scale for -0.04615 and 0.71238 affected households and non-affected households, respectively, imply that maize is produced at a decreasing and close to constant returns to scale, respectively on the same plots. This is lower compared to estimated returns to scale obtained in Chirwa (2007) who found estimated returns to scale at 0.97.

Technical efficiency of smallholder agriculture in Malawi

We present below results on technical efficiency levels of farm households in rural Malawi (Table 2 to 5). We

Table 2. All affected households - time invariant inefficiency model.

ly	Coef.	Std. Err.	z	P>|z|	[95% Conf. Interval]	
lh |	0.1007392	0.102612	0.98	0.326	-0.1003765	0.301855
lb |	0.1085327	0.0901766	1.20	0.229	-0.0682102	0.2852757
lf |	0.4336891	0.04524	9.59	0.000	0.3450203	0.5223579
ls |	0.1577066	0.0912915	1.73	0.084	-0.0212215	0.3366346
t |	0.0893292	0.0980233	0.91	0.362	-0.102793	0.2814514
sex |	0.0694879	0.1141588	0.61	0.543	-0.1542593	0.2932351
age |	0.0013903	0.0027628	0.50	0.615	-0.0040248	0.0068053
edu |	-0.0549349	0.0735401	-0.75	0.455	-0.1990709	0.0892011
_cons |	4.349331	0.4160277	10.45	0.000	3.533932	5.16473
/mu |	-15.52872	103.0175	-0.15	0.880	-217.4393	186.3818
/lnsigma2 |	1.854568	5.608098	0.33	0.741	-9.137102	12.84624
/ilgtgamma |	2.231164	6.199783	0.36	0.719	-9.920186	14.38251
sigma2 |	6.388936	35.82978			0.0001076	379358.7
gamma |	0.9030134	0.5429784		.	0000492	0.9999994
sigma_u2 |	5.769295	35.82332			-64.44312	75.98171
sigma_v2 |	0.6196414	0.0624205			0.4972995	0.7419834

Time-varying decay inefficiency model; number of obs, 410; group variable, id; number of groups, 263; Obs per group min, 1; average, 1.6; maximum, 2; Wald chi^2 (8), 186.90; Log likelihood, -516.93128; Prob > chi^2, 0.0000.

Table 3. Non-affected households – time varying inefficiency model.

ly |	Coef.	Std. Err.	z	P>|z|	[95% Conf. Interval]	
edu |	-0.0277003	0.2596591	-0.11	0.915	-0.5366228	0.4812221
lh |	0.0213395	0.2104826	0.10	0.919	-0.3911989	0.4338779
lb |	0.3961175	0.2981535	1.33	0.184	-0.1882526	0.9804875
lf |	0.1920051	0.0226711	1.57	0.118	-0.048426	0.4324361
ls |	0.1387257	0.1391391	1.00	0.319	-0.133982	0.4114334
t |	-0.0122531	0.3185195	-0.04	0.969	-0.6365399	0.6120338
sex |	0.1434649	0.2675285	0.54	0.592	-0.3808814	0.6678111
_cons |	4.933464	0.8611172	5.73	0.000	3.245705	6.621223
/mu |	-5.591935	119.6798	-0.05	0.963	-240.16	228.9762
/eta |	0.5060411	1.112504	0.45	0.649	-1.674427	2.686509
/lnsigma2 |	1.088286	12.71114	0.09	0.932	-23.82509	26.00166
/ilgtgamma |	0.9810913	17.25328	0.06	0.955	-32.83472	34.79691
sigma2 |	2.969179	37.74165			4.50e-11	1.96e+11
gamma |	0.7273247	3.421731			5.50e-15	1
sigma_u2 |	2.159557	37.60863			-71.55201	75.87112
sigma_v2 |	0.8096218	0.2462347			0.3270107	1.292233

Time-varying decay inefficiency model; number of obs, 120; group variable, id; number of groups, 86; time variable, t; Obs per group min, 1; average, 1.3; maximum, 2; Wald chi^2 (8), 12.13; Log likelihood, -109.91634; Prob > chi^2, 0.0963.

disaggregate the results by gender of the household head, and mortality and morbidity. Results show that technical efficiency levels for non-affected households under time varying and time-invariant models are at 73 and 78%, respectively. This is higher than efficiency levels for affected households, under time varying and time-invariant inefficiency models at 69 and 71%, respectively. These results are in line with findings from Adeoti and Adeoti (2008) and Yusuf et al. (2007). Male headed households are technically more efficient than female headed households for both affected and non-affected households. Similarly, households with morbidity are technically more efficient than households with mortality. In general, Malawian farmers are technically efficient and the mean technical efficiency levels of over 65% are relatively higher than the one obtained in Chirwa

Table 4. Non-affected households –time invariant inefficiency model.

ly	Coef.	Std. Err.	z	P>|z|	[95% Conf. Interval]	
lh |	0.1529719	0.174869	0.87	0.382	-0.1897649	0.4957088
lb |	0.2215137	0.1875177	1.18	0.237	-0.1460143	0.5890417
lf |	0.3705726	0.1187773	3.12	0.002	0.1377733	0.6033719
ls |	-0.0170625	0.1050737	-0.16	0.871	-0.2230032	0.1888781
edu |	0.0839356	0.1743479	0.48	0.630	-0.25778	0.4256513
t |	0.3222047	0.274804	1.17	0.241	-0.2164012	0.8608106
_cons |	3.981378	0.6149625	6.47	0.000	2.776074	5.186683
/mu |	-7.218936	462.2765	-0.02	0.988	-913.2643	898.8264
/lnsigma2 |	0.1975602	3.345199	0.06	0.953	-6.358909	6.75403
/ilgtgamma |	-3.03209	72.74732	-0.04	0.967	-145.6142	139.55
sigma2 |	1.218426	4.075879			0.0017313	857.5072
gamma |	0.045997	3.192246			5.76e-64	1
sigma_u2 |	0.056044	4.076862			-7.934459	8.046547
sigma_v2 |	1.162382	0.1502299			0.8679373	1.456828

Time-varying decay inefficiency model; number of obs, 120; group variable, id; number of groups, 86; time variable, t; Obs per group min, 1; average, 1.4; maximum, 2; Wald chi^2 (8), 36.98; Log likelihood, -179.30621; Prob > chi^2, 0.0000.

Table 5. Technical efficiency levels for affected and non-affected farm households 2004/05-2006/07.

Household	Time-varying model		Time invariant model	
	Affected	Non-affected	Affected	Non-affected
All households	0.693 (0.006)	0.731 (0.119)	0.7129 (0.082)	0.7824 (0.142)
Female headed	0.652 (0.020)	0.701 (0.012)	0.681589 (0.1982)	0.7155 (0.106)
Male headed	0.726 (0.21)	0.758 (0.022)	0.78003 (0.125)	0.8120 (0.0504)
Mortality	0.456 (0.044)	0.626 (0.048)	0.51358 (0.0639)	0.6574 (0.052)
Female headed	0.296 (0.072)	0.613 (0.057)	0.3015 (0.223)	0.620 (0.184)
Male headed	0.521 (0.056)	0.637 (0.073)	0.5562 (0.067)	0.673 (0.085)
Morbidity	0.725 (0.007)	0.733 (0.014)	0.754 (0.092)	0.7684 (0.0671)
Female headed	0.648 (0.025)	0.711 (0.012)	0.70285 (0.0127)	0.7348 (0.027)
Male headed	0.742 (0.006)	0.753 (0.022)	0.7673 (0.008)	0.775 (0.069)

Authors' estimation results from time varying and time invariant inefficiency models.

(2007) of 46% for a cross-section of Malawian farmers in one district in Southern Malawi. Nevertheless, the mean efficiency levels are comparable to those obtained for other African countries whose means range from 55 to 79% (Adeoti and Adeoti.,2009; Yusuf et al., 2007; Obwona, 2006; Seidu, 2008; Ogundele, 2006; Nchare, 2007). The lowest mean technical efficiency recorded is for affected female headed households with mortality at 29%. At this level, theses households are technically inefficient.

Conclusion

This study examines the levels of technical efficiency for affected and non-affected farm households in Malawi. We use time-varying and time invariant inefficiency models of production. The results reveal that government policy of subsidizing hybrid maize seeds and fertilizers since the 2006/06 agricultural season has enhanced technical efficiency of small-scale farmers. Nevertheless, there is still scope for improvement of the productivity of small-scale farmers, as some farm households, particularly female headed households, are still operating at low levels of efficiency.

Three main policy issues emerge from the results of this study. First, all types of obstacles that could limit the use of inputs should be removed. This should include complete liberalisation of purchase and distribution of such inputs and the development of some low-cost technology to reduce labour constraints on the farm. Second, there is need to develop social capital in smallholder farming through the recommencement of farmers' clubs or by setting up agricultural cooperatives. For a long time, from independence to 1992, smallholder agriculture was largely organized around farmers clubs for effective delivery of extension services and agricultural credit. The farmer club system collapsed in 1992 following the collapse of the agricultural credit scheme that worked through the club system and only a few farmers today belong to farmers club or association. Third, the results of this study questions the usefulness of the homogenous conceptualization of the "affected households" and "non-affected households" especially in the context of proposal for targeted assistance and other policies. As has been seen, within the non- affected farm households, there are some vulnerable sections of households that require assistance as much as the AIDS-affected households.

ABBREVIATIONS

TIPs, Targeted input programs; **HIV**, human immunodeficiency virus; **AIDS**, acquired immune deficiency syndrome; **N**, nitrogen; **P,** phosphorus; **K,** potassium; **MoAFS**, Ministry of Agriculture and Food Security; PSU, primary sampling units; **EAs**, enumeration areas; PPS, probability proportional to size.

REFERENCES

Adeoti AI, Adeoti JO (2008). HIV/AIDS and Farm Production Efficiency in Benue State, Nigeria, Afr. J. Biomed. Res. 11:154-153.

Arrehag L, Durevall D, Sjoblom M, De Vylder S (2006). The Impact of HIV/AIDS on the Economy, Livelihoods and Poverty of Malawi," Country Economic Report, Sida, Department for Policy and Methodology.

Battese GE, Coelli TJ (1995). Prediction of farm level technical efficiencies with a generalized frontier production function and panel data. J. Econ. 38:387.

Battese GE, Malik MSJ, Gill MA (1996). An investigation of technical inefficiencies of production of wheat farmers in four districts of Pakistan," J. Agric. Econ. 47(1):37-49.

Chapoto TS, Jayne T (2005).Measuring Impact of Working-Age Adult Mortality on Small-scale Farm Households in Kenya," World Development, 32(1):91-111.

Chirwa EW (2003)."Sources of technical efficiency among smallholder maize farmers in Southern Malawi," Department of Economics, Chancellor College, Zomba, Malawi, Unpublished working paper.

Chirwa EW (2007). Sources of Technical Efficiency among Smallholder Maize Farmers in Southern Malawi. *AERC Research* African Economic Research Consortium, Nairobi. P. 172.Country Economic Report, Sida, Department for Policy and Methodology.

Ministry of Agriculture and Food Security (MoAF). (2008). Evaluation of the 2006/07 Agricultural Input Subsidy Programme. Final Report

National Statistical Office (NSO) and World Bank (2008), "*Integrated Household Survey, 2004/05*," Final Report, Lilongwe, Malawi.

National Statistical Office (NSO) (2005). Integrated Household Survey, 2004/05. Final Report, October, 2005.

Ngom P, Clark S (2003). Adult mortality in the era of HIV/AIDS: Sub-Saharan Africa. ttp://www.un.org/esa/population/publications/adultmort/CLARK_Paper3.pdf.

Obwona M (2006). Determinants of technical efficiency differentials amongst small and medium-scale farmers in Uganda: A case of tobacco growers. AERC Research Paper 152, African Economic Research Consortium, Nairobi, Kenya.

Ogundele O (2006). Technical efficiency differentials in rice production technologies in Nigeria. AERC Research Paper 154, African Economic Research Consortium, Nairobi, Kenya.

UNAIDS. (1998). Report on global AIDS epidemic. Joint United Nations Programme on HIV/AIDS.

World Bank, 2007, "*World Development Report*," World Bank, Washington, DC.

World Health Organization (WHO) (1998). Report on AIDS epidemic.

APPENDIX

Table 1. HIV/AIDS affected female headed households – time varying model.

ly	Coef.	Std. Err.	z	P>|z|	[95% Conf. Interval]	
lh	-0.1464857	.2081148	-0.70	0.482	-0.5543833	0.2614119
lb	-0.0239417	.1408583	-0.17	0.865	-0.3000189	0.2521355
lf	0.291392	.082366	3.54	0.000	0.1299575	0.4528265
ls	0.4253015	.1882309	2.26	0.024	0.0563757	0.7942274
edu	0.4591335	.1752686	2.62	0.009	0.1156133	0.8026537
age	0.0104994	.0048185	2.18	0.029	0.0010554	0.0199435
t	0.5872303	.2333192	2.52	0.012	0.1299331	1.044527
_cons	2.910739	.6584723	4.42	0.000	1.620157	4.201321
/mu |	-13.59856	99.44494	-0.14	0.891	-208.5071	181.3099
/eta |	-0.7416717	.5430449	-1.37	0.172	-1.80602	0.3226768
/lnsigma	2 2.333394	6.278556	0.37	0.710	-9.97235	14.63914
/ilgtgamma	3.396715	6.429311	0.53	0.597	-9.204504	15.99793
sigma2	10.31288	64.74999			0.0000467	2278749
gamma	0.9676017	.2015502			0.0001006	0.9999999
sigma_u2	9.978759	64.72925			-116.8882	136.8457
sigma_v2	0.3341197	.083358			0.1707411	0.4974984

Time-varying decay inefficiency model; number of obs, 84; group variable, id; number of groups, 56; time variable, t; Obs per group min, 1; average, 1.5; maximum, 2; Wald chi2 (8), 69.17; Log likelihood, -95.434398; Prob > chi2, 0.0000.

Table 2. HIV/AIDS affected female headed households – time invariant model.

ly	Coef.	Std. Err.	z	P>|z|	[95% Conf. Interval]	
lh	-0.2125451	0.2221401	-0.96	0.339	-0.6479317	0.2228415
lb	-0.0014876	0.1479815	-0.01	0.992	-0.2915259	0.2885507
lf	0.3612391	0.0828154	4.36	0.000	0.1989238	0.5235543
ls	0.4600556	0.2041268	2.25	0.024	0.0599745	0.8601368
t	0.3615004	0.1868881	1.93	0.053	-.0047934	0.7277943
_cons	3.65666	0.6903172	5.30	0.000	2.303663	5.009657
/mu |	-683.0762	6574.975	-0.10	0.917	-13569.79	12203.64
/lnsigma2 |	5.953372	9.596864	0.62	0.535	-12.85614	24.76288
/ilgtgamma |	6.814653	9.611913	0.71	0.478	-12.02435	25.65366
sigma2 |	385.0496	3695.269			2.61e-06	5.68e+10
gamma |	0.9989036	0.0105267			6.00e-06	1
sigma_u2 |	384.6274	3695.269			-6857.967	7627.222
sigma_v2 |	0.422157	0.0981351			0.2298158	0.6144982

Time-varying decay inefficiency model; number of obs, 84; group variable, id; number of groups, 56; time variable, t; Obs per group min, 1; average, 1.5; maximum, 2; Wald chi2 (8), 47.93; Log likelihood, -101.15863; Prob > chi2, 0.0000.

Table 3. HIV/AIDS affected male headed households – time varying model.

ly |	Coef.	Std. Err.	z	P>|z|	[95% Conf. Interval]	
lh |	0.160703	0.1158756	1.39	0.165	-0.0664089	0.3878149
lb |	0.1718361	0.1086481	1.58	0.114	-0.0411102	0.3847824
lf |	0.4550206	0.0516567	8.81	0.000	0.3537754	0.5562659
ls |	0.1077177	0.1004148	1.07	0.283	-0.0890916	0.304527
age |	-0.0019778	0.0033057	-0.60	0.550	-0.0084569	0.0045014
edu |	-0.1470649	0.0800287	-1.84	0.066	-0.3039183	0.0097884

Table 3. Contd.

t |	0.1629195	0.1686683	0.97	0.334	-0.1676644	0.4935034
_cons |	4.630223	0.4604201	10.06	0.000	3.727816	5.53263
/mu |	-3.727407	19.03826	-0.20	0.845	-41.04172	33.5869
/eta |	-0.400697	0.4228422	-0.95	0.343	-1.229452	0.4280585
/lnsigma2 |	0.9219152	2.965069	0.31	0.756	-4.889513	6.733343
/ilgtgamma |	1.139702	3.884281	0.29	0.769	-6.47335	8.752753
sigma2 |	2.514101	7.454481			0.0075251	839.9505
gamma |	0.7576249	0.7132683			0.0015417	0.999842
sigma_u2	1.904745	7.439822			-12.67704	16.48653
sigma_v2 |	0.6093555	0.0738493			0.4646136	0.7540975

Time-varying decay inefficiency model; number of obs, 326; group variable, id; number of groups, 207; time variable, t; Obs per group min, 1; average, 1.6; maximum, 2; Wald chi2 (8), 154.38; Log likelihood, -410.18861; Prob > chi2, 0.0000.

Table 4. HIV/AIDS affected male headed households – time invariant model.

ly |	Coef.	Std. Err.	z	P>|z|	[95% Conf.	Interval]
lh |	0.155736	0.1149688	1.35	0.176	-0.0695986	0.3810707
lb |	0.1625474	0.1086214	1.50	0.135	-0.0503465	0.3754413
lf |	0.4560497	0.0523809	8.71	0.000	0.3533851	0.5587143
ls |	0.1103033	0.0996142	1.11	0.268	-0.0849369	0.3055436
age |	-0.0017558	0.0033166	-0.53	0.597	-0.0082562	0.0047447
edu |	-0.1401075	0.0793258	-1.77	0.077	-0.2955831	0.0153682
t |	0.0441724	0.1117666	0.40	0.693	-0.1748861	0.2632309
_cons |	4.731005	0.432139	10.95	0.000	3.884028	5.577982
/mu |	-50.21051	483.6436	-0.10	0.917	-998.1345	897.7135
/lnsigma2 |	2.798489	9.083801	0.31	0.758	-15.00543	20.60241
/ilgtgamma |	3.207509	9.448778	0.34	0.734	-15.31176	21.72677
sigma2 |	16.41982	149.1544			3.04e-07	8.86e+08
gamma |	0.9611159	0.353121			2.24e-07	1
sigma_u2 |	15.78135	149.1524			-276.552	308.1147
sigma_v2 |	0.6384701	0.0686842			0.5038515	0.7730886

Time-varying decay inefficiency model; number of obs, 326; group variable, id; number of groups, 207; time variable, t; Obs per group min, 1; average, 1.6; maximum, 2; Wald chi^2 (8), 150.66; Log likelihood, -410.62175; Prob > chi^2, 0.0000.

Table 5. Non-affected households female headed households – time invariant model.

ly |	Coef.	Std. Err.	z	P>|z|	[95% Conf.	Interval]
lh |	0.1516798	0.1748581	0.87	0.386	-0.1910358	0.4943955
lb |	0.232685	0.1886791	1.23	0.217	-0.1371192	0.6024892
lf |	0.3846814	0.1141878	3.37	0.001	0.1608775	0.6084853
ls |	-0.0113513	0.1048879	-0.11	0.914	-0.2169278	0.1942253
age |	0.002679	0.005139	0.52	0.602	-0.0073933	0.0127513
t |	0.3098368	0.2731376	1.13	0.257	-0.2255031	0.8451766
_cons |	3.823813	0.7814863	4.89	0.000	2.292128	5.355498
/mu |	-5.045455	757.306	-0.01	0.995	-1489.338	1479.247
/lnsigma2 |	0.1706477	4.754444	0.04	0.971	-9.147892	9.489187
/ilgtgamma |	-3.881292	235.398	-0.02	0.987	-465.253	457.4904
sigma2 |	1.186073	5.639117	0.0001064			13216.05
gamma |	0.0202074	4.660658	8.8e-203			1
sigma_u2 |	0.0239674	5.641791	-11.03374			11.08167
sigma_v2 |	1.162105	0.1501363	0.8678437			1.456367

Time-varying decay inefficiency model; number of obs, 120; group variable, id; number of groups, 86; time variable, t; Obs per group min, 1; average, 1.4; maximum, 2; Wald chi^2 (8), 37.03; Log likelihood, -179.28624; Prob > chi^2, 0.0000.

Table 6. Non-affected households female headed households –time invariant model.

ly |	Coef.	Std. Err.	z	P>|z|	[95% Conf. Interval]	
lh |	0.1529719	0.174869	0.87	0.382	-0.1897649	0.4957088
lb |	0.2215137	0.1875177	1.18	0.237	-0.1460143	0.5890417
lf |	0.3705726	0.1187773	3.12	0.002	0.1377733	0.6033719
ls |	-0.0170625	0.1050737	-0.16	0.871	-0.2230032	0.1888781
edu |	0.0839356	0.1743479	0.48	0.630	-0.25778	0.4256513
t |	0.3222047	0.274804	1.17	0.241	-0.2164012	0.8608106
_cons |	3.981378	0.6149625	6.47	0.000	2.776074	5.186683
/mu |	-7.218936	462.2765	-0.02	0.988	-913.2643	898.8264
/lnsigma2 |	0.1975602	3.345199	0.06	0.953	-6.358909	6.75403
/ilgtgamma |	-3.03209	72.74732	-0.04	0.967	-145.6142	139.55
sigma2 |	1.218426	4.075879			0.0017313	857.5072
gamma |	0.045997	3.192246			5.76e-64	1
sigma_u2 |	0.056044	4.076862			-7.934459	8.046547
sigma_v2 |	1.162382	1502299			0.8679373	1.456828

Time-varying decay inefficiency model; number of obs, 120; group variable, id; number of groups, 86; time variable, t; Obs per group min, 1; average, 1.4; maximum, 2; Wald chi^2 (8), 36.98; Log likelihood, -179.30621; Prob > chi^2, 0.0000.

Table 7. Non-affected male headed households –time varying model.

ly |	Coef.	Std. Err.	z	P>|z|	[95% Conf. Interval]	
lh |	0.0728867	0.264469	0.28	0.783	-0.445463	0.5912364
lb |	0.5799829	0.2622139	2.21	0.027	0.066053	1.093913
lf |	0.0834944	0.0321394	0.63	0.527	-0.1754941	0.3424829
ls |	0.0328632	0.2235375	0.15	0.883	-0.4052622	0.4709887
age |	-0.0123291	0.0074576	-1.65	0.098	-0.0269458	0.0022876
t |	0.4758284	0.3009817	1.58	0.114	-0.1140849	1.065742
_cons |	5.142039	0.9183292	5.60	0.000	3.342146	6.941931
/mu |	0.1460862	.	.	.	.	.
/eta |	-9.300796	396.1137	-0.02	0.981	-785.6693	767.0677
/lnsigma2 |	0.1831447	0.1467638	1.25	0.212	-0.104507	0.4707963
/ilgtgamma |	-50.22023	0.0009449	-5.3e+04	0.000	-50.22208	-50.21838
sigma2 |	1.200988	0.1762615			0.9007685	1.601269
gamma |	1.55e-22	1.46e-25			1.54e-22	1.55e-22
sigma_u2 |	1.86e-22	2.73e-23			1.32e-22	2.39e-22
sigma_v2 |	1.200988	0.1762615			0.8555219	1.546454

Time-varying decay inefficiency model; number of obs, 89; group variable, id; number of groups, 64; time variable, t; Obs per group min, 1; average, 1.4; maximum, 2; Wald chi^2 (8), 36.98; Log likelihood, 0; Prob > chi^2, 0.0504.

Pedologists' approach to agro-technology transfer: Case study of Ikwuano, Abia State, Nigeria

Godwin O. Chukwu*, Tessy U. Madu, Emilia C. Chinaka and Nkoli H. Anyaegbunam

National Root Crops Research Institute, Umudike, P.M.B. 7006, Umuahia, Abia State, Nigeria.

Soil resource illiteracy and a weak link between agricultural research and farmers are major problems limiting agricultural development in Nigeria. Pedologists provide interpretative soil maps (land capability classification and soil fertility maps) that will enable potential land users to appreciate the advantages and limitations of the soil resource base that is utilized for agricultural production. Such maps are visual aids to facilitate effective communication between extensionists and farmers, in an effort to (1) adopt appropriate land and agronomic management practices, (2) increase productivity and (3) enhance the efficiency of resource use in agricultural production. This approach was illustrated using Ikwuano Local Government Area (LGA) of Abia State, Nigeria, as a case study.

Key words: Pedologists, interpretative soil maps, agro-technology transfer, Ikwuano LGA, Nigeria.

INTRODUCTION

Nigerian agriculture, and indeed African agriculture, has failed to meet the continuously changing needs of the citizenry. Persistent food insecurity and failure of agriculture to supply adequate quantities of raw materials to industries are stark realities. This is attributed to many factors, among which is soil resource illiteracy. OFAR (1984, 1985) identified poor knowledge of soil as a major problem hindering agricultural development in Umuahia Agricultural Zone. This is not a surprise as similar observations have been made in most parts of Nigeria and Africa. Some of the reasons for this situation are related to lack of soil survey reports of most rural communities and Local Government Areas (LGAs) where food and fiber production take place. Also, the scales of most national soil surveys are so coarse (at reconnaissance level) that pedological information about rural communities is virtually non-existent. Similarly, in these communities, soils where most agricultural research and production take place are not characterized. All these reasons and perhaps more contribute to the widespread problems of soil resource illiteracy. This in turn hinders effective agro-technology transfer between areas with similar and/or different soils. There is a clear need for linking soil (pedological) science with extension education.

The term pedology, in its restricted sense, is the science of identification, formation and distribution of soils, which covers soil classification, soil genesis and survey (Dudal, 1987). Pedology has the fundamental task of identifying and characterizing the soil body, determining its distribution, establishing relationships, facilitating international communication and guiding the development of new technologies and their applications.

Agricultural development is hindered by a weak link between agricultural research and farmers (Osborn, 1995) and the extension sector is expected to strengthen this linkage. Extension, though an informal education, has been described by Christoplos (1996), as communication. Visual aids are communication channels and vital tools to facilitate the transfer of ideas from a source to a receiver. Interpretive soil maps could be excellent visual aids for extension education. This is because they could show the productive capacity of the land, its suitability to specific land uses (crop options, fishery, irrigation, etc), and guide resource use (such as quantity of fertilizers to apply) and land management (such as erosion control). These advantages, probably

*Corresponding author. E-mail: goddychukwu2003@yahoo.com.

prompted Fagbemi (1987), to suggest the production of soil fertility capability maps of LGAs as a new dimension towards efficient use of land and fertilizers in Nigeria. In response to this call, FMANR (1990) produced soil fertility maps of North East, North West, South East, South West and Middle Belt agro-ecological zones of Nigeria. Sobulo and Adepetu (1987) had produced similar maps for South West agro-eco zone.

What are interpretive soil maps?

The use of interpretative soil maps at rural level in extension education should be seen as a novel approach to achieving the food security initiative of Nigerian Government. It is also a road map to achieve the Millennium Development Goals on hunger and poverty, as well as reverse natural resources depletion through agriculture. Interpretive soil maps are simplified cartographic representations of the soil or part of it in a specific area or location. Such maps are non-existent in Ikwuano LGA of Abia State, Nigeria. Soil fertility classification maps are attempts to interpret soil test for fertilizer predictions based on critical soil levels of N, P and K as determined from field trials. A good interpretive soil map should be able to address all or some of these problems:

1. It should have a purpose – what does the map show?
2. It should be able to provide basic information about an area in a simplified manner.
3. It should be able to assist potential land users to solve soil-related problems in an area with less difficulty.
4. It should guide input procurement and site selection, in relation to a particular land use, or an agricultural project. Examples include choice of crops, fertilizer type and quantity to apply, erosion control, etc.
5. It should provide information on the nature and distribution of a particular soil problem or sets of problems.
6. It should enhance the predictability of soil attributes in an area covered by the map by a potential user. It should, therefore, facilitate dissemination and adoption of agricultural innovations within a recommendation domain.

The objective of this study was to improve soil resource literacy by providing interpretive soil maps as visual aids for extension education. This is illustrated, using Ikwuano LGA in Abia State, Nigeria, as a case study.

MATERIALS AND METHODS

The study area

Ikwuano LGA is a major food-producing area in Abia State. It is located between 5°24' to 5°30' N latitudes and 7°31'N to 7°37' E longitudes covering an area of 31,000 ha. The area represents the typical degraded humid forest ecology in the southeast agro-ecological zone of Nigeria. The area is characterized by bi-modal annual precipitation totaling over 2000 mm, air temperature ranging from 22 to 31°C, and high relative humidity (77%) during the wet season. The soils are characterized by differences in topography (toposequence) and parent materials (lithosequence) (Chukwu and Ifenkwe, 1996) due to uplands and inland valleys within the rural landscape. As a result, there is an enhanced potential for a diversity of land use options, spanning from arable crop and tree crop production to forestry, animal husbandry and fishery.

Production of interpretive soil maps

As a guide to the production of interpretive soil maps, available literatures on technical soil classification of the area were reviewed. They included studies on soil-plant relations (edaphological) approach to biodiversity conservation rice-fish culture (rizipisciculture) potential of inland valleys, land capability for agricultural development, land resources problems in rural development and agricultural land use (Chukwu and Chinaka, 2001). Results of these reviews were synthesized, simplified and delineated on maps (Figures 1 and 2) that are less technical for extension agents to use as teaching aids (visuals) and literate farmers to understand and use.

Interpretation of soil tests for fertilizer predictions for cassava and yam, which are crops commonly grown in the area, based on critical levels of N, P and K (SPFS, 2004) involved soil fertility classification mapping. The soil fertility classes were rated low, medium and high, as in Sobulo and Adepetu (1987) with their associated fertilizer application recommendations being: (1) full fertilizer dose for low fertility (500 to 600 kg/ha N P K fertilizer 15:15:15); (2) 1/3 to 1/2 dose for medium fertility and (3) no fertilizer for high fertility.

RESULTS AND DISCUSSION

Land capability classification map

The land capability map is a cartographic interpretation of soil survey reports in the area, indicating in practical terms, the potentialities and limitations of the land for alternative uses.

Two land capability classes (Figure 1) represented by Roman numerals (II and VII) and three capability sub-classes or groups of capability units (IIw, IIen and VIIes) were delineated. A capability unit combines the capability class and the limitations (w, e, n and s) inherent in that class. These limitations or problems found in each capability unit are represented by subscripts defined as:

w = wetness or poor drainage; e = soil erosion; n = low nutrient retention; s = slope

Capability unit IIw represents arable land occupying 15.0% of the area (4,650 ha). The physiographic surface is nearly level to concave plain. The area is characterized by medium to fine loamy surfaces (sandy clay loam to clay loam textural classes at the epipedon (topsoil) underlain by clay endopedon (subsoil). The soils are

Figure 1. Land capability classification of Ikwuano. IIw represents nearly leveled poorly drained arable soils (4,650ha), VIIes represents well drained coarse texture non-arable soils prone to gully erosion (3,100 ha), and IIen represents well-drained loamy sand soils with low nutrient retention capacity in undulating to concave and/or nearly level arable land (23,250 ha).

somewhat poorly drained because of high clay contents. Consequently, access could be very difficult at the peak of rainy season. Most of the cocoa producing communities: Umuokwo, Amuro, Iberenta, elemaga, Isiala Ibere and part of Umuariaga, belong to this unit (Chukwu and Chinaka, 2001). Some of the areas like Oburo, Amuro and Nkalanta could be flooded at the peak of the rainy season. It also has a high potential for rice, fishery, rice-fish culture (rizipisculture) and sugarcane (Chukwu and Ifenkwe, 1996).

Capability unit IIen is the most extensive arable land in Ikwuano (23,250 ha), equivalent to 75.0% of the area. The unit represents well-drained coarse loamy to sandy soils. The topography encompasses gently undulating to concave or nearly level plains. It is prone to sheet erosion and has low nutrient retention capacity for sustainable crop production. The sandiness increases as one traverse from the proximal end (Umudike) to the distal end (Ariam). Aggradations of colluvial sand from upper elevation are common. Among the communities in the capability unit IIen are Amaba, Amawom, Ndoro, Aro Iyama, Isiala, Ekebedi, Awomukwu, Oloko, Umuigwu, and Ogbuebule.

Capability unit VIIes comprises well-drained sandy to coarse loamy soils occupying 10.0% of the area or 3100 ha. Generally, the area is a non-arable land because of morpho-erosion (gulling and land sliding) (Chukwu and Ifenkwe, 1996). This unit is a strip of land meandering from Umuariaga community through Ihim, Iyila, Ngwugwo, Obohia Ibere to Ogbuebula. At the Ogbuebula end, clay is extracted at the foot of the gully for industrial purposes. The physiography is a highly undulating plain with wide, deep and frightening gullies that are susceptible to land sliding. However, a minor strip of land

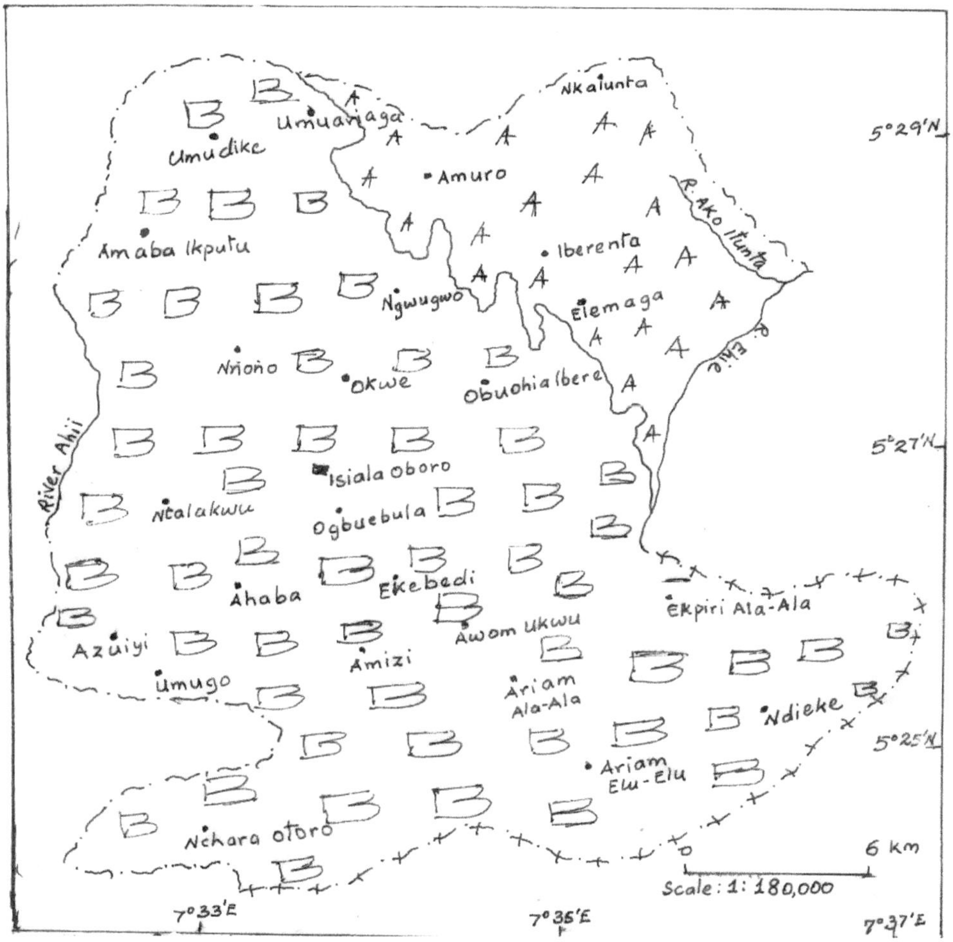

Figure 2. Soil fertility class map of Ikwuano. A represents medium fertility soils that are low-medium in N, P and K (4,650ha); B represents low-fertility soils that are low in N and K and medium-high in P (26,350ha).

at the valley bottom, endowed with perennial streams has potential for crop production. Appropriate land management for this unit is forestry and slopping agricultural land technology (Chukwu and Chinaka, 2001)) to stabilize the soil and improve it's potentialities for arable crop production.

Soil fertility mapping

Figure 2 shows that two soil fertility classes (A and B) could be delineated in the area. Mapping unit A represents soils of medium fertility, where N, P and K could range from low to medium. Total N varies from 0.08 to 0.16%. Available P (Bray 2P) ranges from <5 to 20 mg/kg, while exchangeable K ranges from 0.06 to 0.30 cmol/kg.

The soil fertility mapping unit B represents soils that suffer from multi-nutrient deficiencies (Chukwu and Asawalam, 2001) of two primary nutrients (N and K). Total N, available P and exchangeable K in this unit could vary from 0.05 to 0.12%, 10 to 30 mg/kg and 0.04 to 0.11 cmol/kg.

Fertility map unit A for cassava and yam production based on the recommendations of SPFS (2004) and (Sobulo and Adepetu, 1987), is medium. Application of 200 to 300 kg/ha (4 to 6 bags) of NPK fertilizer 15:15:15 is economic fertilizer rate to maximize cassava and yam production on Unit A. However, unit B has low fertility and requires about 500 to 600 kg/ha (10 to 12 bags) of NPK fertilizer 15:15:15 to sustain cassava and yam production.

Conclusion

A proposed pragmatic approach to effective agro-technology transfer in Nigeria, resulted in delineating Ikwuano LGA, Abia State, into three land capability classification maps and two soil fertility maps. These maps will enhance understanding of the physical environment on which agricultural land utilization type or types will be sited and guide judicious management of resources (labour, fertilizer input, soil, water etc.) within a recommendation domain, to enhance and sustain higher agricultural productivity.

REFERENCES

Chukwu GO, Chinaka C (2001). Agricultural land use in Ikwuano LGA, Abia State: Implications for agro-technology transfer. Proceedings Agricultural Society of Nigeria Conference, Abubakar Tafawa Balewa University, Bauchi, pp. 192 -195.

Christoplos I (1996). Poverty, pluralism, and extension practice. Gatekeeper Series, IIED, UK 64:20.

Dudal R (1987). The role of pedology in meeting increasing demands on soils. Soil Survey Land Evaluation 7:101-110.

Fagbemi A (1987). Land use in agro-industrial development in Nigeria. Proceedings National Fertilizer Seminar, Port-Hacourt. pp. 29-46.

FMANR (Federal Ministry of Agriculture and Natural Resources) (1990). Literature review on soil fertility investigations in Nigeria.5 2:92-95; 5:252-261.

OFAR (On –farm Adaptive Research) (1985). Annual Report in Imo State Unamma RPA, Odurukwe SO, Ene LSO, Oputa CO (eds.). FACU, Ibadan and NRCRI Umudike. p. 20.

OFAR (1984). Annual Report in Umuahia Agricultural Project Zone, Imo State. Unamma RPA, Odurukwe SO, Okoli PSO and Oputa CO (eds.). NRCRI Umudike, FUT., Owerri and FACU, Ibadan. p. 23.

Osborn T (1995). Participatory agricultural extension: Experiences from West African. Gatekeeper Series IIED, UK 48:19.

Sobulo RA, Adepetu JA (1987). Soil testing and fertilizer formulation crop production in Nigeria. Proceedings National Fertilizer Seminar, Port Hacourt. pp. 93-105.

SPFS (Special Programme for Food Security) (2004). Handbook on soil test-based fertilizer recommendations for Extension Workers. National Special Programme for Food Security, Abuja. p. 36.

5

The study of implementation and evaluation of an alternative agricultural extension presentation system

Yuh-Tyng Chen[1] and Lin-Fan Chen[2]

[1]Department of Information Management, Tainan University of Technology, Yung-Kang District, Tainan 710, Taiwan.
[2]Department of Applied English, Tainan University of Technology, Yung-Kang District, Tainan 710, Taiwan.

In Taiwan, to deal with the global competitive pressure, agricultural extension improvement will play an important role. The agricultural extension workers not only need to have professional agricultural knowledge but also have good presentation skills for extension activities, especially for training courses. However, a good presentation for extension activities depended on a well-designed material. This study intends to construct an alternative presentation tool – an interactive agricultural extension presentation tool. It further analyzes effects of training design on agriculture extension workers' cognitive attitude of learning interactive agricultural extension presentation system. In doing this, the participants of agricultural extension workers training camp were surveyed. Results of this study reveal that training course with interactive agricultural extension presentation system do have effects on agricultural extension worker's perception of effectiveness. The interactive agricultural extension presentation system could successfully attract their interests and engagement, best fit their needs and preferences, and help to understand the contents thereby they felt confident and satisfactory with this training activity.

Key words: Agricultural extension, interactive agricultural extension presentation system, cognitive attitude.

INTRODUCTION

In Taiwan, to deal with the seriousness of the agricultural labor shortage problem, the global proneness of free trade and international competitiveness, the traditional model of agriculture business has to be transformed and keeping the young farmers to stay on the farm also needs to be encouraged. Agricultural extension improvement will play an important role at this moment. The major function of extension workers who are employed by Farmers' Associations (FA) is to help farmers to improve their living standards by using the best available techniques (Allo, 1983). The local-level FA is the major agricultural extension system. Its main function is to introduce modern farm management, provide technical training, and offer guidance on the establishment of an entrepreneurial production and distribution throughout local farmers. The agricultural extension worker is often under tremendous pressure but still be depicted as a sort of superman. They not only need to have professional agricultural knowledge but also have good presentation skills for extension activities, especially for training courses which provided the knowledge of new technologies and techniques. However, a good presentation for extension activities depended on a well-designed material (Chen et al., 2011).

There has been little concern about the presentation of agricultural extension courses in the past, but the presentation of extension will affect the comprehension of the courses, changing trainees' attitudes, and altering behaviors (Mulder et al., 2006). Hence the presentation of agricultural extension must be considered. In general, integrating information technologies into the presentation activities is a popular tendency; especially the multimedia-based instruction is being widely applied in the learning environment to improve learning (Liao and Ho, 2010; Mayer, 2001). The multimedia representation includes text, audio, graphs, photographs, animation, or video. PowerPoint, a user-friendly package and popular presentation tool, can be used for the creation of visually clear, dynamic and attention capturing presentations (Holzl, 1997). Highly flexible, audience engagement, great visual tool and simple to use are the benefits of using PowerPoint. Images, audio clips, text and sound effects are used to enrich reports and studies in the presentations and slide shows. Its adoption conducts the speaker to a well-organized path on which the most important points are emphasized (Harknett and Cobane, 1997). Besides, it is easy for a user to combine multimedia with a presentation and even for a novice user to create colorful and easy-to-read slides. Without question, PowerPoint has some very attractive features as mentioned above, but the real question is whether the use of PowerPoint enables teaching/learning effectiveness. the traditional linear PowerPoint presentation is known as a "slide show" (Matheson et al., 2002) which includes a series of screens presented one after another just as slides in an old-fashioned slide projector. It is known that traditional way of teaching discourages active learning, and the slide show presentation simply enlarges the passive nature of the instruction. Hence, the specific objectives of this experiment were to apply the multimedia technologies and modularity concepts into PowerPoint on demand for the training activities of agricultural extension workers and evaluate their learning effects.

MATERIALS AND METHODS

Sample and research question

In order to inspire an agricultural extension worker to get a better understanding of agricultural affairs and learn the knowledge of new technologies and techniques, the Taiwan Council of Agriculture (COA) has implemented serious extension courses from the "Project of Motivation Farmer's Lifelong Learning". The Council of Agriculture is the competent authority on the agricultural, forestry, fishery, animal husbandry and food affairs in Taiwan. Its responsibilities include guiding and supervising provincial and municipal offices in these areas. Due to the limitation of government budget facility, the participants in our study consisted of 38 people recruited from an agricultural extension training camp in southern Taiwan. The unusable surveys which were either incomplete questionnaire or not followed instructions were identified and discarded. As a result, 32 respondents (84% of 38 cases) were used as the basis for data analysis. Of these participants, 62% were males, and 38% were females, 12% were under 30 years old, 45% were 30-50 years old, and 43% were above 50 years old. With regard to the respondents' agricultural backgrounds, more than half had majored in agriculture, and more than half had senior high school diplomas.

Instrument

The presentation tool - interactive PowerPoint

Due to innovations in information technology and telecommunications, the use of multimedia technology is becoming more widespread throughout various sectors including education, business as well as agriculture. This study uses the PowerPoint as the experimental medium because PowerPoint is a widely accepted, easily compiled and maintained presenting medium. Furthermore, previous studies have constantly suggested that learners generally believed that the use of PowerPoint can effectively help their learning (Apperson et al., 2006; Rankin and Hoaas, 2001). The description of the interactive PowerPoint is as follows and is shown in Figures 1 to 3.

(1) Incorporate meaningful visual content and related verbal information together. According to the dual coding theory, brain encodes visual and verbal information simultaneously but differently, in separate areas (Lane and Wright, 2009). The brain clearly handles visual content differently than it does textual information. Text, a coding system, has meaning only in a symbolic sense and viewers must expand a great deal of cognitive resources decoding words and phrases on slides. Under this kind of situation, they have little capacity left to pay attention to the speaker or they pay attention to the speaker and ignore text-heavy slides altogether. Both situations are unfavorable ideal. In contrast, visual processing can occur simultaneously and efficiently along with verbal processing because different brain regions are involved. Thus, pictures and graphics are powerful communication tools if used correctly. Further, considering the knowledge divide of agriculture and educational divide, we used the picture-based visual communication to enable the learners' comprehension of learning contents (Levie and Lentz, 1982).
(2) Separate individual ideas onto their own slides. Based on the cognitive theory of multimedia learning, separating individual topics onto their own slides can resolve the richly visual with performances (Lane and Wright, 2009).
(3) Classify the slides with the concepts of modularity. For preventing the passive learning attitude from the traditional linear presentation, we classified the slides with the concepts of modularity to build the hierarchically organized structures called presentation networks; each module is an entirely independent slide show.
(4) Hyperlink all slides together. We borrowed the navigation technology in the presentation networks to provide presenters/learners with the ability to rapidly find and display whatever content they need, whenever they need it.
(5) Highlight using animation. For keeping audiences' attention, a more engaging approach is to use animation to have each element appear at a time and at the right moment on the slide.

Learning perception survey

Learning motivation is an important factor for inspiring and encouraging the participation of learners during the learning process. This study adopted the ARCS motivation model proposed by Keller (1983). The four constructs in this model include attention, relevance, confidence and satisfaction, they describe the perception procedure: while keeping the learners' attention is critical, instructors will provide an interactive and participative environment to gain and maintain learners' attention. Learners will feel relevant that the course content, activities, and assignments must be related to their personal and professional goals. Learners

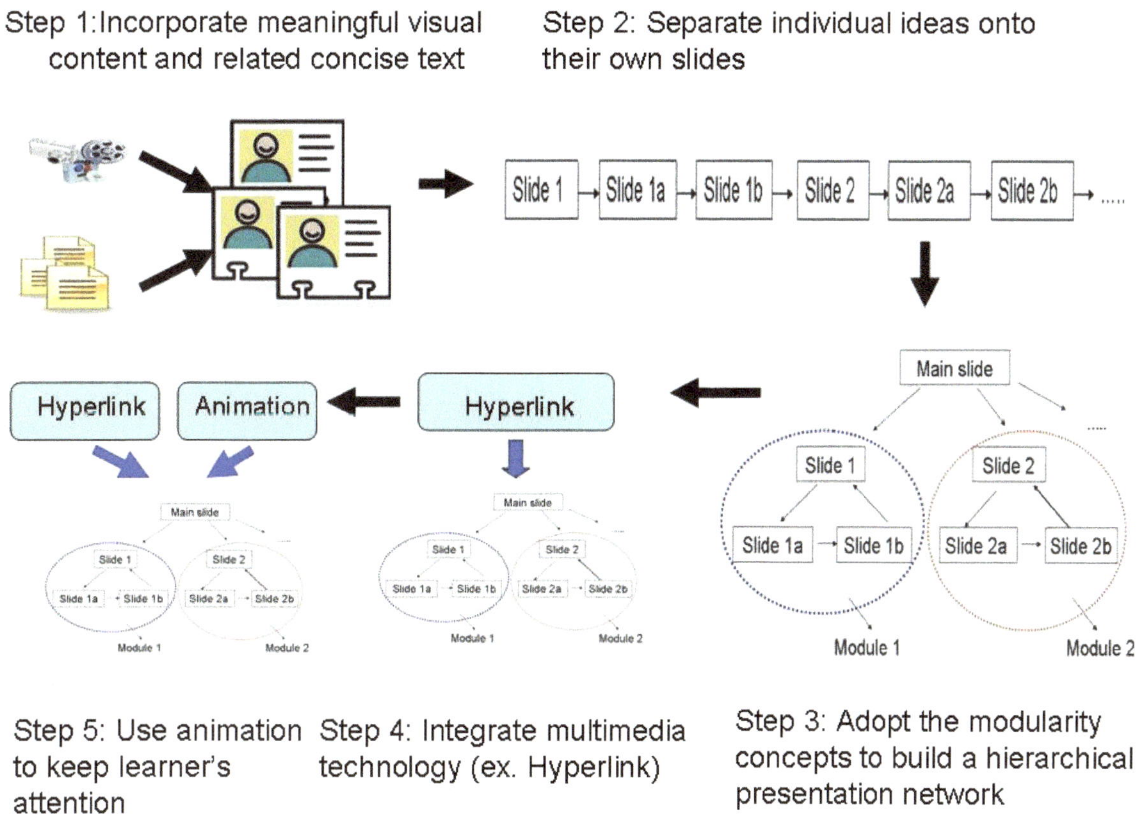

Figure 1. Five steps of implementing interactive agricultural extension materials.

Figure 2. The title page of interactive agricultural extension materials.

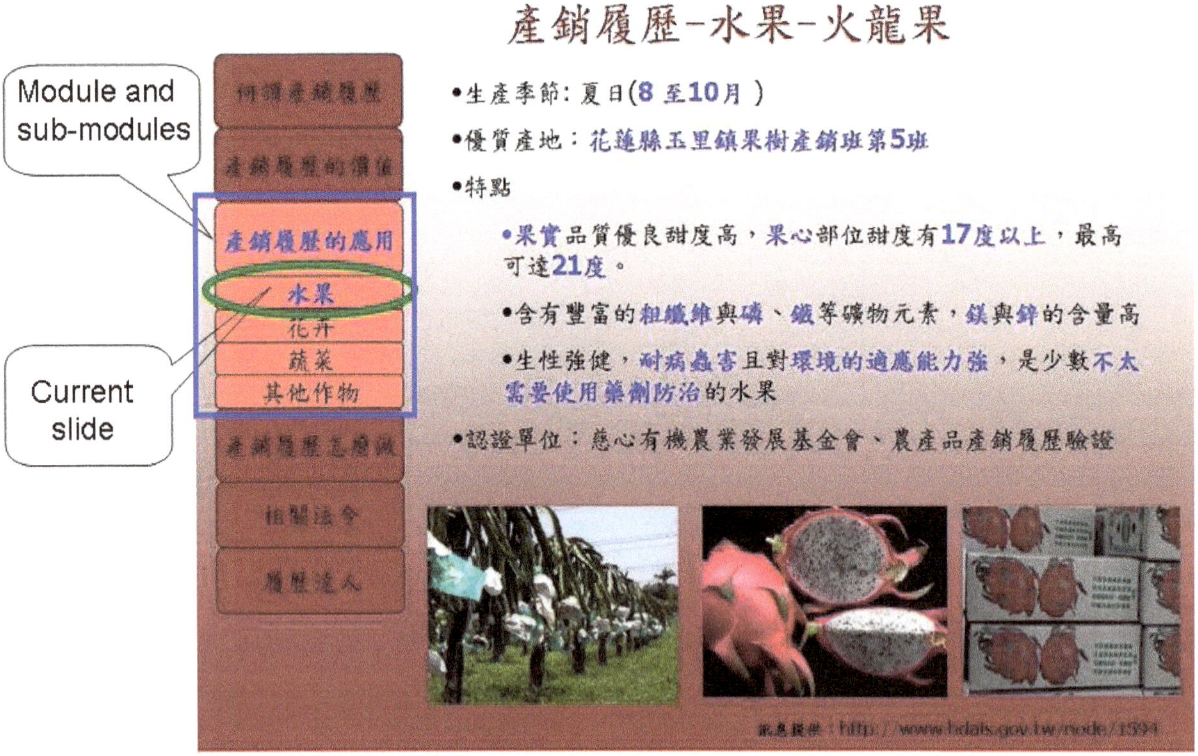

Figure 3. One of the presentations of interactive agricultural extension materials.

will be confident that they can achieve the expected outcomes of the course and satisfaction which derive from the instruction (Johnson and Aragon, 2003).The questionnaire, the IMMS --overall perception to learn was evaluated. IMMS that developed around Keller's ARCS model of perceptive design was designed to evaluate how instructional materials affected motivation to learn. It contained a 36 7-point Likert scale statements, ranging from extremely dissatisfied (1) to extremely satisfied (7), as well as to provide open comments on the system. Each statement measured an individual ARCS component. In order to minimize possible error because of participants' varying levels of English comprehension, a Chinese version of the questionnaire was used, with the Chinese version of IMMS administered by ESL/EFL and translation experts to prevent any translation mistakes.

Procedure

In order to verify the learning perceptions of the PowerPoint presentation, an exploratory study was conducted. In this experimental activity, total workshop duration was 20 h. The participants were randomly assigned to two classes after they registered for the extension course. Then, the participants were randomly assigned as the experimental group and the control group. The experimental group (17 participants) was lectured by a teacher using the interactive PowerPoint presentation, and the control group (15 participants) was lectured by a teacher using a linear PowerPoint presentation. The internal controls, which included teaching hours, instruments, course content, and instructor, were the same for the two groups. At the end of the training activity, IMMS was conducted to assess the participants' general attitudes, interest and efficacy for the PowerPoint presentation.

RESULTS AND ANALYSIS

After distributing the questionnaires, the researcher gathered the responses and used Statistical Package for the Social Sciences (SPSS) for Windows, a statistical program, for data analysis. The data collected was coded and entered into a computer by optical scoring, and analyzed using SPSS. Descriptive statistics, including frequencies, means and standard deviations, were reported in order to understand the learners' cognitive results of the training activity. T-tests were used to determine the effects of experimental activity. The standard for significance in this study was <0.05.

The reliability of the IMMS, as assessed by Cronbach alpha for internal consistency, was .883. For the four components (attention, relevance, confidence, and satisfaction) of IMMS, Cronbach alpha was between 0.818 ~ 0.875. Table 1 shows the significant results of the experimental analysis, which indicated as following: It showed that the interactive PowerPoint presentation provided the rich visual interactivity and the instructor effectively used the instructional technology to present the teaching materials. Hence, the instructor's use of the instructional technology helps participants pay attention in class, $t= 7.999$, $p < 0.001$, $d = 0.814$. It also showed that the participants felt relevant that the course content and activity were closed to their personal and professional works (goals), $t= 9.330$, $p < 0.001$, $d =$

Table 1. Descriptive statistics of learning perception in different groups.

Model constructs	Control (n=15) M(SD)	Experiment (n=17) M(SD)	t-value	Effect size Cohen's d
Attention	21.18 (2.128)	28.53 (3.044)	7.999***	0.814
Relevance	26.88(3.080)	36.60(2.772)	9.330***	0.856
Confidence	26.88(4.608)	36.60(3.832)	6.434***	0.754
Satisfaction	26.18(5.341)	33.47(5.357)	3.848**	0.563

*,**, *** Significant at <0.05, <0.01 and <0.001, respectively.

0.856. According to the cognitive theory of multimedia learning (Mayer, 2001), participants learned better from words and pictures than from printed or spoken words alone. The slides modality concepts and hyperlink technology conduct the interactive presentation which effectively let participants feel that the presentations could promote their understanding of the learning contents. They felt they were more confident, $t = 6.434$, $p < 0.001$, $d = 0.754$. Finally, they satisfied with this extension activity, $t = 3.848$, $p < 0.01$, $d = 0.563$. In summary, the results of IMMS that the participants in experimental group get higher learning motivation than that of in control group.

DISCUSSION

The results of survey pointed out that the interactive PowerPoint is helpful in the extension course. Three statements can be verified from this experiment. First, the interactive PowerPoint could facilitate learning. The interactive PowerPoint presentation provided the rich visual information and technology could successfully attract participants' interests and engagement. Second, considering the cognitive load and knowledge divide, we simplified the contents of the slide such as one slide presented one topic and used the visual information to substitute the complicated texts. And the contents of the instructional materials are more close to participants' life experience. As doing so, participants felt that best fit their needs and preferences and thereby reinforcing their learning (Jones and Plass, 2002). Finally, the interactive PowerPoint presentation incorporated hyperlinks and hierarchical design strategies that provided the instructor with the flexibilities and efficiency in managing the presentation slides. The instructor also easily changed and selected the particular slide to help participants with their questions, thereby helping the cognitive process in the comprehension of the content being taught (Hasler et al., 2007; Wallen et al., 2005). However, in the linear PowerPoint display environment, instructor cannot quickly move and change the desirable slides to explain two successive or separated slides. He/She may lose chances to make important links between more distantly related topics by showing materials in a set order. This may result in decreasing participants' ability to solve a particular problem. Thus, the participants of the experimental group felt more confident and satisfactory than those of the control group.

Conclusions

The results of survey not only showed that the interactive PowerPoint presentation could benefit agricultural extension workers' learning but also showed that the extension course could best fit their needs and preferences. Since the extension workers play the role of facilitator, trainer, and consultant to the farmers that also facilitate them to popularize their jobs. The research findings confirmed that the interactive PowerPoint presentation could contribute effectively to providing the learning environment for agricultural extension.

However, there is still room to improve in this study. For example, we focused the samples who were recruited from only one extension camp in this experiment and thus the results could reflect a bias. In addition, since the training for agricultural extension workers is crucial that a panel study is conducted to track the behavioral changes of participants, specifically their willingness eventually apply the training knowledge in the agricultural extension.

REFERENCES

Allo AV (1983). The Training of Extension Workers. http://www.agnet.org/library/eb/200a/.

Apperson JM, Laws EL, Scepansky JA (2006). The impact of presentation graphics on students' experience in the classroom. Comput Edu. 47(1):116-126.

Chen YT, Chen TJ, Tsai LY (2011). Development and evaluation of multimedia reciprocal representation instructional materials. Int. J. Phys Sci. 6(6):1431-1439.

Harknett RJ, Cobane CT (1997). Introducing instructional technology to international relations.Political Sci. Politics 30:496-500.

Hasler BS, Kersten B, Sweller J (2007). Learner control, cognitive load and instructional animation. Appl. Cogn. Psy. 21:713-729.

Holzl J (1997). Twelve tips for effective PowerPoint presentations for the technologically challenged. Med. Teach. 19(3):175-179.

Johnson SD, Aragon SA (2003). An instructional strategy framework for online learning environments. In S. A. Aragon (Ed.), Facilitating learning in online environments. New Directions for Adult and Continuing Education. San Francisco: Jossey-Bass 10:31-44.

Jones LC, Plass JL (2002). Supporting listening comprehension and

vocabulary acquisition in French with multimedia annotations. Mod.Lan. J. 86(4):546-561.
Keller JM (1983). Motivational design of instruction. In C.M. Reigeluth (Ed.), Instructional Design theories and models. (pp. 383-433). New York: Lawrence Erlbaum Associates.
Lane R, Wright R (2009). And the research says? PowerPoint meets cognitive science, http://office.microsoft.com/en-us/help/and-the-research-says-powerpoint-meets-cognitive-science HA010198311.aspx
Levie WH, Lentz R (1982). Effects of text illustrations: A review of research. Edu. Comm. Tech. J. 30(4):195-232.
Liao SH, Ho SH (2010). Investment project valuation based on a fuzzy binomial approach. Inform. Sci. 180(11):2124-2133.
Matheson VA, Abt-Perkins D, Snedden D (2002). Making PowerPoint Interactive with Hyperlinks. http://www.economicsnetwork.ac.uk/advice/powerpoint.pdf
Mayer RE (2001). Multimedia learning. Cambridge: Cambridge University Press.
Mulder M, Weigel T, Collins K (2006). The concept of competence in the development of vocational education and training in selected EU member states: A critical analysis. J. Voc. Educ. Training 59(1):65-85.
Rankin EL, Hoaas DJ (2001). The use of PowerPoint and student performance. Atl. Econ. J. 29:113.
Wallen E, Plass JL, Brünken R (2005). The function of annotations in the comprehension of scientific texts: Cognitive load effects and the impact of verbal ability. Edu. Tech. Res. Dev. 53(3):59-71.

Factors influencing access to credit: A case study of smallholder farmers in the Capricorn district of South Africa

P. K. Chauke , M. L. Motlhatlhana, T. K. Pfumayaramba and F. D. K. Anim

Department of Agricultural Economics, University of Venda, Thohoyandou, 0950, South Africa.

The study was pursued to ascertain factors that affect smallholder farmer's access to credit sources in the Capricorn District Municipality of Limpopo Province, South Africa. A stratified sampling method was employed to select 250 smallholder farmers in the study area. Structured questionnaires were developed, pre-tested and used for collecting data from the sampled smallholder farmers. The logistic regression model was employed in the analysis. Factors that contributed significantly to credit access were the need for credit, attitude towards risk, distance between lender and borrower, perception on loan repayment, perception on lending procedures and total value of assets. Policy recommendations included the establishment of loans offices close to farmers and operated by officers familiar with farmers to reduce lending procedures, risks and educate them on perceptions on loan repayment.

Key words: Credit, risk, perception, smallholder farmers, logistic regression.

INTRODUCTION

More than 18 years of independence rule in South Africa, little in terms of livelihood generation to smallholder farmers in the country's former homeland rural areas has been achieved, despite government's intention to attain food security in rural areas. One important factor that requires a thorough investigation relates to credit access directed at smallholder farmers. As attested by Okurut et al. (2004), credit is an important instrument for improving the welfare of the poor directly through consumption smoothening that reduces their vulnerability to short-term income. It also enhances productive capacity of the poor through financing investment in their human and physical capital. Manganhele (2010) noted that in many developping countries such as Botswana, Mozambique and Ethiopia, access to credit by smallholder farmers has been burdened for many decades. Despite efforts by some of these governments to provide conducive policies that would ensure credit access to smallholder farmers, actual support has generally not only been static but has even declined in some parts of the developing world. Efforts by most developing countries to channel credit via parastatal institutions have yielded minimal positive results (Machethe, 2004).

The South African government, both during and immediately after independence, focused on financial assistance to emerging rather than smallholder subsistence farmers that are widespread in the rural landscape. Parastatal organisations initiated to achieve such an objective included Ithala Development Finance Corporation in KwaZulu Natal province, Agricultural Development Banks of Ciskei and Transkei, Agribank of the North West Province, Gazankulu Development Finance Corporation, Lebowa Development Finance Corporation both in Limpopo Province, and Uvimba Finance Corporation in the Eastern Cape. Due to lack of investment from government, these institutions have collapsed or merged with other organisations. Consequently access to credit by the small-scale farming sector is still a major challenge that confronts many smallholder farmers in South Africa (Coetzee, 1995). This paper seeks to assess factors that have impact on credit access

*Corresponding author. E-mail: peakay@univen.ac.za.

by smallholder farmers in the Capricorn District Municipality of South Africa.

MATERIALS AND METHODS

Study area and data sources

The study was conducted in the Capricorn District Municipality that is situated at the centre of Limpopo Province in South Africa. The district hosts Polokwane, the capital city and economic hub of Limpopo Province and consists of five local municipalities, that is, Polokwane, Lepelle-Nkumpi, Molemole, Blouberg and Aganang. The Capricorn district covers an area of 2,170,545 ha and has a population of about 1,235,325 (STATSSA, 2007). It therefore ranks number one in the province in terms of population and farming area. Besides Polokwane city in the Polokwane Local Municipality and Lebowakgomo Town in Lepelle-Nkumpi Local Municipality, most of the remaining area is rural or semi-rural. The district therefore hosts a substantial numbers of smallholder farmers as compared to their commercial oriented counterparts (82% vs. 18%), although in terms of area of coverage the latter far outnumbers the former. The district is not only endowed with abundant agricultural resources, it is also one of the country's prime agricultural regions noted for the production of vegetables and field crops (NDA, 2009). Most of the aforementioned commodities are largely produced by white commercial farmers, with the majority of black small-scale farmers cultivating small quantities of field and vegetable crops (Baloyi, 2010).

The study used both qualitative and quantitative approaches in data collection and analyses. Secondary data on farm credit were collected from existing literature while primary data were collected through face-to-face questionnaire based interviews that were administered to smallholder farmers.

Specific predictor variables that were assessed included credit need, attitude towards credit, distance between borrower and lender, smallholder farmers perceptions on loan repayment, lending procedures, total value of assets possessed by the smallholder farmer and the actual time spent per visit to lender.

There were 2500 farmers operating on small-scale farming in the Capricorn District (LDA, 2008). A 10% representative sample, 250 farmers, stratified by the local municipality and status with respect to access to credit was selected based on proportional systematic random selection technique within each of the local municipalities. This technique involved clustering smallholders per local municipality, determining proportional samples based on populations, listing all smallholders in each local municipality, and selecting samples based on a predetermined sampling fraction appropriate for each local municipality. Enumerators administered a pretested questionnaire to ensure validity and reliability of the results of the study.

Data analysis

The data were captured into the SPSS spreadsheet and analyzed through the binary logistic regression modeling technique. The main aim of the analysis was to determine the factors influencing access to credit in the study area. The technique has been used for this kind of a situation (field of social sciences) where prediction of the presence or absence of an outcome based on values of a set of predictor variables is needed. According to Wooldridge (2009), the logistic regression coefficient can be used to estimate odds ratios for each of the independent variables in the model. The logistic regression model (logit model) is applicable to a broader range of research situations than discriminant analysis. The term "logit" refers to the natural logarithm of the odds (log odds) which indicates the probability of falling into one of two categories on some variable of interest (Wooldridge, 2009). According to Harrell (2001), binary logit has only two categories in the response variable, that is, event A and non-event A. Harrell (2001) also asserts that the model shows how a set of predictor (explanatory) variables (X's) are related to a dichotomous response variable Y(ln $(P_i/1 - P_i)$. The dichotomous response variable Y= 0 or 1 with Y=1 denotes the occurrence of the event of interest while Y=0 denotes otherwise. The dummy variables, also known as indicators and bound variables, characterize dichotomous responses.

In this study, since only two options were available, namely "access to credit" or "no access to credit" a binary model was set up to define Y=1 for situation where the farmer accessed credit and Y=0 for situations where the farmer did not access credit from either formal or informal credit sources. Assuming that **X** is a vector of explanatory variables and *p* is the probability that Y=1, two probabilistic relationships as stated by Wooldridge (2009) can be considered as follows:

$$p(Y=1)=\frac{e^{\beta x}}{1+e^{\beta x}} \quad (1)$$

$$p(Y=0)=1-\frac{e^{\beta x}}{1+e^{\beta x}}=\frac{1}{1+e^{\beta x}} \quad (2)$$

Woodridge (2009) concluded that since Equation (2) is the lower response level, that is, the probability that farmers did not access credit from formal and informal credit source, this will be the probability to be modeled by the logistic procedure by convention. Both equations present the outcome of the logit transformation of the odds ratios which can alternatively be represented as:

$$\log it[\theta(x)]=\log\left[\frac{\theta(x)}{1-\theta(x)}\right]=\alpha+\beta_1 x_1+\beta_2 x_2+\ldots\beta_k x_k \quad (3)$$

and thus allowing its estimation as a linear model for which the following definitions apply:

θ = logit transformation of the odds ratio;

∝ = the intercept term of the model;

B = the regression coefficient or slope of the individual predictor (or explanatory) variables modeled and

X_i = the explanatory or predictor variables.

The foregoing operations were feasible within the SPSS package. In relation to Equation (3) the analysis generated the odd ratios using the maximum likelihood procedure (Field, 2005). The logistic regression in this study can be specified as:

$$Yi = \alpha + \beta_1X_1 + \beta_2X_2 + \beta_3X_3 + \beta_4X_4 + \beta_5X_5 + \beta_6X_6 + \beta_7X_7 + \beta_8X_8 \ldots + u_k \quad (4)$$

where, Y_i (SHAC) = the dependent variable defined as the access to credit by smallholder farmers = 1 and 0 otherwise; α = constant and intercept of the equation; X_1 (CRE) = need for credit by the respondent, 1 = had a need and 0 = otherwise; X_2 (RIT) = attitude towards risks by the respondents, 1 = they do not fear risk and 0 = otherwise; X_3 (DIS) = distance between the lender and borrower, 1= 15km and above and 0 = <15 km; X_4 (SHO) = farmers' perception of loan repayment, 1= perceive it as a constraint and 0 = otherwise; X_5 (LEP) = farmers perception of lending procedures, 1 = perceive it as a constraint and 0 = otherwise; X_6 (EXE) = actual time spent with extension officer per visit in hours, 1 = > 1 h and 0 = < 1 h; X_7 (TIN) = total income from farming per year (Rand); X_8 (TAS) = total value of assets possessed by the respondent, 1 = >R 50000 and 0 = < R 50000, and U_k = correction error term. Prior to running the

Table 1. Contingency coefficients of association of included variables.

	CRE	RIT	DIS	SHO	LEP	EXE	TIN	TAS
CRE	1.00							
RIT	0.21	1.00						
DIS	0.095	0.85	1.00					
SHO	-0.153*	-0.153*	-0.231**	1.00				
LEP	-0.109	-0.90**	-0.078	0.364**	1.00			
EXE	-0.136*	-0.151*	0.068	0.071	0.174**	1.00		
TIN	0.980	0.096	0.165**	0.111	0.030	-0.29	1.00	
TAS	-0.165**	0.041	0.002	0.232**	0.444**	0.36	-0.16	1.00

** P<5%; * P<10%; n = 250; Source: Field Survey (2012).

Table 2. Estimate regression model of access to credit.

Explanatory variable	β	Std error	Wald	Exp (β)	P-value
Constant	18.25	4.89	13.92		
CRE	1.67**	0.84	3.98	5.299	0.046
RIT	- 1.20*	0.68	3.10	0.301	0.078
DIS	- 1.62*	0.88	3.41	0.197	0.065
SHO	- 2.27***	0.58	15.48	0.103	0.000
LEP	- 2.55***	0.64	15.79	0.078	0.000
TAS	- 4.75**	0.78	37.33	0.009	0.035
EXE	1.53***	0.73	4.42	4.603	0.000
TIN	-1.94	1.20	2.61	0.143	0.106

***p<0.01; **p<0.05; *p<0.10; N=250; Source; Survey data (2012). Marginal effects could not be calculated since SPSS V.16 was used in the analysis

logistic regression model, the independent variables were checked for the existence of multico-llinearity problems. The result proved negative and non- significance.

RESULTS AND DISCUSSION

The correlation matrix of variables included in the analysis is presented in Table 1, while Table 2 shows the estimated regression model.

The credit need and extension contact had significant positive influences on the respondent's access to credit. The positive signs of credit need and extension contact implied that access to credit increased with increases in the variables. On the contrary, a negative sign of attitude towards risk, repayment period, lending procedures and total value of assets implied that access to credit decreases with an increase in these variables.

As indicated by the Exp (β) values, a value less than 1 would indicate the opposite. Thus as the odds of a credit need increases by one unit, that of actual access to credit increases by more than 5 times. A similar explanation pertains to the need for extension services in which actual need for credit increases by more than 4 times. However as the odds of needs for the other variables increase by 1 unit, there are progressive decreases in the odds for the actual need to access credit. The findings of this study are consistent with that of Beck (2007), whereby it was indicated that extension services play a crucial role in empowering farmers with farming techniques, knowledge and management skills. In addition, extension services provide essential information to farmers regarding agricultural interventions such as farm production technologies, facilitating farm management, marketing and processing equipment. Machethe (2004) further indicated that growth in smallholder farming is indescribable without support services. While logistic results for this study depict decreasing need for accessing credit with unitary increases for the remaining predictor variables, a study by Yehuala (2008) in Ethiopia confirmed that repayment period was a critical factor in influencing access to credit. Although risks and uncertainties are not only unique to agricultural production, they are much more conspicuous in farming than most non- farming activities. The type and severity of risks faced by farmers in general vary with the type of farming system, physical situations, economic conditions and prevailing policies (Klein et al., 1999). Hussien (2007) has affirmed that farm households are discouraged to borrow when credit sources are located further away from their farming operations. The results of the total value assets are contradictory with a study conducted by Atieno (2001) which indicates that among other factors value of farm assets owned is a significant variable that explain the participation of smallholder farmers in formal credit markets. According to Fleisig (1999), farmers experience is a critical factor in adopting modern technologies and accumulation of assets. The finding that increased asset accumulation will result in decreased dependence on credit is understandable and reflective of realities. Contrary to this finding, a study by Kgowedi et al. (2003) associated increased credit needs with increased income generation.

SUMMARY AND CONCLUSIONS

The intention of the paper was to assess the access to credit by smallholder farmers in the Capricorn District Municipality of the Limpopo Province, South Africa. Stratified, clustering and proportional systematic random sampling techniques were used to select the respondents in the study area. The logistic regression was used to analyze the data. The results showed that as the odds of a credit need increased by one unit that of actual access to credit increased substantially (more than 5 times). A similar explanation pertained to the need for extension services in which actual need for credit increased by more than 4 times. Access to credit decreased with unitary increases of the other variables, notably repayment period, risk and uncertainty, distance between borrower and lender, farmer experience and asset accumulation.

RECOMMENDATIONS

The study recommends that since farmers were found to be risk averse and dependent on own assets, there is the need for effective training programmes that would include access to insurance and use of modern technologies such as cell phones, to close the gap between lender and borrower. The study also recommends the establishment of loans offices close to farmers and which will be operated by bank officials who may be familiar with farmers in the area to reduce lending procedures, risks and educate them on perceptions on loan repayment.

ACKNOWLEDGEMENTS

The authors would like to express their special thanks to the University of Venda, National Research Foundation and the Land Bank for the fund provided for carrying out the study.

REFERENCES

Atieno R (2001). Formal and informal institutions' lending policies and access to credit by small-scale enterprises in Kenya: An empirical assessment. African Economic Research Consortium, Nairobi.

Baloyi JB (2010). An analysis of constraints facing smallholder farmers in the Agribusiness value chain: A case study of farmers in Limpopo Province. Master's thesis Department of Agricultural Economics, Extension and Rural Development. Faculty of Natural and Agricultural Science. University of Pretoria, Pretoria.

Beck T (2007). Financing constraints of SME's in developing countries. Evidence, determinants and solutions. J. Int. Money Finan. 31(2):401-441

Coetzee GK (1995). Credit. In R Singini and J van Rooyen (eds). Serving smallscale farmers: evaluation of the DBSA's farmer support programmes, Midrand. Development Bank of Southern Africa. pp. 227-250

Field A (2005). Discovering statistics using SPSS. (2nd Ed.). London:Sage

Fleisig H (1999). The power of collateral. View point. Washington D.C., The World Bank.

Harrell Jr. FE (2001). Regression Modeling Strategies with Applications to Linear models, Logistic Regression and Survival Analysis. Springer-Verlag, New York.

Hussien H (2007). Farm Household Economic Behavior in Imperfect Financial Markets, Doctoral Thesis, Swedish University of Agricultural Sciences, Uppsala.

Kgowedi MJ, Makhura MN, Coetzee GK (2003). Factors distinguishing the choice of money lenders and non-money lenders in Moletjie District (Limpopo Province). Working paper: 2002 -07. University of Pretoria.

Klein B, Mey R, Hanning A, Burnett J, Fiebig M (1999). Better practices in agricultural lending. Agricultural finance revisited. Food and Agriculture Organization of the United Nations (FAO). Deutsche Gesellschft Fur Technisch Zusammenarbeit, p. 3.

Limpopo Department of Agriculture (LDA) (2008). Financial report for April 2007 – March 2008.

Machethe CL (2004). Agriculture and poverty in South Africa: can agriculture reduce poverty? Paper presented at the Overcoming Underdevelopment Conference, Pretoria, South Africa.

Manganhele AT (2010). Improving access to credit for smallholder farmers in Mozambique. Lessons from government effort in developing countries of Africa and Asia. MSc Thesis. Department of Agricultural Economics, Extension and Rural Development. University of Pretoria. South Africa.

National Department of Agriculture (NDA) (2009). Directorate Economic Analysis, Resource centre, Directorate Agricultural Information Services, Pretoria. Available online at: <http:www.nda.agric.za> [Accessed 04 June 2011].

Okurut N, Schoombee A, Van der Berg S (2004). Credit demand and credit rationing in the informal financial sector in Uganda. Paper to the DPRU/Tips/Cornell conference on African Development and Poverty reduction: the Macro-Micro linkage October 2004.

Statistics South Africa (STATSSA) (2007). Community survey 2007 by municipality, type of main dwelling and population group of household.

Wooldridge JM (2009). Introductory Econometrics, A Modern Approach, Fourth edition. Michigan State University.

Yehuala S (2008). Determinants of smallholder farmers access to formal credit. A case of Metema Woreda, North Gonder, Ethiopia. Unpublished Master's Thesis. Haramaya University, Ethiopia.

Effect of root restriction on vegetative growth and leaf anatomy of 'Kyoho' grapevines cultivar

ZhaoSen Xie[1], Xijun Guo[2] and Hongmei Cao[1]

[1]College of Horticulture Science and Technology, Hebei Normal University of Technology and Science, Qinhuangdao 066600, Hebei, China.
[2]Liaocheng Municipal Forestry Bureau, Liaocheng, 252000, Shandong, China.

The effects of root restriction on leaf and shoot growth and leaf anatomy of grape were investigated. Root restriction delayed leaf appearance, decreased leaf growth and reduced the rate of leaf expansion during all stage of berry development. Shoot growth of root-restricted vines was also reduced, with the length and thickness of shoots being lower than those of the control. Root restriction also induced adaptive changes in the leaf anatomy. Somata density of leaves from root-restricted vines was higher than that of controls, but the size of the stomata was smaller. The thickness of the leaf, cuticle, palisade layer, spongy mesophyll, and leaf structure tense ratio (CTR) were increased, however, the leaf tissue loose ratio (SR) and the size of guard cells were decreased. In conclusion, our results suggest that root restriction not only reduces growth of vegetative organ, but also alters their anatomy. These changes may be one of the reasons that change the physiological functions of the leaf.

Key words: Root restriction, shoot and leaf growth, leaf anatomy.

INTRODUCTION

Leaves are the main source of CO_2 assimilation in plants. Root restriction can control the size of the shoot and partitioning of assimilates between vegetative and reproductive organs (Carmi, 1986). Under root restriction, the plant displays distinctive differences in growth habit compared with that under normal field cultivation. Root restriction increased root mass, the amount of fibrous roots, reduce shoot growth, and improve both fruit set and fruit quality (Bar-Tal et al., 1995; Wang et al., 1998, 2001). Many investigators report that root restriction decrease vegetative growth, and alter the metabolism of vegetative organs (Wang et al., 2001; Zhu et al., 2006). Root restriction decreased leaf nitrogen content of *Euonymus* (Dubik et al., 1990), grapevine (Zhu et al., 2006), and peach (Mark and Marra, 1994; Boland et al., 2000), which may reduce the rate of leaf expansion (Radin, 1983; Gastal and Nelson, 1994).

Although, many studies have reported that root restriction affects leaf photosynthesis and nutrient content, information on its effects on leaf growth and anatomy is limited. The importance of leaf anatomy for the main physiological functions of photosynthesis and transpiration, and its response to the environment, has long been recognized (Pachepsky et al., 1999). The objective of this study was to examine the effects of root restriction on vegetative growth and leaf anatomy of'Kyoho' (*Vitis vinifera* × *Vitis labrasca*) grape cultivar.

MATERIAL AND METHODS

Plant material

The experiment was carried out during the 2008 growing season at the experimental farm at Shanghai Jiao tong University, Shanghai,

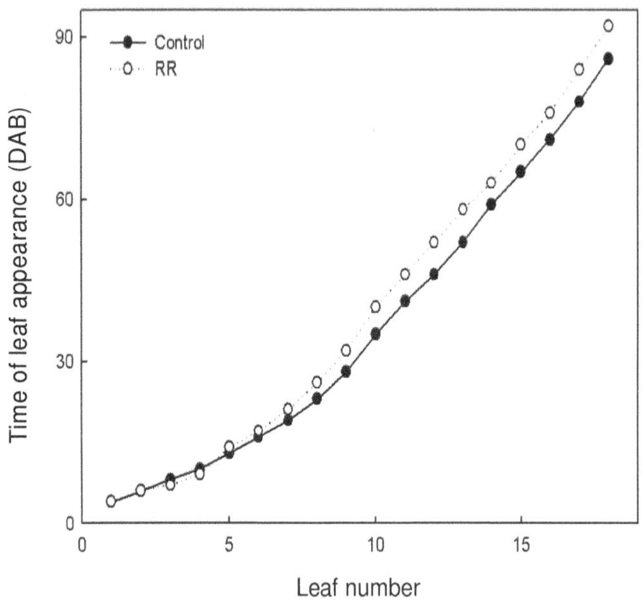

Figure 1. The time of leaf appearance in plants subjected to root restriction (RR) and control. Points represent the mean of 5 shoots.

China (3111N, 12129E).Two groups of 20 uniform 3-years-old 'Kyoho' grapevines (*V. vinifera* × *V. labrasca* cv.) were selected. The first group was subjected to root restriction by planting in 10 L plastic pots, however, the second one, served as the control, was planted in a raised bed 50 cm in deep in a medium composed of a mixture of sand, loam, and perlite (1:1:1) by volume and served as the control. In late winter of 2007, each vine was pruned to 3 to 5 buds. In spring of 2008, one vigorous shoot on each vine was left and the other shoots were removed. Only the basal cluster was retained on the remaining shoot after flowering. The number of berries in those clusters was thinned to about 50. The space between each vine was 60 cm; vines were maintained in a ventilated greenhouse under natural light at the experimental farm from February to September. The greenhouse was covered with a plastic sheet with a transmission of about 87%. Natural ventilation was used to maintain the air temperature inside the greenhouse near from outside air temperature. After budburst, each vine was fertilized with 1 L of complete liquid fertilizer (Hydro Co. Ltd., Israel), containing 32.7 mg NO_3^-, 22.0 mg NH_4, 120 mg P_2O_5, 120 mg K_2O, 20.0 mg MgO, 0.167 mg B, 0.067 mg Cu, 0.467 mg EDTA-Fe, 2.667 mg Mn, 0.027 mg Mo, and 0.167 mg Zn, once a week. Tensiometers were placed at a 15 cm depth in the rooting-zone to monitor soil moisture. Drip irrigation system was used to maintain the soil moisture ≥ -3.0 kPa from replanting to veraison, and ≥ -5.0 kPa from veraison to harvest.

Leaf and shoot growth measurements

Five grapevines in each treatment were marked and all major features of development were monitored on the marked shoots, including the time of new leaf appearance, Leaf area (cm^2) of the second leaf from the shoot apex was measured during different stages of berry development. The leaf area for each leaf was calculated from the non-destructive measurements of leaf length using the following equation, $Y = 0.6933X^2 + 5.5069X - 15.1966$ ($R^2 = 0.9419$) where the X is the length of leaf midvein. The length and thickness of tagged shoots were measured every 3 days after 20 days of anthesis.

Observation of the stomata density and anatomic structure of leaf

Clear nail polish was applied to the adaxial and abaxial leaf surfaces on opposite sides of the leaf to measure the stomata density. The transparent of the leaf surface impression was removed and attached to a glass slide (Abraham et al., 2009).

To measure the anatomical features of the leaves, the lower surface of one fully-expanded leaf from six different individuals for each treatment was assessed. 3 × 3 mm segments were cut from the central area of the cleared leaf blade tissue, then were fixed by FAA, and cut into 10 μm thickness by paraffin sectioning technology (Nii et al., 1995).

Statistical analysis

The comparison of pairs of values was analyzed by t-test and levels of significance are represented by $^*P < 0.05$, $^{**}P < 0.01$ (Xie et al., 2012).

RESULTS

New leaf appearance

The time of leaf appearance was recorded after bud break. The second leaf appeared 1 to 2 days later than the first leaf, the third and the fourth leaves appeared at interval of 2 to 3 days (Figure 1). The difference in time of leaf appearance was not significant from the first to the fourth leaf between root restriction and control treatments. After the fifth leaf, the rate of leaf appearance was at an interval of 4 to 5 days. The time of leaf appearance in plants subjected to root restriction was delayed 2 to 3 days when compared with the control after the fifth leaf (Figure 1).

Leaf growth

The expansion of new leaves was observed during different stages of berry growth, including inflorescence, the first rapid growth stage, the lag stage and the second rapid growth stage (Figure 2). During the first rapid growth stage (15 to 50 days after anthesis), the new leaf had the maximum rate of expansion compared with the other stages (Figure 2B). During all stages of berry growth, the leaf expansion was reduced in plants subjected to root restriction than in control plants (Figure 2).

Shoot growth

Shoot growth was monitored following 20 days after anthesis (Figure 3). The shoot growth of grape vines, including the length and thickness, was reduced in plants subjected to root restriction compared to controls (Figure 3). The increase of shoot length followed a single sigmoid curve. The shoots of root restriction vines were significantly shorter than control following 50 days after

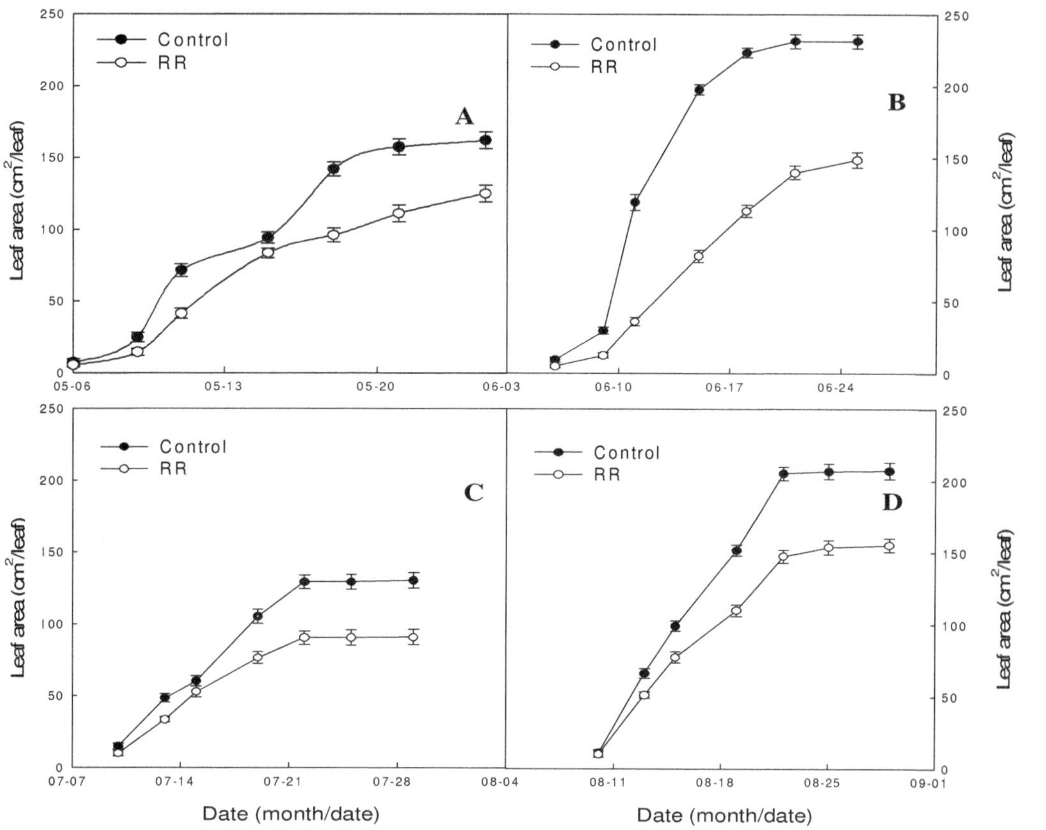

Figure 2. The changes in leaf area of the second leaf from the shoot apex in plants subjected to root restriction (RR) and control during anthesis (A), the first rapid stage of berry growth (B), the lag stage of berry growth (C), and the second rapid stage of berry growth (D).

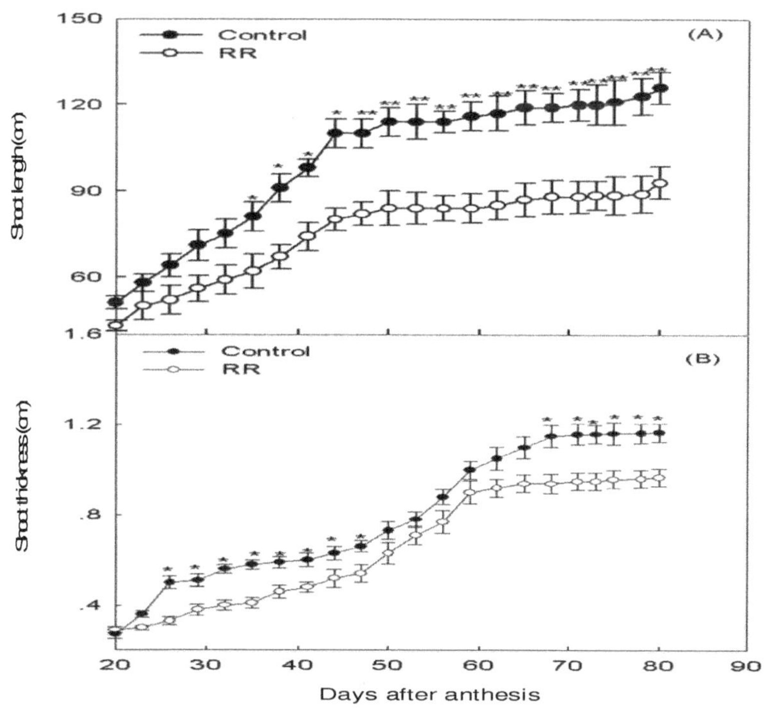

Figure 3. The length (A) and thickness (B) of shoots in plants subjected to root restriction (RR) and control during fruit development.

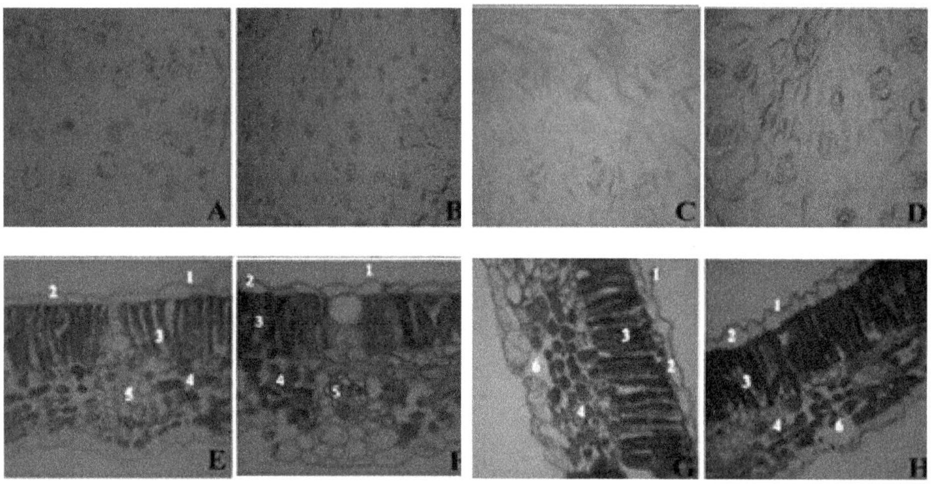

Figure 4. The stomata (A-D) and comparative anatomical structure (E-H) of grape leaves from plants subjected to root restriction (B, D, F, H) and control (A, C, E, G). 1: Cuticle, 2: epidermis, 3: palisad layer, 4: spongy tissue, 5: vascular tissue; 6: stomata.

Table 1. Indices of anatomical structure of grape leaves from plants subjected to root restriction (RR) and control.

Treatment	Control	RR	Significance
Leaf thickness(μm)	319.4	456.8	**
Cuticle thickness(μm)	1.5	2.1	**
Thickness of palisade layer (μm)	176.84	236.13	**
Thickness of spongy mesophyll (μm)	78.35	85.14	*
Thickness of upper epidermis cell (μm)	41.12	52.25	*
Thickness of lower epidermis cell (μm)	16.84	22.7	*
CTR^Z (%)	42.84	51.69	**
SR^y (%)	24.53	18.64	**
Stomata density (number/mm^2)	72	101	**
Stomata longitudinal diameter (μm)	26.6	21.3	*
Stomata horizontal diameter (μm)	4.9	3.2	*

Values represent the means of six replicates. *. Significant at $p < 0.05$; **, significant at $p < 0.01$ based on a paired t-test. ^{z}CTR (leaf tissue structure tense ratio): the thickness of (palisade + lower tight tissue) / mesophyll. ^{y}SR (leaf tissue structure loose ratio): the thickness of spongy/mesophyll.

anthesis (Figure 3A). The change of shoot diameter followed a double sigmoid curve; the maximum significant difference between the two treatments was achieved 67 days after anthesis (Figure 3B).

Leaf anatomy

Grape leaves are comprised of a cuticle, epidermis, mesophyll, and vascular tissue (Figure 4E and F). The thickness of the leaf and cuticle in plants subjected to root restriction was greater than that of control (Table 1). Root restriction increased the leaf tissue structure tense ratio (CTR) and decreased the leaf structure loose ratio (SR) compared with those of control (Table 1, Figure 4G and H). The leaves of plants subjected to root restriction had better uniformity epidermis cells than that of control. The thickness of the upper epidermis of leaves subjected to root restriction and control were 52.25 and 23.12 μm, respectively, however, the thickness of lower epidermis of leaves were 22.7 and 16.84 μm, respectively (Table 1, Figure 4G and H). The palisade layer and spongy mesophyll of leaves subjected to root restriction were thicker than those of controls. Leaves subjected to root restriction had thicker palisade layer, thicker spongy mesophyll, and thicker layers of smaller and more closely-arranged diachyma cells than control (Table 1, Figure 4G and H). Stomata and guard cells of leaves from root restriction plants were significantly smaller than those of control; however, the density of stomata was higher (Table 1, Figure 4A-D).

DISCUSSION

Root restriction, as one type of physical stress, can induce changes in the development of grapevines and other plants (Wang et al., 2001; Xie et al., 2009). Reduced vegetative growth as a result of root restriction is a well recognized phenomenon (Peterson et al., 1991a). In agreement with previous reports, our results showed that leaf and shoot growth was reduced by root restriction. This decline in vegetative growth in plants subjected to root restriction was often interpreted as a feedback inhibition caused by the stress on the roots (Shi et al., 2008).

Leaf structure in plants is determined by both genetic and environmental factors. The total thickness of leaves from root restricted plants was higher than in control (Table 1). Furthermore, an increase in the density of palisade and spongy mesophyll cells and a decrease of intercellular spaces were also observed (Table 1, Figure 4G and H). The arrangement of spongy mesophyll cells in root-restricted leaves may result in a reduction of the CO_2 conductance. These results proved that root restriction, environmental stress, can induce changes in leaf structure.

Wang et al. (2001) reported that water stress occurred almost every day in plants subjected to root restriction, because the amount of available water was reduced. Several reports have suggested that the reduction in shoot growth due to root restriction is partly attributed to disturbance of the internal water relations (Tschaplinski and Blake, 1985; Arp, 1991). Garg et al. (2001) reported that increasing water stress progressively decreased plant growth potential, leaf area, net photosynthetic rate and nitrate reductase activity. It has been reported that root restriction reduced the number of leaves, total leaf area, and shoot initiation in peach (Richards and Rowe, 1977). Leaf growth was the most sensitive process to water stress (Dry and Loveys, 1998).

The growth of leaf and shoot can be mediated by chemical signals synthesized in the roots. The effect of root-sourced signals on leaf growth has been reported for apple, tomato and grape plants. Ternesi et al. (1994) indicated that ABA in xylem of sunflower plants is subjected to root restriction increased seven-fold, which may be associated with a reduction of leaf growth. Similar evidence was found in watermelon (Liu and Latimer, 1995). Root restriction also affected the capability of the root to absorb nutrients. Some reports have revealed that root restriction decreased the concentration of nitrogen in shoots, flower clusters, trunks and canes of grapevines (Wang et al., 1998; Zhu et al., 2006), and of leaves of *Euonymus* (Dubik et al., 1990) and peach (Mark and Marra, 1994; Boland et al., 2000). Reduction of nitrogen concentrations in root-restricted plants is believed to be one of the important reasons why root restriction reduced shoot growth in 'Kyoho' grapevines (Wang et al., 1998).

In conclusion, this study has shown that the dynamic process of leaf and shoot growth was affected by root restriction. The decrease in leaf growth also affected leaf anatomy, which may affect its function.

ACKNOWLEDGEMENT

This study was supported by the National Natural Science of China (31101508).

REFERENCES

Abraham J, Richard BP, Pamela H, Templer S, Sharda M (2009). Long-term relationships among atmospheric CO_2, stomata and intrinsic water use efficiency in individual trees. Am. J. Bot. 96(10):1779-1786.

Arp WJ (1991). Effects of source-sink relations on photosynthetic acclimation to elevated CO_2. Plant Cell Environ. 14:869-875.

Bar-Tal A, Feigin S, Sheinfel D (1995). Root restriction and N-NO_3 solution concentration effects on nutrient uptake, transpiration and dry matter production of tomato. Scientia Hort. 63:195-208.

Boland AM, Jerie PH, Mitchell PD, Goodwin I, Connor DJ (2000). Long-term effects of restricted root volume and regulated deficit irrigation on peach. 1. Growth and mineral nutrition. J. Am. Soc. Hort. Sci. 125:135-142.

Carmi A (1986). Effects of root zone volume and plant density on the vegetative and reproductive development of cotton. Field Crops Res. 13:25-32.

Dry PR, Loveys BR (1998). Factors influencing grapevine vigour and the potential for control with partial root zone drying. Aust. J. Grape Wine Res. 4:140-148.

Dubik SP, Krizek DT, Stimart DP (1990). Influence of root zone restriction on mineral element concentration, water potential, chlorophyll concentration and partitioning of assimilate in spreading Euonymus (E. *Kiautschovica* Loe. '*Sieboldiana*'). J. Plant Nutr. 13:677-699.

Gastal F, Nelson CJ (1994). Nitrogen use within the growing leaf blade of tall fescue. Plant Physiol. 105:191-197.

Garg BK, Kathju S, Burman U (2001). Influence of water stress on water relations, photosynthetic parameters and nitrogen metabolism of moth bean. Biologia. Plantarum. 44:289-292.

Liu A, Latimer JG (1995). Water relations and abscisic acid levels of water melon as affected by rooting volume restriction. J. Exp. Bot. 46:1011-1015.

Mark R, Marra F (1994). Response of young peach trees to root confinement. J. Am. Soc. Hort. Sci. 119:223-228.

Nii N, Watanabe T, Yamaguchi K, Nishimura M (1995). Changes of anatomical features, photosynthesis and ribulose bisphosphate carboxylase-oxygenase content of mango leaves. Ann. Bot. 76:649-656.

Pachepsky YA, Rawls WJ, Timlin DJ (1999). The current status of pedotransfer functions: their accuracy, reliability, and utility in field- and regional-scale modeling. pp. 223-234. In Corwin DL, Loague K, Ellsworth TR (ed.) Assessment of non-point source pollution in the vadose zone. Geophysical Monogr. 108. American Geophysical Union, Washington, DC.

Radin JW (1983). Control of plant growth by nitrogen: differences between cereals and broadleaf species. Plant Cell Environ. 6:65-68.

Richards D, Rowe RN (1977). Effect of root restriction, root pruning and 6-benzylaminopurine on the growth of peach seedlings. Ann. Bot. 41:729-740.

Peterson TA, Reinsel MD, Krizek DT (1991a). Tomato plant response to root restriction Alteration to plant morphology. J. Exp. Bot. 42:1233-1240.

Shi K, Ding XT, Dong DK, Zhou YH, Yu JQ (2008). Putrescine enhancement of tolerance to root-zone hypoxia in *Cucumis sativus*: a role for increased nitrate reduction. Funct. Plant Biol. 35:337-345.

Ternesi MA, Andrad AP, Jorrin J, Benlloch M (1994). Root-shoot signaling in sunflower plants with confined root system. Plant Soil 166:31-36.

Tschaplinski TJ, Blake TJ (1985). Effects of root restriction on growth correlation, water relations and senescence of alder seedlings. Physiol. Plant. 64:167-176.

Wang S, Okamoto G, Ken H (1998). Effects of rooting-zone restriction on the changes in carbohydrates and nitrogenous compounds in 'Kyoho' grapevines during winter dormancy and early shoot growth. J. Jpn. Soc. Hort. Sci. 67:577-582.

Wang S, Okamoto G, Ken H, Lu J, Zhang C (2001). Effect of restricted rooting volume on vine growth and berry development of Kyoho grapevines. Am. J. Enol. Viticult. 52:248-253.

Xie Z, Li B, Forney CF, Xu W, Wang S (2009). Changes in sugar content and relative enzymes activities in grape berry in response to root restriction. Sci. Hort. 123:39-45.

Xie ZS, Cao HM, Li B, Li WF, Wang SP (2012). Changes of water transportation in berry vascular bundle at different dvelopmental phases of Kyoho grape berry. Sci. Agric. Sinica. 45(1):111-117.

Zhu LN, Wang SP, Yang TY, Zhang CX, Xu WP (2006). Vine growth and nitrogen metabolism of 'Fujiminori' grapevines in response to root restriction. Sci. Hort. 107:143-149.

8

Preferences and constraints of maize farmers in the development and adoption of improved varieties in the mid-altitude, sub-humid agro-ecology of Western Ethiopia

W. Abera[1], S. Hussein[1] , J. Derera[1], M. Worku[2] and M. D. Laing[1]

[1]African Centre for Crop Improvement, University of KwaZulu-Natal, Private Bag X01, Scottsville 3209, Pietermaritzburg, South Africa.
[2]CIMMYT Kenya ICRAF House, P. O. Box 1041, Nairobi Kenya.

Understanding farmers' production constraints and preferences is important in maize breeding, especially underlying successful adoption of improved varieties and their production packages. This study was conducted to assess the present importance, and productivity constraints of maize in the mid-altitude, sub-humid agro-ecology of Western Ethiopia. Data was collected through a semi-structured questionnaire and focus group discussions, using 240 randomly selected respondent farmers, in 12 sub-districts, within three administrative zones. Maize was ranked number one as both food and cash crop by 82.9% of respondents. Most farmers (59%) use hybrids, while 24% grow landrace varieties. Unavailability of improved seed and lack of production inputs were the two major constraints reducing maize productivity, as reported by 62 and 60% of respondents, respectively. A high proportion of respondents (80%) indicated that, unpredictable grain prices are the major market constraint as 97% of the respondents sell their maize crop in the local market. Northern corn leaf blight (NCLB) was reported to be important by 46% of respondents. Breeding for improved disease resistance and grain yield, enhancing the availability of crop input and stabilizing market price during harvest time are the most important strategies to increase maize production by small-scale farmers in Western Ethiopia.

Key words: Farmers' constraints, maize, participatory research, production packages, *Zea mays*.

INTRODUCTION

Maize (*Zea mays* L.) is the third important cereal crop globally after wheat and rice (FAO, 2011). The crop occupies a pivotal role in the world economy and is traded widely. Maize demand is projected to increase by 50% worldwide and by 93% in sub-Saharan Africa between 1995 and 2020 (FAO, 2007). In the past, much of the global use of maize has been for animal feed. However, maize is increasingly used for human consumption and accounts for 70% of the food consumed in sub-Saharan Africa (FAO, 2007). The recent volatile food market and rising prices for most food crops may increase the importance of maize production. In addition, because of its productivity and wide adaptation, maize remains an important source of food with great potential to improve the livelihoods of most poor famers in developing countries (FAO, 2011).

In developing countries, most studies on agricultural technology development and diffusion are based on ex-post analysis of intervention programs (Bandiera and Rasul, 2006). Farmers are rarely consulted, *a* priori, about their specific circumstances, priority constraints, and their preferences with regards to production packages. Often the level of adoption is determined after costs are incurred and technologies are developed and diffused. Such top-down interventions usually result in a low level of acceptance by end-users and minimal success for variety development programs (Feder et al., 1981). Prior identification of farmers' constraints and preferences is required in order to design more appropriate, acceptable and cost effective development intervention programs (Walter and Zewdie, 2008). In addition, the likely extent of future adoption of research results has a strong influence on the efficiency of research and on the results of research priority settings (Batz et al., 2003). Complementary contributions of farmers and scientists are essential for effective agricultural research and technology development, and their adoption (Chimdo et al., 2002).

Farmers' participation in setting research priorities and technology evaluation is crucial to scientists in order to design, test and recommend appropriate new production technologies (Ashby, 1991). This can be achieved through participatory research that allows incorporation of farmers' indigenous technical knowledge, identification of farmers' criteria and priorities, and the definition of the research agenda, among others (Girma et al., 2005). Participatory rural appraisal (PRA) is a research tool to capture farmers' perceptions, constraints and preferences (De Groote and Bellon, 2000). These authors emphasized the effectiveness of PRA through seeking insights and promoting dialogue between scientists and farmers on production constraints, technology development and adoption. By integrating farmers' constraints and preferences into agricultural research, technologies can be developed that become widely adopted, resulting in more productive, stable, equitable and sustainable agricultural systems. This has led to the emergence of participatory plant breeding approaches in conventional plant breeding programs which integrates farmers' preferences and constraints (Ceccarelli et al., 2001). Participatory plant breeding approaches are reported in cassava (Witcombe and Virk, 2001), barley technology adoption in Syria, Morocco and Tunisia (Ceccerali et al., 2001), and maize in India (Joshi and Witcombe, 1996) and Ethiopia (Abebe et al., 2005).

In Ethiopia, maize is the second largest food security crop after tef [*Eragrostis tef* (Zucc.) Trotter. It is predominantly grown by small-scale farmers in the mid- and low-altitude, sub-humid agro-ecologies. It is primarily produced and consumed by the small-scale farmers that comprise about 80% of Ethiopia's population (Dawit et al., 2008). Maize has increasingly become a popular crop in the country with steady growth in production area and yield (Doss et al., 2003). The use of improved maize varieties has increased from 5% in 1997 to 20% in 2006 (CSA, 2001, 2004, 2006; Byerlee et al., 2007). The mid-altitude, sub-humid agro-ecology is the most important maize producing environment in Ethiopia (Birhane and Bantayehu, 1989; Kebede et al., 1993). This region is considered to be the major maize growing zone in the country. The region lies at an altitudes of between 1000 to 1800 m above sea level and receives a fairly reliable average annual rainfall (1000 to 1500 mm/year), rendering it a region of high potential for maize production. However, maize production has remained low, with the estimated national average yield of 2.5 t/ha due to several constraints: biotic (inadequate improved varieties, pests and diseases), abiotic (low soil fertility, land and water degradation, and drought) and socio-economic (input unavailability, lack of storage facility, poor access to markets) (CSA, 2010). Therefore, maize production components such as farmer-preferred, improved varieties, farming technologies, farm inputs, and access to markets should be developed and made available to enhance maize production, in order to achieve food security.

Bako National Maize Research Project was established under the Ethiopian Institute of Agricultural Research (EIAR) to coordinate maize research and development nationally. The project is situated in the region of mid-altitude, sub-humid maize growing agro-ecology. It is aimed at development and popularization of improved maize varieties, together with their management technologies. In the past, there was a limited participatory research approach in maize technology development and/or improvement by the project. Consequently, there is little formal documentation of farmers' production constraints and preferences which are needed to guide maize varietal development in the mid-altitude, sub-humid agro-ecology of Western Ethiopia. Therefore, the objective of the study was to assess the present importance, and production and productivity constraints, of maize in the mid-altitude, sub-humid agro-ecology of Western Ethiopia. The study would serve as the basis for formulating research and development strategies to increase maize production and productivity by resource poor famers in these and similar environments.

MATERIALS AND METHODS

Study sites and sampling

A survey study was conducted in three representative maize growing zones of Western Ethiopia; West Shoa, East Wollega and West Wollega. In each zone, two districts were selected. Two sub-districts were selected per district providing a total of 12 study sites. The study zones are situated in the mid-altitude, sub-humid agro-ecology where maize is grown as the predominant cereal crop. The zones receive the main rain from May to September in most years. Most farming activities follow the main rainy season which has a

Table 1. Area, population demography and population density of the study zones and districts (CSA, 2010).

Zone and district	Area (Km^2)	Males (000)	Females (000)	Total (000)	Density per km^2
West Shoa					
Ilu-Galaan	332.04	40.834	40.251	81.085	244.20
Bako-Tibe	1044.52	60.217	66.674	126.891	121.48
East Wollega					
Gobu Sayyo	337.53	22.013	23.874	45.887	135.95
Sibu-Sire	1132.50	61.32	54.00	115.313	101.82
West Wollega					
Najo	958	68.715	71.146	139.861	145.99
Manasibu	1668.1	69.776	74.608	144.384	86.56

uni-modal distribution. Soils are varied and generally of low fertility. This is regarded as one of the major problems for cereal production in the zones. As a result of the high intensity of rainfall in the study areas, the soils are acidic (Wakene et al., 2012).

The study sites were selected based on the importance of maize in the livelihoods of small-scale farmers, and the prevalence of major maize production constraints, including diseases, during production seasons. The survey was conducted during January, February and March, 2011. According to the recent population census (CSA, 2010), the population demography and area of the study districts are shown in Table 1. Overall, Ilu Galaan is the most densely populated (244 per km^2) and Manasibu (86 per km^2), the least.

Sampling procedures

Multi-stage random sampling techniques were applied to select the study sites that represent the diverse ecological and socio-economic environments, and varying maize production systems, in the mid-altitude, sub-humid agro-ecological zones of Western Ethiopia. The major criteria used during selection of the study sites were the relative importance of maize, and the severity of maize diseases.

Data collection

Data sources

Both primary and secondary data were utilized in the study. Primary data was collected through interview of male and female farmers, key informants and focus group discussions. The key informants included maize researchers, experienced farmers in the villages, local leaders and agricultural agents. The facilitators used pictures showing disease symptoms and cards that had drawings representing various traits and constraints to assist the responding farmers during the discussions. The secondary data was obtained from the zonal and district agricultural offices of the respective districts included in the study areas. The offices of agriculture in the respective districts were the main sources for the secondary data. Eighty (80) respondents were available per zone, per two districts (Table 2). Data was collected using a total of 86 variables. Consequently, principal component analyses (PCA) was used for data reduction and to identify the most important variables based on the magnitude of contributions to the total variation in the data set.

Administering questionnaire

A semi-structured questionnaire was administered to 240 respondent farmers from 12 sub-districts of three administrative zones. The sub-districts and respondent farmers were randomly selected in the major maize producing areas of each zone. Both male and female maize farmers were mobilized for focus group discussion through the local administrators, development agents and extension staff of the respective sub-districts. Checklists were developed and used to guide discussions with farmer groups and individual key informants. The objectives of the project and the significance of contributions from various actors were explained to both groups and communication procedures established.

The farmers were encouraged to use a language they were most familiar with. A member of the research team most versed with Afan Oromo, the local language, facilitated the group discussions. For ease of focusing the discussions and reaching a consensus, farmers were allowed to form discussion groups. Sex and age were the important criteria farmers used in categorizing themselves into discussion groups. Amongst other variables, farmers were asked to list maize varieties that they grow and provide the relative proportions of the varieties, to list and rank their criteria for variety selection in terms of their relative importance, and the main constraints to maize production. The interviews in all the study areas were held by five researchers, selected from different disciplines of the Bako Agricultural Research Centre.

Data analysis

Data from questionnaires of individual interviews was coded, captured and analyzed using the SPSS computer package (SPSS 2005). Descriptive statistics, chi-square tests, and PCA analyses were performed using the same computer package.

RESULTS

Demography, socio-economic characters and main food crops grown

Of the total interviewed farmers, 93.8% were males, while 6.2% were females. Of all the respondents 98.8% were household heads and almost all the respondents (99.6%) were farmers in occupation. Among total respondents

Table 2. Study zones, districts and sub-districts and number of sampled maize farmers in the mid-altitudes sub-humid areas of Western Ethiopia.

Zone	District	Sub-district	Female	Male	Total
West Shoa	Ilu-Galan	Siba-Biche	2	18	20
		Jato dirqi	2	18	20
	Bako-Tibe	Tulu Sangota	1	19	20
		Bacharra	2	18	20
East Wollega	Anno	Qeejo	1	19	20
		Anno Bakanisa	2	18	20
	Sibu-Sire	Biqila	1	19	20
		Chari	1	19	20
West Wollega	Manasibu	Teyiba	2	18	20
		Bengua	-	20	20
	Nejo	Goori	1	19	20
		Dilla	-	20	20
Total			15	225	240

Table 3. Major crops grown, area of production and rank in the study zones of the mid-altitude sub-humid areas of Western Ethiopia.

Zone	Crop	Production	
		Area (ha)	Rank
West Shoa	Maize	118,344	2
	Teff	181,018	1
	Pepper	36,818	5
	Wheat	57,425	4
	Sorghum	77,758	3
East Wellega	Maize	121,854	1
	Teff	74,497	2
	Wheat	9,840	5
	Sorghum	51,173	3
	Pearl millet	15,471	4
West Wellega	Teff	34,040	3
	Wheat	2,927	5
	Maize	73,960	1
	Sorghum	55,926	2
	Pearl millet	33,294	4

53.8% were educated up to primary level, 25% secondary level and 19.6% without any education. The mean household age in the study areas was 41, with the maximum and minimum ages of 73 and 20, respectively. The major five crops grown in the study zones are summarized in Table 3. Most of the respondents (82.9%) ranked maize as the number one food crop. Other important crops in the zones included teff, sorghum, pepper, wheat and pearl millet (Table 3).

Maize varieties grown

Table 4 summarizes maize varieties grown across the three zones. Farmers grow an assortment of maize types [hybrids, landraces and improved open pollinated varieties (OPVs), either on the same or different fields. Farmers grow both landraces (often referred to as local varieties) or improved varieties, to meet multiple objectives. Many of the respondents (60%) grow hybrid varieties, whilst 24% grow landraces (Table 4). Among the respondents, 39% grow the maize hybrid BH660 and 15% grow BH540. BH660, BH540, Pioneer hybrids and BH543 are the most widely grown hybrid varieties, in their order of importance in the study areas (data not shown). In some instances, the same landrace varieties are known by different names. The names of the landrace maize varieties are often descriptive, referring to certain key identifiable characteristics, especially grain color, appearance, growth habit and the perceived place of origin.

Farmers' preferences of maize varieties

Farmers said that they use many criteria in selecting maize varieties for production. Overall, the major and common selection criteria of maize varieties and corresponding number of farmers are presented in Table 5. Farmers' perceptions and rating of the different criteria varied across the study sites. The most important criteria across the studied districts were high yield, resistance to disease and insect pests, lodging resistance, ability to perform well under low soil fertility, and a combination of these. There were highly significant ($P < 0.01$) differences among the respondents in their preference to maize variety traits in all the 12 sub-districts (Table 5).

Table 4. Widely grown maize types and names with corresponding frequency and percentage of farmers in the mid-altitude sub-humid zones of Western Ethiopia.

Type	Name	Frequency	Percent
Hybrid	Different hybrids	143	59.6
Landrace	Landraces	58	24.2
Hybrid and OPV	BH-660 and OPV	2	0.8
Hybrid and Landrace	BH-660 and Landrace	37	15.4
Total		240	100

Table 5. The important farmers preferred traits of maize varieties in the mid-altitude sub-humid zones of Western Ethiopia.

Farmers preferred traits	Farmers	
	Number	Percent
Improved yield	54	22.5
Disease resistance	2	0.80
Insect resistance	6	2.50
Yield and disease resistance	41	17.10
Yield, insect and disease resistance	21	8.80
Yield, disease, insect and lodging resistance	57	23.80
Yield, disease and lodging resistance	21	8.80
Yield, insect and lodging resistance	2	0.80
Yield and lodging resistance	12	5.00
Disease and insect resistance	24	10.00
Total	240	100
Chi-square	32.65	
Significance level	0.000	

Major constraints to maize production

Table 6 summarizes the main constraints to maize production in the studied zones. Prioritization of the constraints was based on number of households affected, severity of the constraint, importance of the constraint in attainment of household objectives, frequency of occurrence of the constraint and the likelihood of a solution to be provided by the research team. The most important farmers' constraints in maize farming, as revealed by semi-structured questionnaire, were low soil fertility, high input prices (especially fertilizers and seed), leaf diseases, and early cessation of rain. Other important constraints were insect pests (termite and stalk borer), volatile grain prices, and unavailability of fertilizer (Tables 7 and 8).

Northern corn leaf blight (NCLB) and gray leaf spot (GLS) are the most important reported maize diseases. Farmers in the study area (46.7%) indicated NCLB as the major leaf disease on maize (Table 7). GLS is ranked as the second most important leaf disease in the area, as reported by 17.9% of the respondents. Price fluctuation of maize grain is also one of the major constraints in the maize market, impacting on the production of maize in the study areas.

Farmers in the study areas (80.8%) indicated that the unpredictable grain price was the major financial problem (Table 8). Furthermore, a widespread lack of storage facilities and high transportation costs affected the marketing of maize. During harvest times (November/December) the farm price of a 100 kg of maize is at Birr200 (≈12 USD), that increases to birr 500 to 600 during June/July. In the present study, the major market outlets for the sale of maize were local market as indicated by 97.1% of the respondent farmers. Only a small proportion of farmers (<3%) sold their products to other market outlets, such as unions, grain trade enterprises and farmer to farmer.

Focus group discussion

Focus group discussion was conducted to identify locally preferred maize varieties, and the traits associated with the respective varieties. Farmers listed the varieties they grew, ranked them and identified traits they preferred in

Table 6. Major maize production constraints in the mid-altitude, sub-humid zones of Western Ethiopia as revealed by five randomly selected farmers averaged over 12 sites.

Constraints	Farmers group across 12 sites					Total	Importance
	A	B	C	D	E		
Early cessation of rain	3.25	2.50	2.50	2.75	2.37	13.37	4
Diseases	3.50	3.00	3.37	3.50	3.62	16.99	3
Insect pests	2.25	2.37	3.12	2.75	2.50	12.99	5
Weeds	2.37	2.50	1.62	1.75	2.00	10.24	5
Input costs	4.87	4.62	5.00	4.87	4.87	24.25	2
Labour shortage	2.00	1.37	1.37	2.00	1.12	7.87	5
Low soil fertility	4.90	4.77	4.87	5.00	5.00	24.54	1

1 = most important, 5 = least important; A, B, C, D, E = five different farmers pooled across 12 sites.

Table 7. Major diseases and insect pests in the study area.

Name	Farmers	
	Frequency	Percent
NCLB	112	46.7
GLS	43	17.9
MSV	5	2.1
Termite	1	0.4
Stalk borer	4	1.7
NCLB,GLS, termite and stalk borer	1	0.4
NCLB,GLS and stalk borer	2	0.8
NCLB and GLS	49	20.4
NCLB, termite and stalk borer	3	1.2
NCLB , GLS and MSV	2	0.8
NCLB , GLS and termite	1	0.4
GLS and rust	1	0.4
NCLB and MSV	7	2.9
GLS and MSV	1	0.4
NCLB, GLS and *Phaosphaeria* leaf spot	5	2.1
NCLB and termite	2	0.8
NCLB, MSV and stalk borer	1	0.4
Total	240	100.0

maize, giving reasons for their choice of the selected varieties. Further, the focus group discussion identified important constraints in maize production. Farmers identified, listed and ranked the problems based on their priority of the constraint in the study areas.

A maximum of 20 and a minimum of 15 farmers group participated in group discussion across all the study areas. The gender composition of the participating farmers was considered but most respondents were male farmers. Household leaders in almost all of the study areas were male farmers. Maize was the major cereal crop in terms of area under production in all of the study areas, except in West Shoa, where it was surpassed by teff. Maize was also number one in yield per hectare and total production in the mid-altitude, sub-humid areas of Western Ethiopia (CSA, 2010, 2011). The responding farmers preferred maize because of its adaptability and also for food security. Most of the farmers in the study areas utilized improved maize seed, except in their homestead where they use own seeds of local variety called 'Burre'.

The farmers in the study areas recognize the value of improved maize varieties because of higher yield, lodging resistance, flour quality, and market preferences. Farmers in majority of the study areas indicated that some improved maize varieties are more susceptible to diseases and insect pests than the landrace varieties, which they considered to be more adapted to the production

Table 8. Input and marketing constraints affecting maize production.

Constraint	Farmers	
	Frequency	Percent
Market		
Grain price fluctuation and low price	193	80.8
Storage facility	3	1.2
Price fluctuation and transport cost	14	6.0
Price fluctuation, transport cost and storage facility	12	5.0
Price fluctuation and storage facility	8	3.3
No sales of produce	7	2.9
Transportation cost	2	0.8
Total	240	100.0
Fertilizer		
Readily available	25	10.4
Moderately available	93	38.8
Poorly available	100	42.0
Not available	21	8.8
Do not use	1	0.4
Total	240	100.0

environments. Often the main maize growing season coincides with the severe outbreaks of leaf diseases, including NCLB, GLS and maize streak virus (MSV). Farmers indicated that there were little traditional or modern control methods applied for the control of leaf diseases on maize, especially NCLB and GLS.

Principal component analysis

PCA is most useful in data reduction and projection, and allows maintenance of smaller number of principal components than original variables. PCA is a powerful tool for pattern recognition, classification, modeling, and other aspects of data evaluation (Heberger et al., 2003). The current study used 86 variables to establish maize production constraints, using 240 respondent farmers. According to the PCA, 21 principal components (PC) were identified with Eigen values >1 (Sneath and Sokal, 1973) and contributed to 69% of the total variation (Table 10). As a result, these components and variables that were highly correlated were identified as important contributors to the variation presented in the data sets of this study (data not shown).

The most important variables well-correlated with PC1 ($r \geq 0.5$) were: which constraints limit your maize production, rank of maize among cereals produced for the last three years; maize productivity; availability of fertilizer; and widely grown maize varieties. Considering the PC2 and PC3, the important variables were: Accessibility to inputs; improved quality required from maize varieties; availability of improved seed for the production of maize; and criteria for selecting sites for maize (Table 9). Though PC1, PC2 and PC3 were the most important components and made a significant contribution to the variance (≈70%) in the data set, other components among the 21 were also relevant. Some of the important variables well-correlated with the rests of the PCs were: market for maize produce; trends of maize production; land utilization; area for maize; types of cereals grown, and preference for maize (data not shown). These were also important variables identified and retained through the PCA in the study. These variables could be used in the future studies on the preferences of farmers in similar maize growing environments.

DISCUSSION

In the present study, farmers in all the study areas grow an assemblage of different varieties in order to avoid the risk of crop failure caused by growing the same variety. Girma and Tanto (2008) reported that most farmers in Ethiopia grow a diverse assemblage of genotypes (landraces) to minimize risks of crop failure and to increase food security because the landraces are well-adapted to production environments, and carry farmers preferred traits. Further, in the Hararghe highlands of Eastern Ethiopia, Mulatu and Zelleke (2002) found that farmers' preferred landrace varieties over improved cultivars despite the higher yield potential of the latter. Both studies contended that landraces possess farmers preferred attributes, resulting from long term human selection.

Farmers considered high input costs, as compared to

Table 9. Important principal components with variables having r > 0.6.

Variable	Principal component	r > 0.6
Which constraints limit your maize production	PC1	0.825
How is the accessibility to inputs for maize production?	PCA2	0.877
what improved qualities would you like the improved maize cultivars to have	PCA3	0.676
Total area cultivated last production year	PCA4	0.868
Which cereal crops do you grow?	PCA5	0.814
What is the origin of maize in your community	PCA8	0.747
For how long have you been cultivating and consuming maize	PCA10	0.61
Where do you market your maize producer?	PCA14	0.806
What was the trend of maize production,	PCA18	0.664
Do you like maize production	PC20	0.846

Table 10. The Eigen values, percent of variance and cumulative variance explained by the 21 principal components.

PC	Initial Eigen values		
	Total	% of variance	Cumulative %
1	6.380	10.999	10.999
2	3.776	6.510	17.509
3	2.652	4.572	22.081
4	2.406	4.148	26.229
5	2.311	3.985	30.214
6	2.007	3.460	33.674
7	1.937	3.340	37.014
8	1.839	3.171	40.185
9	1.573	2.713	42.898
10	1.532	2.642	45.540
11	1.511	2.605	48.145
12	1.406	2.424	50.570
13	1.357	2.340	52.910
14	1.345	2.318	55.228
15	1.295	2.233	57.461
16	1.262	2.176	59.637
17	1.190	2.052	61.689
18	1.114	1.922	63.610
19	1.110	1.913	65.523
20	1.058	1.824	67.348
21	1.045	1.802	69.149

grain prices, as a key constraint because they believed that lower inputs costs would lead to alleviation of many other constraints. Dawit et al. (2008) found similar constraints in their studies of maize seed systems in the drought prone areas of Ethiopia. In some instances, farmers did not explicitly indicate the constraints they face. However, they said that they lack cash to buy the optimum crop input. Hailu (1992) reported the unavailability of agricultural inputs as the main bottleneck in maize production and productivity in Ethiopia. The prevalence of low quality agricultural inputs in the market is also one of the major constraints that farmers face. Maize seeds, for instance, are often adulterated or not true to type. Walter and Zewdie (2008) found that unavailability and untimely supply of seed are the major production constraints in maize growing areas of sub-Saharan Africa.

In this study, NCLB was reported as the number one disease followed by GLS. This contradicts Dagne et al. (2004) who reported GLS to be the number one leaf disease, followed by NCLB in the mid-altitude, sub-humid maize growing areas. The present result suggests that a shift in disease levels has occurred making NCLB a research priority. Declining soil fertility was also identified as a major limitation in the production of maize in these zones. Farmers indicated that a decline in soil fertility was exacerbated by mono-cropping and limited soil conservation practices.

The present study indicated that the unpredicted grain price was the major bottleneck in the maize grain market, which was further aggravated by a lack of, or poor infrastructure, including postharvest storage facilities and transport systems. In the study by Dawit et al. (2008), the volatility of the maize price was seen as one of the major challenges for maize production. Furthermore, this affected the allocation of land for maize seed production and the total quantity of maize seed produced over the last decade.

The present findings showed that maize production in the mid-altitude, sub-humid areas of Western Ethiopia is constrained by a myriad of related factors. The most important diseases, in decreasing order of importance, are NCLB, GLS and MSV. This study indicated a shift in the order of importance of major maize diseases in the study areas when compared with previous studies (Dagne et al., 2004). Recurrent future studies are required every 4 to 5 years in order to collate information on the relative importance of maize diseases and other relevant constraints in the area. Maize breeding cannot incorporate all the desired attributes instantaneously. But the key attributes should be included in a particular variety, and other candidate varieties should be bred focusing on the preferences of different groups of farmers. Equally, farmers should have access to improved seed and other inputs with reasonable price. To overcome the setback of low grain prices following maize harvests, farmers should be provided with technical and financial support to store their maize product till market prices improve.

Farmers have diverse perceptions and complex combinations of criteria to select maize varieties. From this study, farmers' preferred attributes included high yields, resistance to disease and insect pests, fair seed costs and the ability of a variety to give a reasonable yield with little application of external inputs, especially fertilizer and pesticides. To increase maize production, research should be inclusive and take into consideration the farmers' circumstances and preferences. Maize varieties and crop management packages developed should meet farmer's needs. Incorporation of farmers' preferences and production constraints in selection of maize varieties in any breeding project would increase the likelihood of adoption of the varieties. The findings from this study will form the basis for farmer-oriented maize breeding in the mid-altitude, sub-humid zone of Western Ethiopia and similar areas in the sub-Saharan Africa.

ACKNOWLEDGEMENTS

The Alliance for a Green Revolution in Africa (AGRA) is sincerely thanked for financial support to initiate the study. The Ethiopian Institute of Agricultural Research (EIAR) is acknowledged for providing leave of absence and hosting the field research to the first author. The all round support provided by the National Maize Research Project of Ethiopia is highly appreciated.

REFERENCES

Abebe G, Assefa T, Harun H, Mesfine T, Al-Tawaha AM (2005). Participatory selection of drought tolerant maize varieties using mother and baby methodology: A case study in the rift valley of Ethiopia. World J. Agric. Sci. 7 (1):22-27.

Ashby J (1991). Adopters and adapters. Participation of farmers in on-farm research. *In*: D.B. Thripp, ed. Planned Change in Farming Systems, Progress in On-Farm Research. New York: John Wiley and Sons. pp. 59-91.

Bandiera O, Rasul I (2006). Social networks and technology adoption in northern Mozambique. Econ. J. 116 (2):869-902.

Batz F, Janssen W, Peters K (2003). Predicting technology adoption to improve research priority setting. Agric. Econ. 28(3):151-164.

Birhane G, Bentayehu G (1989). The maize mega-environments of Eastern and Southern Africa and germplasm development. *In*: J.K. Ransom, ed. Maize Productivity Gains Through Research and Technology Dissemination, 3-5 December 1988 Nairobi. Nairobi: CIMMYT-Nairobi. pp.197-211.

Byerlee D, Spielman DJ, Alemu D, Gautam M (2007). Policies to promote cereal intensification in Ethiopia: a review of evidence and experience. International Food Policy Research Institute (IFPRI), discussion paper P. 707. IFPRI, Washington, D.C, U.S.A.

Ceccarelli S, Grando S, Bailey E, Amri A, El-Felah M, Nassif F, Rezgui S, Yahyaoui A (2001). Farmer participation in barley breeding in Syria, Morocco and Tunisia. Euphytica 122(2):521-536.

Chimdo A, Aberra D, Shemelis D, Fekadu B, Nigusse E, Belete G, Akalu T, Maikel T (2002). Research center based maize technology transfer: efforts and achievements. *In*: Mandefro N. and Tanner D. (Eds). Enhancing the Contribution of Maize to Food Security in Ethiopia, 11-13 November 2001, Addis Abeba. Addis Abeba: CIMMYT-Ethiopia.

CSA (2011). Agricultural sample survey. Report on farm management practices. Private peasant holdings, Meher season. Stat. Bull. Addis Ababa, Ethiopia. pp. 6-9.

CSA (2010). Agricultural sample survey. Report on farm management practices. Private peasant holdings, Meher season. Stat. Bull. Addis Ababa, Ethiopia. pp. 5-8.

CSA (2006). Agricultural sample survey. Report on farm management practices. Private peasant holdings, Meher season. Stat. Bull. Addis Ababa, Ethiopia. pp. 4-7.

CSA (2004). Agricultural sample survey. Report on farm management practices. Private peasant holdings, Meher season. Stat. Bull. Addis Ababa, Ethiopia. pp. 4-8.

CSA (2001). Agricultural sample survey. Report on farm management practices. Private peasant holdings, Meher season. Stat. Bull. Addis Ababa, Ethiopia. pp. 5-9.

Dagne W, Demisesew K, Girma D (2004). Assessment of losses in yield and yield components of maize varieties due to gray leaf spot. Pest Manag. J. Ethiop. 8(1):59-69.

Dawit A, Wilfred M, Nigussie M, Spielman DJ (2008). The maize seed system in Ethiopia: challenges and opportunities in drought prone areas. Afr. J. Agric. Res. 3 (1):305-314.

De Groote H, Bellon M (2000). Farmers' Participatory Research: Application for Breeders. Paper presented at a workshop "PRA for IRMA" Nov. 9-11, 2000– KARI HQ, Nairobi.

Doss CR, Mwangi W, Verkuijl H, de Groote H (2003). Adoption of maize and wheat technologies in eastern Africa: a synthesis of the findings of 22 case studies. CIMMYT Economics Working Paper 03-06, CIMMYT, Mexico, D.F.

FAO (2011). World Agricultural Production. [online] Available at: http://faostat.fao.org/default.aspx. Accessed 14 April. 2012.

FAO (2007). FAO Statistical Databases.[online]. Available at: http://faostat.fao.org/default.aspx. Accessed 1 July, 2007.

Feder G, Just R, Sibeman D (1981). Adoption of agricultural innovations

in developing countries. World Bank, Staff Working Paper P. 444, Washington, D.C., U.S.A.
Girma A, Teshale A, Hussen H, Tewodrose M and Abdel-Rahman Al-Tawaha M (2005). Participatory selection of drought tolerant maize varieties using mother and baby methodology: A case study in the semi arid zones of the central rift valley of Ethiopia. World J. Agric. Sci. 7(1):22-27.
Girma B, Tanto T (2008). Conservation of genetic diversity and supporting informal seed supply in Ethiopia. *In*: MH Thijssen, Z Bishaw, A Beshir and WS de Boef (Eds). Farmers, Seeds and Varieties: Supporting Informal Seed Supply in Ethiopia, 5-7 November 2007. Wageningen, The Netherlands. pp. 141-149.
Hailu G (1992). Availability and use of seed in Ethiopia. Program support unit, Canadian International Development Agency, Addis Ababa, Ethiopia.
Heberger K, Csomos E, Simon-Sarkadi L (2003). Principal component and linear discriminant analysis of free amino acids and biogenic amines in Hungarian wines. J. Agric. Food Chem. 51(1):8055-8060.
Joshi A, Witcombe JR (1996). Farmer participatory crop improvement. II. Participatory varietal selection, a case study in India. Exper. Agric. 32(2):461-477.
Mulatu E, Zelleke H (2002). Farmer's highland maize (*Zea mays L.*) selection criteria: implication for maize breeding for the Hararghe highlands of eastern Ethiopia. Euphytica 127(1):11-30.
Kebede M, Gezahegne B, Benti T, Mosisa W, Yigzew D, Asefa A (1993). Maize production trends and research in Ethiopia. *In:* T Benti and JK Ransom (Eds). Proceedings of the first national maize workshop of Ethiopia, 10-12 November 1992 Addis Abeba. Addis Abeba: CIMMYT. pp. 4-12.
Sneath PHA, Sokal RR (1973). Numerical Taxonomy 2nd ed. San Francisco: Freeman.SPSS, Institute. 2005. Statistical package for social sciences - Users guide. Chicago.
Wakene N, Tolera A, Minale L, Tolessa D, Tenaw W, Assefa M, Zerihun A (2012). Soil fertility management technologies for sustainable maize production in Ethiopia. In: W Mosisa, A Twumasi, W Legesse, T Berhanu, D Girma, B Gezahagn, W Dagne and M Prasanna (Eds). Ethiopia Meeting the Challenges of Global Climate Change and Food Security through Innovative Maize Research. Proceedings of the third national maize workshop of Ethiopia, 18-20 April 2011 Addis Abeba. Addis Abeba: EIAR/CIMMYT, Pp.123-127.
Walter SB, Zewdie B (2008). A system perspective for linking farmers and professionals supporting farmers' practices in seed supply. In: M.H. Thijssen, Z. Bishaw, A. Beshir and W.S. de Boef (Eds). Farmers, Seeds and Varieties: Supporting Informal Seed Supply in Ethiopia. Wageningen, The Netherlands. pp. 47-53.
Witcombe JR, Virk DS (2001). Number of crosses and population size for participatory and classical plant breeding. Euphytica 122 (1):451-462.

An econometric approach to ascertain sorghum supply response in Zimbabwe

Munyati Vincent[1], Mugabe Douglas[2], Chipunza Nyasha[2], Mafuse Never[1], Chagwiza Godfrey[1] and Musara Joseph

[1]Department of Agricultural Economics, Education and Extension, Bindura University of Science Education, P. Bag 1020, Bindura, Zimbabwe.
[2]Department of Agricultural Economics and Development, Midlands State University P. Bag 9055, Gweru, Zimbabwe.

Despite failures of government policy to stimulate sustained growth in sorghum production in the face of increased climatic shocks on maize, there have been very little efforts to understand sorghum response to policy incentives. The main purpose of the paper was to determine how sorghum farmers responded to changes in price and non-pricing policies. The major sources of the data were the ZIMSTAT, FAOSTAT, Meteorological department and the Ministry of Agriculture. The data on the area planted sorghum, capital expenditure and that of area of maize which was used as a substitute crop were obtained from Ministry of Agriculture. The data on price of sorghum, price of maize, exchange rate and inflation was obtained from ZIMSTAT. The data on the weather variable was obtained from the Meteorological Department. The international price of sorghum and maize were obtained from the FAOSTAT. The Consumer Price Index and inflation figures were obtained from the Reserve Bank of Zimbabwe (R.B.Z). The Nerlovian partial adjustment model was used to determine the responsiveness of sorghum farmers to price and non-price. It was found that sorghum supply is inelastic to own price both in the long run and short run. In the long run the own price elasticity was found to be 0.51 whilst in the short run was 0.24. This result means that agricultural price policy alone cannot guarantee sorghum production growth targets, but a policy mix that goes beyond factor and product markets and acknowledges the structural and institutional constraints faced by sorghum farmers is likely to achieve a substantial growth in sorghum output in both the short run and long run.

Key words: Econometric approach, sorghum supply response, Zimbabwe.

INTRODUCTION

In large parts of Africa, sorghum remains critically important for rural food security. Most of the output is consumed by the households producing the crop, and only a small proportion of harvests enter the commercial market. Since many sorghum-producing areas still experience periodic food deficits, production must be increased in order to improve household food security. Sorghum area will continue to expand over the foreseeable future as climatic risk increase and rural populations expand and crop land is extended into drier and more fragile ecosystems. Average sorghum yields, which have been falling by 1.0% per annum since the early 1980s, will need to grow in order for food production to keep pace with population growth.

The increase in the demand of small grains, particularly sorghum, is attributable to their growing importance in economies of both developed and developing countries. The growing importance of small grains is also a result of their adaptability to rainfall variability. With the advent of climate change, there has been increasing risks of crop failures due to frequent droughts and dry spells. As such there has been an enormous pressure among developing nations, particularly Sub-Saharan Africa whose majority of agriculture occurs under rain-fed conditions, to diversify into small grains that are less susceptible to moisture stress. Sorghum is therefore increasingly used as a substitute for maize in most parts of the region in order to reduce the problems of food insecurity.

It seems that the pricing and non-pricing policy are the major factors at the heart of Zimbabwe's sorghum activity stagnation in terms of output and these have contributed to the current starvation in the country. However, the return to grain marketing monopoly and price controls since 2000 has been a source of great concern to farmers. In light of these developments the pricing and non-pricing policy are the major factors contributing to Zimbabwe's sorghum activity stagnation in terms of output and these have contributed to the current starvation in the country.Policy makers need an applied understanding of how sorghum production responds to input and factor prices. This knowledge will enable them to meaningfully explore market and policy strategies to stimulate food and non-food production growth trends in the economy. Knowledge of elasticities of sorghum product supply and associated factor demand functions would improve precision of government policy makers to make better informed short run and long run planning decisions. The main objective of the paper is to determine the supply response of sorghum to different policy initiative by the government in-order to boost the production of sorghum in Zimbabwe.

METHODOLOGY

The data for the study was obtained from a number of sources. The major sources of the data were the ZIMSTAT, FAOSTAT, Meteorological department and the Ministry of Agriculture. The data on the area planted sorghum, capital expenditure and that of area of maize which was used as a substitute crop were obtained from Ministry of Agriculture. The data on price of sorghum, price of maize, exchange rate and inflation was obtained from ZIMSTAT. The data on the weather was obtained from the Meteorological Department. The international price of sorghum and maize were obtained from the FAOSTAT. The Consumer Price Index and inflation figures were obtained from the Reserve Bank of Zimbabwe (R.B.Z) publication. The CPI was used to deflate nominal prices and real prices which were used in the analysis. The analysis was conducted for the period 1980 to 2011 where more consistent data was readily available.

The Nerlovian model is a dynamic model, stating that area is a function of expected price, output adjustment, and some exogenous variables. The reduced form of the Nerlovian model is an autoregressive model because it includes lagged values of the dependent variable among its explanatory variables (Nkang et al., 2007).

Sadoulet and Janvry (1995) noted that a central problem in the estimation of supply response equation is that producers respond to expected as opposed to actual prices. In addition, they argued that the observed quantities may differ from the desired ones because of adjustment lags in the reallocation of variable factors.

$$Y_t - Y_{t-1}=\beta(Y^*_t - Y_{t-1}) \quad (1)$$

Thus

$$Y_t = \beta Y^*_t -+ (1-\beta)\ Yt_{-1} \quad (2)$$

Since farmers base their production plans on expected prices, the change in price expectation is specified as some proportion of the error made in formulating expectations in last year (Nkang et al., 2007; De Janvry and Sadoulet, 1995).

$$P^*_t - P^*_{t-1}=\delta\ (P_t - P^*_{t-1}) \quad (3)$$

Therefore

$$P^*_t= \delta P_t(1-\delta)P^*_{t-1} \quad (4)$$

However, the desired output can be specified as a function of expected price and other exogenous variables which influence supply, hence (Nkang et al., 2007; De Janvry and Sadoulet, 1995):

$$Y^*_t = a_0 + a_1\ P^*_t + a_2 Z_t + U_t \quad (5)$$

Since P^*_t is unobservable, we assume that farmers make their planting decisions based on their knowledge about prices that prevailed immediately in the preceding period that is ($P^*_t = P_{t-1}$). Hence, P^*_t is taken to be lagged price and when the other variables which affect supply response are included in Equation (5), we have

$$Y^*_t=a_0+a_1P_{t-1}+a_2PA_{t-1}+a_3ER+a_4CE_{t-1}+a_5\ \pi_{t-1}\ +a_6W_t+a_7D+a_8T+Ut \quad (6)$$

Substituting Equation (6) into (2), the following estimation equation is obtained (Nkang et al., 2007; De Janvry and Sadoulet, 1995):

$$Y_t = a_o\beta+ a_1\beta P_{t-1}+ a_2\beta PA_{t-1} + a_3\beta ER+ a_4\beta CE_{t-1} + a_5\beta\ \pi_{t-1} + a_6\beta W_t+ a_7\beta D + a_8\beta T + (1-\beta)Y_{t-1}+ e_t.......... \quad (7)$$

So that the reduced form of Equation (7) becomes the following equation which in the Nerlovian supply response function (Nkang et al., 2007; De Janvry and Sadoulet 1995):

$$\ln\ (Y_t) = b_0+b_1\ln\ (P_{t-1}) + b_2\ln\ (PAt_{-1}) + b_3\ln(ER) + b_4\ln\ (CE_{t-1}) + b_5\ln \pi_{t-1} +b_6\ln\ (W_t) + b_7D+b_8T+b_9\ln\ (Y_{t-1})+\varepsilon+... \quad (8)$$

Y_t = Actual hectarage of sorghum in year t; P_{t-1} = Lagged price of sorghum in year t; PA_{t-1} = lagged price of maize in year t; ER = Real exchange rate; CE_{t-1} = Government capital expenditure on agriculture lagged one year; π_{t-1} = Rate of inflation lagged one year; W_t = Weather variable (rainfall used as a proxy); D = SAP

Table 1. Factors affecting sorghum production in Zimbabwe.

Variable (area as the dependent variable)	Aggregate famers (whole sector)	Communal farmers	Commercial famers
Real price of sorghum(t-1)	0.24(1.01)*	0.38(0.94)*	0.46(1.95)**
Real price of maize(t-1)	-0.93(-4.79)***	-0.27(0.67)*	-0.29(-1.16)*
International price (t-1)	0.39(2.03)**	-1.2(-2.85)**	0.40(1.32)*
Exchange rate(t-1)	-0.2(3.22)***	0.11(0.87)*	0.18(2.41)***
Inflation rate(t-1)	-0.95(-1.02)*	-0.45(-2.41)**	-0.32(-2.72)***
Weather(rainfall as proxy)(t-1)	-0.65(-3.37)***	-0.50(-1.29)**	-0.47(2.17)***
Credit(agriculture) (t-1)	-0.52 (-3.72)***	0.34(1.41)**	-0.62(-2.99)***
Yields (Y_{t-1})(t-1)	0.12(0.58)	-0.307(-1.67)**	-0.323(-1.58)**
Area (A_{t-1})(t-1)	0.53(2.88)**	0.44(1.58)**	0.29(-2.16)**
Structural adjustment program dummy	-0.67(-4.18)***	0.68(2.07)**	-0.47(-2.02)**
Constant	15.79(4.00)***	9.1(1.56)*	23.64(7.14)***
R-squared	0.9971	0.9652	0.88
Adjusted R-Squared	0.9947	0.9357	0.7833
Durbin Watson test	2.5	2.14	1.817

***Significant at 0.01 level; ** significant at 0.05 level; * significant at 0.1 level.

dummy – to capture government policy shifts; T = Trend variable- to capture technological changes; Y_{t-1} = Area planted previous year, and Є = Error term (satisfying the normal classical regression assumption).

The supply model chosen uses area as the dependent variable. This approach is particularly applicable to Zimbabwe where the extreme variability of yields due to the unreliability of rainfall means that farmers have limited control over actual output. Area planted is therefore used to indicate farmers planned output.

The area of the crop planted the previous year has been included on the basis of Nerlove's Partial adjustment model which states that the achieved agricultural output by a farmer in any one period is only a fraction of the desired change. This means that the adjustment of farmers' crop plans to a change in price is unlikely to take place in full in one year but will probably persist and be distributed over several years. Nerlove's partial adjustment model shows that the partial adjustment in any one period can be measured with observed output (or as in this case observed area planted A_t) are written as a function of lagged observed output (lagged observed area planted A_{t-1}) and other exogenous variables (Nerlove, 1971).

The producer price variable has been lagged one year because farmers do not usually know what the price for their crop will be at the time of planting. The assumption is that they base their price expectations on the previous year's price. There are other factors which play a role in price expectations but it was considered reasonable to assume that farmers are most influenced by the previous year's price.

In deciding how much of a crop to plant farmers also take into account the opportunity cost of producing that crop which means the price of alternative crops will influence their decisions. These prices have also been lagged a year on the assumption that the previous year's price is the indication for expected price.

Elasticities of supply response

Elasticities, rather than estimated structural coefficients, are more appropriate for purposes of any comparisons that can be made. The short run and the long run elasticities were derived from the Nerlovian supply response model. Specifically the short run elasticities for the log form of the Nerlovian supply response function were taken as b_1. Long-run elasticities are calculated as follows:

$$E_i = \frac{E^{sr}}{1-\beta_1}, \tag{9}$$

E^{sr} = short-run elasticity and, β_1 is the coefficient for the lagged dependent variable (De Janvry and Sadoulet, 1995).

RESULTS AND DISCUSSION

The logarithmic form of the Nerlovian model was estimated using ordinary least squares. In this paper the supply response of commercial and smallholder sorghum farmers and aggregate supply response was estimated. The reason for choosing area was because it shows the decisions of farmers to plant more of the crop. It is a good indicator of farmers' response to policies, as farmers will decide how many hectares to plant sorghum given the policy environment.

From the results, it can be observed that the model fits the observed data fairly well as indicated by the R^2 values of 0.99, 0.96 and 0.88. The short run price elasticity of acreage response with respect to the lagged market price was computed as 0.24 (Table 1) for the whole sector of the sorghum enterprise. The coefficient of the deflated price in a paper by Alwan and El-Habbab was also found to be positive and the results are similar to that of the paper by Alwan and El-Habbab which were also positive. The sorghum sector is price inelastic to the changes in the price of sorghum in the country. The farmers are not responding to changes in the market price of sorghum. The communal farmer's responsiveness to the changes in the market price of sorghum is 0.38 (Table 1).The communal farmers are price inelastic to changes in the market prices of sorghum. The communal farmers usually produce for home consumption and not for the market. The commercial farmers produce for the market, the commercial farmers responsiveness to changes in the market price of sorghum is 0.46 (Table 1). The short-run market price elasticity with respect to the lagged price variable is inelastic for the commercial farmers. Whenever there is an increase in the price of sorghum the increase in the area of sorghum will not be proportional to the increase in the price of sorghum.

The area of the area planted in the previous year was found positive and its value is 0.52 which is between zero and one showing that its inelastic and this was also similar to a research by Alwan and El-Habbab which were also positive. This means that the farmers are slow in changing the planted areas. It took more than one year to do this change (Alwan and El-Habbab, 2002).

If the exchange rate is overvalued this will reduce the amount of area allocated to sorghum in Zimbabwe. The coefficient for the exchange rate is -0.2 (Table 1). For every rise in the exchange rate by 10% the area allocated to sorghum in the country will reduce by 2%. In order to increase the amount of area allocated to sorghum the country should devalue the currency and this will increase the area allocated to sorghum in the country. If the currency is overvalued the sorghum exports from the country will be expensive relative to other sorghum producing countries in the region.

The sorghum sector is highly sensitive to the changes in the maize prices this is due to the fact that maize and sorghum are substitutes and they compete for land so an increase in the price of maize will led to farmers switching to the production of maize. Before farmers grow a crop they will look at the opportunity cost of growing that the crop. The cross price elasticity of sorghum is -0.93 (Table 1), this means that for every increase in the price of maize by 10%, the acreage of sorghum will reduce by 9.3%.

In view of the low responsiveness of aggregate sorghum supply to price and sorghum international price it should be noted that farming uses land which is fixed in the short term. The low aggregate supply response is attributable to lack of technical progress and the slow rate of mechanization of agriculture by sorghum farmers. This does not, however, mean that positive sorghum prices can be neglected for output growth (though undoubtedly they are essential); it means that less can be expected from changing the general sorghum price level alone A package of changes may elicit a better response from farmers than a price change alone.

The elasticities of weather are all negative for both the communal and commercial farmers. The coefficient of weather for the sorghum subsector is -0.65 (Table 1), this is attributed to the fact that sorghum is a drought resistant crop and whenever there is an increase in the mean annual rainfall this will result in farmers growing other crops which requires a lot of rainfall. The farmers will grow other crops and reduce the area under sorghum in the country. Whenever the country experience above normal rainfall most farmers will grow other crops and leave sorghum which is a drought resistant crop. Usually sorghum is grown in times of drought because of its character as a drought resistant crop.

Long run elasticities

Long-run elasticities are calculated as follows:

$$E_i = \frac{E^{sr}}{1-\beta_1},$$

E^{sr} = short-run elasticity and, β_1 is the coefficient for the lagged dependent variable (De Janvry and Sadoulet, 1995).

The long run own price elasticity for the sorghum sector is 0.51 (Table 2). In the view of the low responsiveness of aggregate supply to price, it should be noted that agriculture uses land which is usually fixed both in the long term and short term and this is why the sorghum is inelastic to changes in the producer price of sorghum. For every rise in the producer price of sorghum by 10% (from Table 2) the area of sorghum will expand by 5.1% (Table 2) which shows that sorghum does not respond well to changes in own price. In a study by Nkang (2007) the long run price elasticity was also found out to be positive. The results are in line with previous studies on supply response which shows that if the price of an agriculture commodity increases the farmers are likely to respond to the changes in prices. If the price increases the acreage under the commodity will also increases.

In the long run the macro-economic variables a significant role in determining the level of production of most of agricultural commodities. A rise in the level of inflation will result in a reduction in the area of sorghum. This may due to the fact that inputs become relatively expensive and in the marketing side the personal disposable income is reduced and this will reduce the demand of traditional crops such as sorghum and consumers' will be buying basic commodities only.

Policy insights and implications of findings for policy

Agricultural price policy alone cannot guarantee sorghum production growth targets, but a policy mix that goes

Table 2. Long run elasticities.

Variable (Area as the dependent variable)	Aggregate famers
Real price of sorghum(t-1)	0.51
Real price of maize(t-1)	-1.98
Sorghum international price(t-1)	0.83
Exchange rate	-0.43
Inflation rate	-2.02
Weather (rainfall used as a proxy)	-1.38
Credit to the agricultural sector	-1.11
Yields(Y_{t-1})	0.26
Structural adjustment program dummy	-1.43

beyond factor and product markets and acknowledges the structural and institutional constraints faced by sorghum farmers is likely to achieve a substantial growth in sorghum output in both the short run and long run.

The most competitive niche for sorghum is in the rural market. Most semi-arid production zones where sorghum is widely grown are consistently net food importers. The immediate priority for improving the productivity of sorghum and its use must be to improve the food security of rural households. In most semi-arid parts of the country, major production gains will first serve to improve the level and stability of food consumption among rural households. Productivity gains must also be backed by improvements in grain storage and in increasing grain transfer directly from surplus to deficit households.

Efforts to promote sorghum production should not be based simply on desire for equity or concern about the welfare of those producing insufficient food. The development of the small grains food system should be viewed as a contribution to national economic growth. Policies and investment strategies should be designed to exploit the competitive advantages of these small grains- a basis for improving the productivity of the extensive semi-arid regions of the country and of their rural labour force. Gains to the economy will also accrue from improving rural food security, reducing the need for drought relief, lowering the level of subsidies underlying grain markets, and, at least in the short run, stemming migration from rural to urban areas.

Sorghum production is more sensitive to the rate of inflation in the country. Whenever there is a rise in the level of inflation this will result in the farmers reducing the area planted to sorghum. There is need for the country to implement policies that will reduce the level of inflation in the country because if the levels of inflation continue to increase this result in the country being food insufficient, leading to problems of food insecurity and the country will not be able to achieve some of the millennium goals of complete eradication of hunger.

Conclusion

The production of sorghum in the country was found to be more sensitive to the cross price elasticity of maize. Area under sorghum was also found to be sensitive to the producer price. An increase in the price of sorghum will result in more farmers growing the crop. The commercial farmers are responding to both price and non-price incentives more than the smallholder farmers in the country.

REFERENCES

De Janvry A, Sadoulet E (1995). Quantitative Development Policy Analysis. The Johns Hopkins University Press, London.

Nkang NM, Ndifan HN, Edet EO (2007). Maize Supply Response To Changes In Real Prices in Nigeria. A Vector Error Correction Approach. Agric. J. 2(3)419-425.

Evaluation of fuel wood marketing in Adamawa State, Nigeria

Taru, V. B.[1] and Ndaghu, A. A.[2]

[1]Department of Agricultural Economics, Federal University of Technology, P. M. B. 2076 Yola, Adamawa State, Nigeria.
[2]Department of Agricultural Technology, The Federal Polytechnic, P. M. B. 35 Mubi Adamawa State, Nigeria.

The falling of trees for fuel wood (energy) consumption cuts across most rural and urban communities in sub-Saharan Africa. However, this activity denies the existence of forest trees and wood is burnt to produce carbon dioxide which is linked to climate change. In this paper, fuel wood marketing in Adamawa State, Nigeria is evaluated. Both descriptive statistics and gross margin (GM) analysis were used to evaluate marketing activities of 120 fuel wood marketers selected using a simple random sampling technique. Specifically, the objectives were to investigate the marketing channel and as well to examine the GM in fuel wood marketing in the study area. The GM was N2,132.91 with margin of 28.00%. The distribution channel indicates that the urban retailers who buy from urban whole sellers sell the product to home users, bakeries, breweries, food sellers, kebab joints and local food processing units, while the rural whole sellers hewn down the product from the nearby joint.

Key words: Fuel wood, marketing channel, gross margin.

INTRODUCTION

The word marketing extends beyond dictionary definition and is a widely interconnected subject with numerous definitions. The American Marketing Association (AMA) revisits its definition every 5 years and in 2008 defined marketing as the activity, set of institution and processes for creating, communicating, delivering and exchange offering that have value for customers, clients, partners and society at large.

Fuel wood is one of the most important sources of domestic cooking and heating (energy) especially in the rural areas despite the fact that it is difficult to flame, smoky and hazardous for children. In the opinion of Nzeh and Eboh (2007), forests contribute directly and indirectly to rural household livelihoods through the generation of income and employment from the sale and exchange of gathered and unprocessed non-timber forest products such as fuel wood.

World Bank (2000) reported that in sub-Saharan African countries, especially Nigeria, fuel wood is the dominant source of energy for cooking and other activities. About 73% of the rural areas and urban households rely on fuel wood as a major source of energy for cooking. In urban areas, most people who use fuel wood buy in small bundles of– wood or charcoal, while in rural areas, people often gather their own fuel wood, sometimes rural people earn a small income by gathering fuel wood for those in urban areas. Despite these opportunities, fuel wood collection has series of environmental implications. It leads to deforestation and erosion which contribute to global climatic change. The demand for fuel wood is directly related to population. Adamawa State Economic Empowerment and Development Strategy (ADSEED, 2004) noted that the State used to be like a "Garden of Eden, with trees and forests everywhere" but with an increase in human population and more people now there is serious exploitation of natural resources, including cutting down trees for crop cultivation, shelter, cooking and exploitation of animals and fish. No one would ever have thought that these resources were exhaustible.

Furthermore, Agroforestry today (1993) showed that about 25% of the global warming effect is attributed to the clearing of tropical rainforest at the rate of 17 million hectare per annum. Also, CTA (2007) reported that Africa lost over 9% of its trees between 1990 and 2005 and this represented half of global loss. Bigger losers in this regard are countries like Angola, Cameroon, Democratic Republic of Congo, Nigeria, Sudan, Tanzania, Zambia and Zimbabwe.

The use of fuel wood has been on the increase due to increase in cost and scarcity of alternative sources, particularly Kerosene (Paul, 2008). In addition, fuel wood is consumed in large quantities in most parts of rural Africa.

In fact much has been known about the use, effect and exploitation of fuel wood but barely little is known about its marketing in Northern Nigeria in general and Adamawa State in particular. Salisu (2008) showed that the efficient functioning of the market system is based on a number of conditions. These include the existence and efficiency of certain legal and institutional foundations that guarantee private property, a well developed infrastructure that ensures reliable access to transportation and communication at minimum cost, and ready information about quality that must be symmetric.

More often than not, marketing of fuel wood has been facing the problem of lack of uniform measures, high transport cost and storage problems among other problems; it is against this background that that study evaluates the cost and returns, marketing margin and marketing channels of fuel wood in Adamawa State, Nigeria.

MATERIALS AND METHODS

The data for this study was predominately from primary sources. This was obtained through the use of structured questionnaires administered to 120 fuel wood marketers selected using a simple random sampling technique in Yola South and Yola North Local Government Areas of Adamawa State.

Analytical techniques

Budgetary techniques involving gross margin (GM) was used to determine the cost and returns associated with fuel wood marketing. GM was used for analysis because of the negligibility of fixed costs. GM is the differences between the gross income (GI) and the total variable cost (TVC) (Afolabi, 2007; Ehirim el al., 2007; Taru et al., 2010).

$$GM = GR - TVC$$

Where;
GM = gross margin (Naira/Bundle), TR = total revenue (N), TVC = total variable cost, *150 N = IUS $.

Marketing margin

The marketing margin is the differences between the price paid by purchaser (consumer) and that received by seller is expressed as (Olukosi et al., 2007):

$$MM = \frac{CP - SP}{CP} \times 100$$

RESULTS AND DISCUSSION

Marketing margin

Table 1 reveals the analysis of cost and returns associated with fuel wood marketing. The market tax constituted 4.6% of the TVC, while cost of ropes constituted 0.31% of the TVC. The average monthly cost of labour and cost of transportation were N121.35 and N113.28 and contributed 1.1 and 1%, respectively. The average monthly sales revenue is estimated to be N13,266.08, while the average monthly TVC is estimated to be N11,133.17. Therefore, the GM being the difference between total revenue and TVC is N 13, 26608 - N 11,133.17. The retailers make an average profit of N2,132.91 on fuel wood per month as revealed by the GMs, indicating that fuel wood marketing is profitable. It also suggests that the price is not detrimental for the consumers.

Marketing channel

According to Olukosi et al. (2007), market channel is the path of a commodity from its raw form to finished form. Figure 1 shows the distribution channel of fuel wood. It shows that the urban retailers who buy from urban wholesalers sell the product to home users, bakeries, breweries, food sellers, Kebab (suya) joints, and local food processing units, while the rural wholesalers hewn down the product from the nearby forest.

The retailers display the fuel wood in time and in a form convenient for the consumers. For bulk purchasers, the consumers can deal directly with wholesaler, thereby by-passing the retailers.

Marketing margin

The cost of fuel wood constituted 93% of the TVCs, while the average monthly cost of fuel wood was purchased at N10,35.6 and sold for N13,266.08 on average. The figures were extracted from Table 1, the margin is analyzed as follows:

$$\text{Market margin} = \frac{\text{Cost price - Seller's price}}{\text{Cost price}} \times 100$$

Table 1. Average costs and returns of fuel wood per month.

Item	Cost (N)	Percentage of total
Cost		
Total revenue		
Sales of fuel wood/month	13,266.08	
Variable cost		
Market tax	506.95	4.6
Cost of ropes	33.99	0.31
Labour	121.35	1.1
Cost of transportation	113.28	1.0
Cost of fuel wood	10,357.6	93.0
Total variable cost (TVC)	11,133.17	
Gross margin TR-TVC		2,132.91
Total		100

Source: Field Fancy (2010).

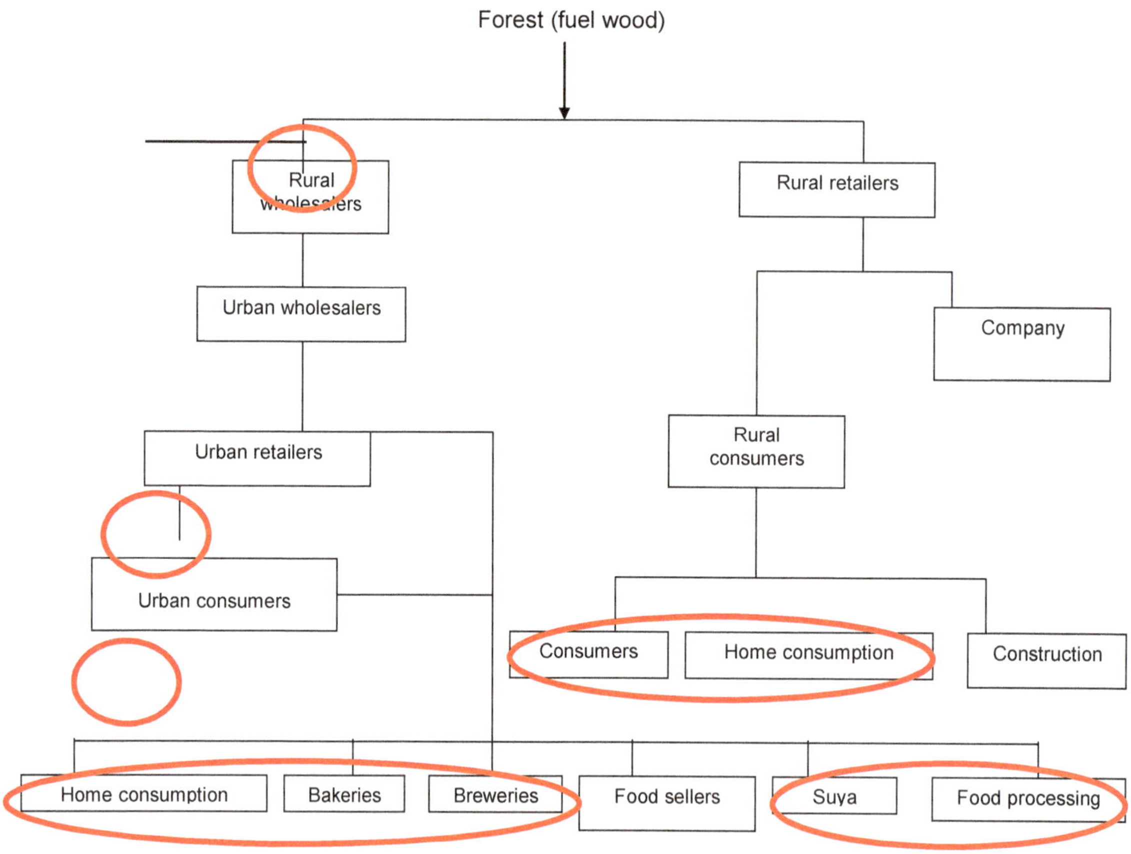

Figure 1. Marketing channel of fuel wood.

$$= \frac{10{,}357.6 - 13{,}266.08}{10{,}337.6} \times 100$$

= 28.08%

Analysis of fuel wood marketing margin shows 28.08% as the final selling price received by the retailers, while the remaining 71.92% goes to wholesalers and gatherers of fuel wood in the industry. The high percentage (71.92%)

captured by the gathers and wholesalers could well be the reason why they remain in the business.

Conclusion

The study revealed that marketing of fuel wood involves different agents namely gatherers, rural and urban wholesalers, rural and urban retailers. The sale of fuel wood that generates income to the marketers is in the study area especially the dry season. However, the business contributes significantly to climate change.

RECOMMENDATION

1) The demand for fuel wood is usually high due to inadequate supply and high cost of kerosene. Therefore, the government should ensure adequate supply of kerosene at much subsidized rate. This would ensure accessibility to kerosene by households and reduce the demand for fuel wood.
2) The government should intensify their campaign through sensitization programmes on the effect of deforestation to the environment, encouraging afforestation to replace felled trees

REFERENCES

Adamawa state Economic Empowerment and Development Strategy (ADSEED) (2004). Adamawa State Planning Commission. P. 65.

Afolabi JA (2007). Evaluation of Poultry Egg Marketing in South-Western Nigeria. Int. J. Poultry Sci. 6(5):362-336.

Agroforestry today (July-September 1993). The World Wide Problem of slash and burn Agriculture. 2:4-6.

Center for Tropical Agriculture CTA (2007). Sharing knowledge, improving rural Liveli-hood. pp. 1-6.

Ehirim NLC, Onyemauwa CS, Ikheloa E, Umezurumba IE (2007). Economics of yam Marketing in Umuahia, Abia state, Nigeria. Int. J. Agric. Food Syst. 1(91):51-56.

Olukosi JO, Isitor SU, Moses OO (2007). Introduction to Agricultural Marketing and Prices. Principle and application 3rd edition living 3rd edition Living Books series publication Abuja, Nigeria. pp. 1(27):51866.

Nzeh CEP, Eboh EC (2007). Analysis Of Income Effects Of Forest Products Activities Among Rural Households In Enugu State, Nigeria. J. Agric. Soc. Res. 7(1):23-33.

Paul J (2008). Comparative Analysis of household Energy use in Yola Metropolitan Area. An unpublished M.Sc Thesis Department of Agricultural Economics and Extension, Federal University of Technology, Yola Adamawa state Nigeria.

Salisu M (2008). Structure and performance of cowpea in Yola Metropolis of Adamawa state unpublished M.Sc Thesis Department of agricultural Economics and extension federal university of technology Yola Adamawa State Nigeria.

Taru VB, Kyagya IZ, Mshehia IS (2010). Profitability of Groundnut Production In Michika local government area of Adamawa state. World J. Agric. Sci. 1(1):25-29.

World Bank (2000). Solution for sustainable charcoal use leaflet.

11

Characteristics of market garden in Anambra State, Nigeria

Ofoka C. I., Chah J. M. and Madukwe M. C.

Department of Agricultural Extension, University of Nigeria Nsukka, Nigeria.

The study examined the characteristics of market gardens in Anambra State Nigeria. Questionnaire was used to collect data from 48 respondents using snow ball sampling procedure. Descriptive statistics were used to analyze data. Majority (79.2%) of the respondents were sole owners of the gardens. About 38% of the respondents used personal savings as initial investment in their operations. A greater proportion (31%) of the respondents invested N111,000 - N140,000 in establishing gardens. The output of the gardens include *Citrus aurantifolia* (100%), hedge seedlings such as *Ixora coccinea* (100%), *Solanum melongena (*87.5%), medicinal seedlings (e.g. *Ocimium viridus*) and grasses (52.1%). All the respondents sold their produce at farm gates. Majority (98.0%) of the operators used wheel barrow for transportation of the produce for sale. Watering, planting in polythene bags, and manure application were practiced by all the respondents. All the respondents used rain harvested water in their gardens. However, 60.4% of the gardeners dug shallow wells. Some (29.2%) of the respondents had pruning saw, 25.0% raffia rope and 22.9% budding knife as tools used for budding. Constraints encountered by farmers include difficulty in securing land, lack of capital and ability to control fire outbreak.

Key words: Market garden, characteristics, fruits, vegetables, grasses, Nigeria.

INTRODUCTION

Market garden is the name given to commercial horticulture and is defined as the science and art of cultivating, processing and marketing of fruits, vegetables, nuts and ornamental plants (Adebisi-Oyedele et al., 2010). According to Ojeifor et al. (2006) market gardening is the growing of vegetables, mostly for commercial purposes and for domestic use. The economic potentials and production contribution of fruits, vegetables and other horticultural crops to the agricultural economy of Nigeria were estimated at 20 billion naira (FAOSTAT, 2007) which was more than that of forestry (1.4 billion) and fishery (1.25 billion). On the other hand, the country was only able to export 9 metric tons (MT) of fresh/dried horticultural produce/products in 2002 and 3.2 MT in 2003, even though international trade in fresh horticultural and floricultural products (FHFP) was growing at a rate of 7% per year, compared with only 2% for staple crops (Gioè, 2006). Therefore, the export sector faces major challenges to maintain growth of exporting horticultural products. Any growth in the domestic horticultural industry is likely to reduce poverty in different ways in rural and urban areas. Bachmann (2002) reports that market gardening entails the intense production of high-value crops and gives farmers the potential to increase their income from a few acres.

The government of Anambra State established market gardens in Awka, Onitsha and Nnewi only in 2005 (Erinne, 2005). The delay in setting up government market gardens gave private individuals opportunities to establish small scale gardens in cities and towns to satisfy the ever increasing demand for horticultural products. New regulations in international horticultural trade and intense competition are redefining the marketing system which may challenge gardeners. In other to curtail challenges face by market gardeners in Anambra State, and therefore improve on their production scale, there is need to examine the characteristics of market garden operations in the State; describe the socioeconomic characteristics of market gardeners and identify constraints encountered in market garden enterprises in Anambra State.

METHODOLOGY

The study was conducted in Anambra State of Nigeria. The state lies between latitudes 5°4[1] and 7°05[1] North and longitude 6°3[1] East and occupies an area of approximately 5025 km^2 (NPC, 2006). The inhabitants of the rural communities are mainly farmers. Within urban towns and cities, there is the cultivation of market gardens. Anambra State is divided into four agricultural zones (namely: Aguata, Anambra, Awka and Onitsha). The agricultural zones are comprised of 21 Local Governments Areas (LGAs).

Three out of the four agricultural zones were purposively selected (Onitsha, Aguata and Awka) due to high concentration of market garden operators. Two LGAs were similarly purposively selected from each of the agricultural zones to give a total of six LGAs used for the study. Selected LGAs were Onitsha North and South, Awka North and South, Nnewi North and Aguata. Each LGA has on the average 15 market gardens. Considering the manageable number and strong network among gardeners in the state, snowball sampling procedure was used to select eight operators from each of the six LGAs giving a total of 48 respondents. Questionnaire was used to collect data for the study. The questionnaire contained questions on the socio-economic characteristics of respondents, characteristics of market garden operations and constraints encountered in market garden enterprises

To examine the characteristics of market garden operations, the respondents were asked to indicate their sources of fund for investments, the amount invested on equipment, tools and planting materials, and production practices. Respondents were asked to indicate their marketing channels, sources of water and methods of transportation of their products. To seek information on the constraints confronting the gardeners, 16 possible constraints were listed and respondents were asked to indicate the level of their perceived seriousness on a 3-point Likert type scale. Their response categories were as follows: Not serious (NS) = 0, serious (S)=1 and very serious (VS)=2. The cutoff point was determined as follows; 0+1+2=3/3=1. Any constraint that had a mean score of one and above was regarded as being serious while any mean score below one was regarded as not being serious. Descriptive statistics was used to analysed data collected.

RESULTS AND DISCUSSION

Socio-economic characteristics of respondents

The percentage distribution of respondents according to their socio-economic characteristics is shown in Table 1. Majority (79.2%) of the respondents were sole owners/proprietor of their operation. Others were managers (14.6%) and workers (6.3%). About 42% were within the age bracket of 41 to 50 years. Majority (70.8%) of the respondents were males. All the respondents had some form of formal education and 72.9% practiced market gardening on full time bases. Sixty two percent had no formal training on market garden and 83.3% were willing to undergo training. The areas of training indicated by respondents were pest control (75.0%), propagation by budding (52.1%), herbicide use (75.0%), marketing strategies (47.9%) and wreath making and marketing (41.7%). Most (52.0%) used government land for their operations.

Characteristics of market garden operations

Source of capital and amount invested

A greater proportion (37.5%) of the respondents used personal savings as initial investment in their operations (Table 2). Other sources of initial capital invested included: family sources (33.3%), loan from friends (25.0%) and bank loan (4.2%). High interest rates and requirement for collateral demanded by banks may be responsible for the poor utilization of banks as source of capital for investment in market gardening in the state. Thus, to boost market garden activities in the state, state government should provide financial assistance to market gardeners. About 31% of the respondents invested N111,000 - N140,000 while 25.0, 8.3 and 4.1% invested N171,000 and above, N51,000 - N80.000 and 20,000-50.000, respectively. It was observed that market garden operators adopted working tools used in other farming activities in their enterprises. For this reason, capital was invested only for purchasing those items not owned, replacing damaged ones and planting materials.

Tools used in market gardening

Polythene bags were used by all the respondents for their garden operations as shown in Table 3. This is because it enhances the ease of handling and conveyance of seedlings for transplanting and to the market. Water storage containers owned by 92% of the respondents, was found to be important to the extent that even those located very close to perennial source of water still deemed it necessary to own water storage containers. Water storage containers are very important for market gardening activities especially during the dry season. Other tools used for production include hoes (98%), spade (93.8%), machetes (81.3%), hand trowel (83.3%) and digging folk (83.3%). Although a pump is very necessary for distributing water irrespective of the source of water, only 12.5% of the respondents owned one. This may be because farmers could not afford the cost. The use of these implements suggests that, market gardeners in the study area were small scale operators. The Kenyan Horticultural Crops Development Authority estimated that 40% of exported fruit and 70% of exported vegetables

Table 1. Percentage distribution of respondents by socio-economic characteristic.

Variable	Percentage
Sole owner/proprietor	79.2
Manager	14.6
Worker	14.6
Sex	
Male	70.8
Female	29.2
Age	
20-30	14.6
31-40	29.2
41-50	41.7
51-60	10.4
61 and above	4.2
Educational qualification	
Primary school incomplete	25.0
Primary school completed	6.3
Secondary school incomplete	20.8
Secondary school completed	33.3
NCE, OND, HSC	8.3
HND, Degree	6.3
Mode of farming	
Full-time	72.9
Part-time	27.1
Type of training on horticulture received	
Formal	25.0
Non-formal	62.5
No training	12.5
Willingness to undergo further horticultural training	
Yes	83.2
No	16.8
Areas of training needs*	
Pest control	75.0
Propagation by budding	52.1
Herbicide weed control	75.0
Marketing strategies	47.9
Wreath making	41.7
Land acquisition	
Government land (government land)	
Rentage	52.0
Purchase	26.0
Inheritance	16.0

*Multiple response, NCE = National Certificate of Education, OND- Ordinary National Diploma, HSC- Higher School Certificate, HND- Higher National Diploma.

Table 2. Percentage distribution of respondents based on source of capital and amount invested.

Variable	Percentage (n=45)
Source of capital*	
Personal savings	37.5
Family sources	33.3
Loan (from friends)	25.0
Bank Loan	4.2
Capital invested (₦)**	
20,000-50.000	4.2
51,000-80.000	8.3
81,000-110,000	0.0
111,000-140,000	31.3
141,000-170,000	0.00
171,000 and above	25.0

*Multiple responses ₦: Naira, **₦160 = US Dollar 1.

Table 3. Percentage distribution of respondents according to equipment and tools purchased and use for market gardening.

Equipment	Percentage
Polythene bags	100.0
Spade	93.8
Water tank	91.7
Plastic basins	85.4
Trowel	83.3
Machetes	81.3
Wheel barrow	68.8
Head pan	62.5
Hoe	97.9
Knapsack sprayer	35.4
Pruning saw	29.2
Raffia rope	25.00
Budding knife	22.9
Budding stool	20.8
Twine	18.8
Wire netting	16.7
watering can	14.6
Water pump	12.5
Motorcycle	8.3
Digging fork	8.3

are produced by smallholders (Harris, 2001). Some (68.0%) of the respondents had wheel barrow while 62.0% had head pan which were tools used for carrying seedlings and other materials within the farm. Others had pruning saw (29.2%), raffia rope (25.0%), budding knife (22.9%) and twine (20.2%) as tools used for budding practices in the garden.

Most important market garden seedlings/crops produced

A relatively restricted number of crops dominated market garden activities in the study area. Table 4 shows the seedlings and market garden products grown in the study area. All the respondents produced citrus seedlings/fruits. Majority (83.3%) of the respondents produced each of mango and guava seedlings/fruits. About 73% produced paw-paw seedlings/fruits, 62.5% avocado, 62.5% *Irvinga gabonesis*, 52.1% plantain and bannana, 52.1% apple, 31.3% cashew and 20.8% melon seedlings. Fruit from these plants rarely existed in Anambra State before the year 2000 (Olufolaji, 2000).

Majority (87.9%) of the respondents were involved in the production of egg-plant (*Solanum melongena*), pepper, pumpkin, cury leaves and bitter leave (*Vernonia amygdalina*) (Table 4). The market garden activities undertaken by majority of the respondents studied were vegetable production. This may be attributed to the fact that there is more demand for vegetables than fruits as they are used for everyday cooking. Fruit and vegetables play an important role in human nutrition and health (Gioè, 2006). World Bank (2004) found that the vegetable sector presents more market opportunities for sub Saharan Africa (SSA) producers than the fruit sector. All the respondents produced hedge seedlings and spot beautification seedlings while 81.3% produced grasses and 52.1% produced medicinal seedlings. This implies that ornamental plants are of immense importance in the study area. Lawson (2000) opined that the economic importance of ornamentals has been increasing in many countries, and international demand has rapidly expanded.

Production practices in market garden enterprise

Watering, planting in polythene bags, and manuring were practiced by all the respondents (Table 5). This indicates the importance of these practices to the success of market garden operations. Sixty percent of the operators practiced mulching during dry season to conserve water. Budding was carried out by 41.7% of the respondents while 21.0% practiced grafting to enhance early maturity and fruiting of crops. The use of fencing sticks (16.7%) and wire gauze (35.4%) around the seedlings was to protect seedlings from damage by predators. A small proportion of respondents employed chemicals in the control of weed (10.4%) and pest (16.7%) in their gardens. This may be attributed to the gardeners'' lack of knowledge on the availability or method of application of chemicals in weed and pest control. Despite the low use of chemicals for pest control it has been noted that,

Table 4. Percentage distribution of respondents according to types of market garden products.

Types of seedling/produce*	Common name	Percentage
Fruit seedlings		
Citrus aurantifolia		100
Irvinga gabonesis	Bush mango	62.5
Carica papaya.	Pawpaw	72.9
Annona muricata	Soursop	41.7
Mangifera Indica	Mango	83.3
Musa species	Plantain/banana	52.1
Anacardium Occidentale	Cashew	31.3
Cycamine		20.8
Psidium guajava	Guava	83.3
Malus domestica	Apple	52.1
Persea Americana	Avocado pear	62.5
Pyrus communis	Pear	62.5
Vegetable seedlings/produce*		
Abelmoschus Esculentus	Okra	18.8
cucurbitaceae spp	Melon	20.8
Amaranthus viridus L.	Green	62.5
Talinum Triangulare,	Water leaf	41.7
Vernonia amygdalina	Bitter leaf	72.9
Cucurbita moschata Decne	Pumpkin	81.3
Corchoris olitorius		41.7
Pegularial extensa seedling and leaf		62.5
Murraya Koenigii leaves and seedling	Cury leaves	72.9
Lactuca sativa leave	Lettuce leaves	10.4
Capsicum annuum seedling and fruit	Pepper	83.3
Solanum melongena	Egg plant	87.9
Ornamental seedlings*		
Hedge seedlings: cherry (*Ixora coccinea, Prunus spp),* yellow fiscus (*Flagellaria indica*) and yellow bush (*Lupinus arboreus*), *Hibiscus* spp. etc)		100
Spot beautification: seedlings pines (*Pinus aphremphous*), *Acasia* spp, ashoka tree (*Saraca indica* etc)		100
Grasses: (*Chasmistes cujus* (cui-cui), carpet grass (*Axonopus fissifolius*), peanut grass *(Arachis hypogaea)* etc)		81.3
Medicinal seedlings: nchuanwu (*ocinium viridus*), lemon grass (*Cymbopogon* spp), neem (*Azadirachta indica*), ginger (*Zingiber officinale*), garlic (*Allium sativum*) etc.		52.1

*Multiple responses.

pesticide use will continue to be a significant strategy for dealing with arthropod pests so that ornamental producers can stay competitive in both national and international markets (Bethke and Cloyd, 2009).

Market channels used by respondents

All the respondents sold their produce at farm gates (Table 6). Sale at farm gates results in reduction of cost of production and a resultant increase in profit margin. In addition to farm gate sales, 53.2% of the respondents indicated that they also sold their products along the streets. Over 37% of the respondents sold their products by decorating entertainment places, while 35.0% made wreath during funerals and 31.3% sold in super markets. This shows that respondents had multiple market channels to dispose of their produce. Bachmann (2002) indicated that farmers' market, roadside stands, pick-your-own, subscription marketing arrangements, and sales to restaurants are common marketing strategies in market gardening. According to Erinne (2005), effective

Table 5. Percentage distribution of respondents according to production practices in market garden enterprise.

Production practices*	Percentage
Planting in polythene bags	100.00
Manuring	100.0
Watering	100.0
Weeding (hand weeding)	87.5
Provision of light shade over the seedlings	83.3
Rogueing out diseased plants	81.3
Clay slurring of seedling exposed roots of seedlings	81.3
Mulching	60.4
Budding	41.7
Wire guaze	35.4
Planting insitu	20.8
Grafting	20.8
Use of fencing sticks around the seedlings	16.7
Chemical control of pest	16.7
Chemical control of weed	10.4

*Multiple responses.

Table 6. Percentage distribution of respondents base on marketing channels.

Market channels*	Percentage
Farm gate	100.0
Along the streets (carried in barrows or motor cycles)	52.1
Landscaping of clientele houses	52.1
Decorations of entertainment places	37.5
Making of wreath during funerals	35.0
local daily markets	33.3
Super markets	31.3
Primary and secondary schools	16.7
Photographers studios	16.7

*Multiple responses.

marketing is a function of marketing strategies and channels adopted to distributed and sell products. Farmers need to develop a clearly focused marketing plan before any crops are planted (Bachmann, 2002). Having marketing channels in mind before planting helps to ensure but will not guarantee that most of what you plant will be sold, thus eliminating wasted time, space, produce, and money.

Sources of water used in market garden enterprises in Anambra State

All the respondents used harvested rain water in their gardens (Table 7), because according to them it was cheaper. It was observed that most (70.8%) of the market garden enterprises were located near perennial streams. However, 60.4% of the gardeners dug shallow wells while 41.7% used piped borne water and another 48.0% used tanker vendors. About 52% of the respondents used water from boreholes. The water sources used by most of the respondents depended on accessibility and cost. However, the danger associated with the use of perennial streams and other small surface sources is that the period of maximum need will probably coincide with the period of lowest water levels.

Hence, particularly with resource-poor small farmers, it is essential that sustainable and cost-effective methods of applying and managing irrigation water are adopted, and that the quality of the water applied and its impact on soil and crop water balances is carefully monitored.

Means of transporting produce

Majority (98.0%) of the operators used wheel barrow for

Table 7. Percentage distribution of respondents based on sources of water used in the enterprises.

Sources of water	Percentage
Rain	100.0
Perennial streams	70.8
Shallow well	60.4
Pipe borne water	41.7
Bore hole	52.1
Buy from tanker vendors	48.0

Table 8. Percentage distribution of respondents based on methods of transportation.

Means of transportation*	Percentage
Personal car	60.4
Taxi	54.2
Bus	41.7
Commercial motorcycle	12.5
Personal motorcycle	50.0
Wheel barrow	97.9

*Multiple response.

transportation of the produce for sale as shown in Table 8. About 60% of the respondents used personal car, 54.2% used taxi, and 50.0% used personal motorcycle. Means of transportation was dependent on the affordability and quantity of produce to be sold or carried.

Constraints encountered by market garden operators

Out of 15 possible constraints listed, 12 were considered serious in market garden enterprises in Anambra State (Table 9). These include: lack of capital (M=1.5), poor sales during dry season (M=1.5), high cost of transportation (M=1.4), no operators association for co-operate actions (M=1.4), poor access roads (M=1.3), high cost of inputs (M=1.3), low extension visit (M=1.3) wild fire outbreak (M=1.2), inadequate land for expansion, (M=1.2), and high cost of labour (M=1.2). High cost of implements (M=1.1) and lack of credit facilities (M=1.1) were also identified as constraints by operators. These factors may constitute a barrier for respondents to expand their operations and as well as discourage individuals taking up market garden enterprise in the state.

Marketing can be hampered by poor access roads. This may be the reason why most of the farmers sale at farm gates. This can also act as barrier to participation in the value chain for small farms and firms. Small producers and processing firms are frequently eliminated from markets for failure to understand market dynamics

Table 9. Mean score of constraints encountered by market garden operators.

Constraint	Mean (M)
Poor sales during dry season	1.5*
Lack of capital	1.5*
High cost of transportation	1.4*
No gardeners' association or co-operate action	1.4*
Low extension workers visits	1.3*
High cost of input	1.3*
lack of credit facilities	1.1*
Poor access roads	1.3*
Fire outbreak	1.2*
Inadequate land space for expansion	1.2*
High cost of labour	1.2*
Lack of credit facilities	1.1*
High cost of implements (tools)	1.1*
Lack of production skill	0.9
Lack of water	0.9
Government interference	0.8

or, because of their inability to meet new production, sanitary and quality standards. Because market garden development and marketing are strongly dependent upon knowledge, human capital and technical inputs must be provided if interventions and growth are to be sustainable.

Compared to cereal crops, most market garden crops demand high levels of inputs. Producers in developing regions often lack access to appropriate inputs and the necessary technical production skills due to inadequate input and credit markets as well as weak extension systems (USAID, 2005). Improving access to appropriate inputs and information resources can help farmers raise productivity and contribute to sound natural resource management. USAID (2005) reported that, inadequate government regulation and organization, coupled with insufficient input markets and infrastructure, severely limit rural producers' access to planting materials.

The absence of an effective, well-trained extension network is a significant constraint to the development of the market garden industry and the capacity of small producers in particular. Ineffective and inaccessible extension and education networks have resulted in inadequate human technical capacity and expertise throughout the market garden chain in developing countries. The market garden industry changes rapidly, and extension and education networks must have effective mechanisms to ensure that all parts of the market chain are trained in currently required methodologies and practices. Inadequate land, lack of labour and access to credit are of core importance to all aspects of market garden development as these may hinder expansion. However, high labour demands of market garden production have the benefit of local employment generation. Per hectare, the production of market garden crops creates more than twice the number of jobs that cereal production generates (Ali and Farooq, 2002).

Gardeners association is also an important factor to increase productivity among producers as this may increase their bargaining power and information sharing/exchange. This did not exist in the study area and consequently, new production knowledge, standards, regulations and innovations did not reach the producers. In the absence of this information, small producers are at an economic disadvantage and often find their product excluded from the marketplace for lack of compliance with poorly understood regulations. Given the high local demand for market garden produce and the potential for exports, it is important to address these challenges for producers to benefit from market opportunities and to increase their incomes. Dealing with these challenges will require a whole value chain approach.

Conclusion

Market garden enterprise in Anambra State is operated on a small scale. All the respondents used harvested rain water in their gardens as well as sold their produce at farm gates. Also, watering, planting in polythene bags and manuring were practiced by all the respondents. However, operators lack proper knowledge about chemicals for pest and disease control and encountered a number of problems such as lack of capital, fire outbreak, difficulty in securing land space for expansion, inadequate extension visit etc. nonetheless, it is a lucrative venture and therefore a veritable source of livelihood for the operators. Young people, especially those in the job market and retired civil servants could be encouraged to take up full-time market gardening as a profession by providing funds, land and ensured market of produce. There is need for extension education on the use of chemical for pest and disease control through seminars, demonstrations and regular visits.

REFERENCES

Adebisi-Adelani O, Adeoye IB, Usman IB, Olajide-Taiwo FB, Usman JM, Agbarevoh P, Oyedele OO (2010). Gender analysis of production, potentials and constraints of *Thaumatococcus danielli* in Ekiti State. Cont. J. Agric. Sci. 4:54-59.

Ali M, Farooq US (2002). Vegetable research and development in the ASEAN region: a guideline for setting priorities. In: C G Kuo eds. "Perspective in ASEAN cooperation in vegetable in research and development Ashanhua, Taiwan" Asian Veg. Res. Dev. Center. pp. 20-64.

Bachmann J (2002). Market gardening: A start-up guide. Horticultural system guide. Appropriate Technology Transfer for Rural Area (ATTRA). www.attra.ncat.org accessed 29/08/12.

Bethke JA, Cloyd RA (2009). Pesticide use in ornamental production: what are the benefits. Pest Manag. Sci. 65:345-50.

Erinne EC (2005). Need for establishment of functional market gardens at Awka Capital city and other agricultural zonal offices of Anambra State MOA/Comm/148/1/1.

FAOSTAT (2007). Statistical Yearbooks (http://www.faostat.org)

Gioè M (2006). Can Horticultural Production Help AfricanSmallholders to Escape Dependence on Exportof Tropical Agricultural Commodities? Crossroads 6:16-65.

NPC (2006). National population census news, Federal Republic of Nigeria.

Ojeifor IM., Emuh FU, Denton OA (2006). Crop production systems of market gardens in Nigeria. J. Food Agric. Environ. 4:246-250.

Olufolaji A (2000). Horticulture in Nigeria. Report presented to Korea International Cooperation Agency (KOICA) and the International Agricultural Training Center (IATC), Kyung Pook.

USAID (2005). Global horticultural assessment. The world vegetable center. June. Available http://pdf.usaid.gov/pdf_docs/pnadh769.pdf. Accessed 26/12/12.

World Bank (2004). Global Economic Prospects 2005. Washington D.C.: The World Bank.

12

Role of wetlands resource utilisation on community livelihoods: The case of Songwe River Basin, Tanzania

D. Kalisa[1], A. Majule[2] and J. G. Lyimo[2]

[1]Ministry of Agriculture, Box 9192, Dar es Salaam, Tanzania.
[2]Institute of Resource Assessment, University of Dar es salaam, Box 35097 Dar es Salaam, Tanzania.

The role of wetland resources on community livelihoods was examined in Katumbasongwe and Mpunguti villages in Songwe River Basin (SRB), Tanzania. The study focused on wetland resource utilization patterns and its significance in supporting rural livelihoods. Various methods were used to complement each other in addressing the study, namely structured questionnaires, participatory methods and remote sensing techniques. The study shows that 81.7 and 85.6% of the sample population in Katumbasongwe and Mpunguti village respectively depended largely on wetland resources for their livelihoods. These include rice production, livestock keeping, fishing and harvesting of macrophytes for weaving. Wetland resource utilisation patterns have been towards expansion of rice production and settlement at the expense of other land cover types. This has been driven largely by high market demand for rice and increased population. The study observed that the nature and extent of wetland resource utilization patterns vary considerably from one socio-economic group to another depending on their flexibility and access to livelihood assets. Agricultural intensification and input support to farmers is expected to increase agricultural productivity. However, such intervention should take into consideration the socio-economic differentiation of the households. Moreover, proper land use plan and management are crucial for enhanced sustainability of wetland resources and other potentials in supporting livelihoods.

Key words: Agriculture, wetland resources, land use, rural livelihood, Songwe River basin.

INTRODUCTION

In many sub-Saharan African (SSA) countries, agriculture is the main source of livelihood to most of the rural communities (Majule and Omolo, 2008). However, agricultural production is constrained by several factors including scarcity of arable land, declining soil fertility and environmental changes particularly climate. The need to address some of these constraints as well as increasing market demands of agricultural products has compelled communities to increase utilization of wetlands for agriculture (Majule and Mwalyosi, 2005; Lyimo, 2005). Wetlands in Tanzania are among the world's most biologically productive ecosystems and are rich in species diversity (Mwakaje, 2009). It play significant role in supporting livelihoods through provision of a variety of goods and services to diverse users. For example in Tanzania some of the wetlands have been used for production of crops such as large scale rice production and vegetables, direct resource extractions and good source of pasture for livestock production especially during the dry season (Lyimo, 2005; Majule and Mwalyosi, 2005; Mwakaje, 2009). Small-scale fisheries in the wetlands are source of food security, employment, and generate additional income to rural communities living in such areas.

This study examined wetland resource utilization patterns and their implications to rural livelihoods in Songwe River Basin (SRB). Wetlands are defined by the Ramsar Convention of 1971 as "areas of marsh, fen, peat land or water, whether natural or artificial, permanent or temporary, with water that is static or flowing, fresh, brackish or salt, including areas of marine water, the depth of which at low tide does not exceed six meters" (Davis, 1994). Wetlands in the SRB contains natural resources such as land, forest, water, wildlife and other biodiversity, which provide a wide range of goods and services such as, food, income, medicine and recreation to the adjacent communities. Such goods and services are crucial in supporting rural livelihoods. However, socio-economic and natural factors have increased pressure on the wetland resources leading to changes in land use, consequently threatening wetland resource sustainability (Majule and Mwalyosi, 2005; Majule and Kalonga, 2009). Unsustainable land use practices such as cultivation in river banks and marginal lands, extensive clearing of vegetation and uncontrolled use of agrochemicals, usually result in land degradation, water pollution, declining soil fertility and loss of biodiversity that consequently affect the livelihoods of local communities and their environment. As demand of arable land increases with limited access to extra land, changes in agricultural land use have been towards intensification. This entails use of more input such as labour and fertilizer/manure to increase land productivity and expansion of cultivation to marginal areas or change the livelihood portfolio to non-farm income activities (Boserup, 1965; Netting and Stone, 1993; Majule and Mwalyosi, 2005).

In the exploitation of the wetlands resources various households pursue different livelihood strategies as conditioned by various socio-economic and natural factors (Lema and Majule, 2009). Livelihood strategies frequently vary among households and individuals depending on the ownership of livelihood assets, which to a large extent reflect the flexibility and capability of optimizing livelihood (Ellis, 1998; Majule and Kalonga, 2009). Households with more livelihood assets can utilize opportunities such as market to produce highly marketable crops such as rice and keep livestock (Lyimo, 2005, Yanda et al., 2005). Households with limited assets may depend more on natural resource -based livelihood. Such differences in the utilization of wetland resources may contribute differently to income and food security as well as to the environment.

Due to the importance of the wetlands ecosystems to rural livelihoods the sustainability of wetlands resources needs to be ensured through understanding current utilization patterns and major driving force. Assessment of the level of dependence on wetland resources by different socio-economic groups also need to be made. This will contribute among others in informing policy intervention. The overall objective of this study was to investigate the contribution of wetland resource on community livelihoods with reference to agriculture.

MATERIALS AND METHODS

Description of the study area

Songwe River basin covers an area of approximately 4,214 km^2. This area includes Kyela, Ileje, Mbozi and Mbeya Districts in Tanzania and Karonga and Chitipa Districts in Malawi. About 54.6% of the area is located in Tanzania and 45.4% in Malawi. It has an estimated human population of 288,000, with the highest population density in Kyela (254 persons/ km^2) and the lowest in Mbozi district (43 persons/km^2) (NORPLAN, 2003). The basin has been subdivided into 3 zones, the lower, middle and upper basin (Yanda et al., 2004). This study concentrated on the lower basin on the Tanzania side, covering the floodplain in Kyela District (Figure 1).

The sample population was obtained from two villages namely Katumbasongwe and Mpunguti villages situated in Katumbasongwe ward in the southern part of Kyela District, about 18 km from Kyela Town. The area comprises two major rivers, namely, Songwe and Nsesi Stream, both pouring their water into Lake Nyasa.

Data collection methods

Documentary search

Secondary data on socio-economic and physical aspects relevant to the study area were obtained from various sources, both published and unpublished records, including reports obtained within and outside the Kyela District headquarters.

Satellite image interpretation

In this study image interpretation from Landsat TM taken in 1991 and 2001, at the scale of 1:80,000, was done. According to Lillesand and Kiefer (2000), this is a medium-range scale which is widely used to map vegetation cover and agricultural land use. The main purpose of satellite image interpretation was to come up with general land cover/land use maps so as to gain an insight into the extent of land use/cover changes in relation to wetland resource utilization. Through image interpretation land use/cover maps of the study area for the 1991 and 2001 were produced. A comparison of the two land use maps was made to establish the extent of land use/cover changes.

Participatory methods

Participatory methods including focus group discussion, key informants interviews at district and village level guided by checklists were used. This was aimed at capturing livelihood diversity, type of wetland resources and other uses. The focus group discussions comprised of 12 representatives from each sub-village in both villages. The level of their dependence on the wetland resources and major socio-economic groups in the study areas were also discussed. Also household categorization into different socio-economic groups in the study villages was undertaken based on the concept of wealth and poverty as understood by the villagers, as well as interpretation made by the authors during interviews. Such socio-economic groups reflect diversity in endowment of livelihood assets as well as resource utilization patterns by the local community.

Field observations were made in accessible areas of the wetland

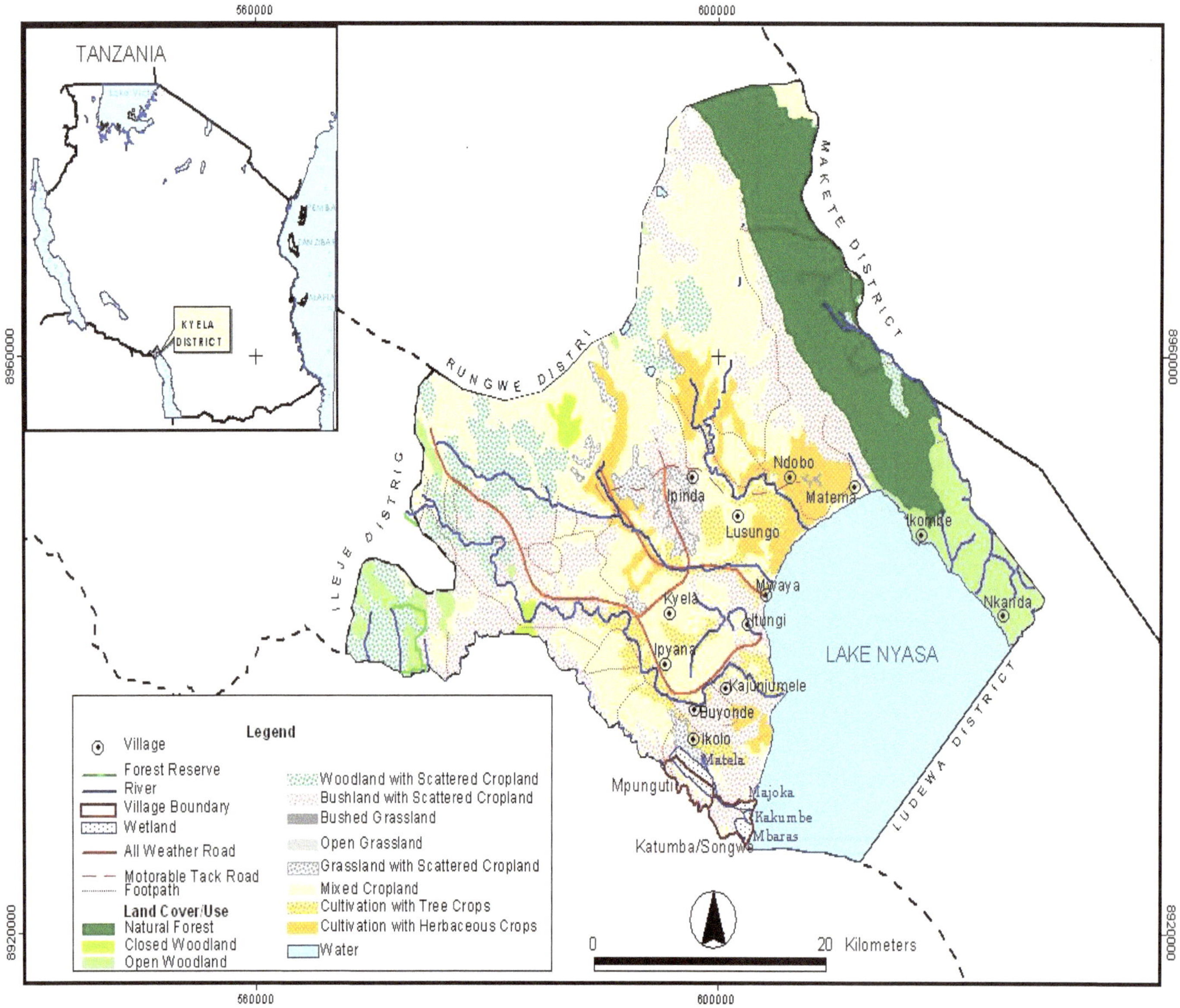

Figure 1. Location of Kyela district in Tanzania.

such as Majoka, Kakumbe and Mbalasi in Katumba Songwe village. Also Matela floodplain in Mpunguti village was visited and photographs were taken in subsequent areas covering features of interest to the study.

Household surveys

A sample size of 10% of all households in each village was randomly selected for interviews using a structured questionnaire. Questionnaires were administered to 44 and 60 respondents from Mpunguti and Katumbasongwe villages respectively.

Data analysis

The data collected were analyzed both qualitatively and quantitatively. The Statistical Package for Social Sciences Program (SPSS) software version 10 for Windows was used for data coding and for descriptive statistics. Descriptive statistics included frequency and percentages of the sample population. Qualitative data analysis was done through assessment of narratives of the focus group discussion and evidence from field observations.

RESULTS AND DISCUSSION

Land cover/ land use changes and resource utilization patterns

Changes in the land cover/land use types in the study villages included bush-land, grassland, permanent swamps and cultivated land. The cultivated land mainly consists of rice production in the lowland areas. The

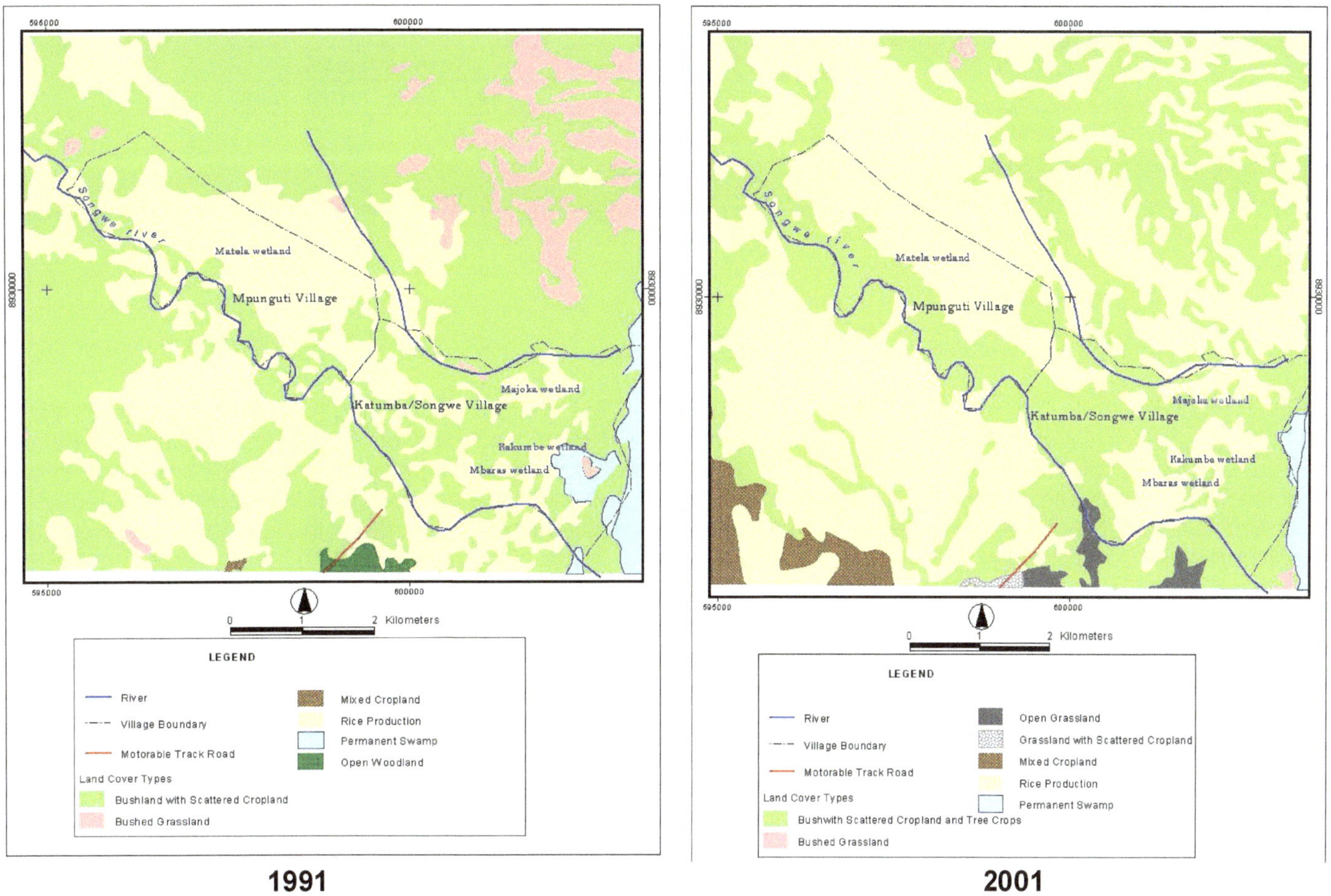

Figure 2. Land use/land cover maps of the study villages in 1991 and 2001.

upland crops included maize, banana and tree crops, and are to a large extent also found in areas with settlements. Analysis of land use changes in the study area has shown that there has been an overall increase in land under rice production in both villages (Figure 2).

Figure 2 shows that agricultural land and settlements has increased at the expense of other land use types from 59.5% in 1991 to 66.5% in 2001 in Katumbasongwe and from 31.6% in 1991 to 42.3% in 2001 in Mpunguti village. Based on the land use maps of 2001 it was observed that rice cultivation especially in Mpunguti village has expanded mainly to the north parts in Matela wetland. Due to increasing land scarcity for rice production households especially in Mpunguti has been compelled to cultivate land outside their village. Such land was not formally under cultivation. Focus group discussions revealed that increased demand of land for rice production has largely contributed to the observed changes. Such changes have significant effects on socio-economic activities and livelihoods of the local people. Increased market demand for rice has compelled farmers to expand their agricultural land as well as intensification to generate more income and food (Lyimo, 2005). Nevertheless, agricultural expansion within such ecological systems proceeds at the expenses of other important land uses. This has resulted into undermining other sources of livelihoods to the communities and loss of biodiversity. For instance, according to key informants, conversions of permanent swamps into rice cultivation were reported to have caused loss of about four fish species namely, *Barbus* spp. (Swahili, Sila), *Labeo* spp. (Swahili, Ningu), *Rhamphocromis* spp. (Swahili, Ndilwale) and *Opsalidium microcephalus* (Swahili, Mbelele), as well as a reduction in macrophytes. Furthermore, other activities that were traditionally undertaken in this area such as hunting were greatly affected.

Most people in Mpunguti village were reported to cultivate rice in Matela wetland outside their village (Figure 2). This is because a large part of the study villages are under high pressure to crop production as well as from settlements,. The overall process of land use changes in Songwe River basin has been towards agricultural intensification and expansion where land is available, which is associated with commercialised rice cultivation.

Table 1. Population trends in the study area.

Village	Year			Growth rate (%)	
	1978	1988	2002	1978-1988	1988-2002
Katumbasongwe	2222	2266	2667	0.2	1.6
Mpunguti	1438	1576	2001	0.9	2.4
Average growth rate				0.6	2.0

Source: URT (1978, 1988, 2003).

Population dynamics and wetland resource utilization

Findings from the study have shown that from 1978 to 2002 the study area is experiencing increasing population size as well as population growth rates (Table 1). High population has contributed to the increased demand of land for food production, leading to land expansion and intensification.

During the inter-census periods 1978 – 1988 and 1988-2002 the population growth rate has increased from 0.2 to 1.6% and 0.9 to 2.4% in Katumbasongwe and Mpunguti village respectively, though the growth rate was less than the national growth rate of 2.9 (URT, 2003). High population has contributed to the increased demand of land for food production as well as other domestic needs.

This has impact on the extent of wetland resource utilisation as well as the sustainability of the wetland ecosystems. There is a strong relationship between population growth and land use changes (Malthus, 1960; Boserup, 1965; Nettings, 1993). Population dynamics normally exert pressure on environmental resources as a result of human being striving to meet their basic needs. For instance, Kangalawe and Liwenga (2005) observed in Kilombero wetlands that over the last two decades the use of wetlands for agriculture has increased because of increasing population and the resultant need to produce more food. However, the main concern was deforestation and the trampling of soils by increasing numbers of livestock. Also, Palela (2000) noted a reduction in area coverage of the swamp in Msimbazi wetland from 5.71 km² in 1975 to 2.90 km² in 1992 due to establishment of settlements.

The study has also shown that most of the natural vegetation has been cleared, while permanent swamps no longer exist in the study area due to over-exploitation of the resources through human activity. Similar observations were also reported by NORPLAN (2003) which show that the lower plain of Songwe River has been intensively cultivated over the years and that over 95% is used for crop cultivation. This problem is magnified by the customary land tenure system operating in the study area. Customary ownership of land which exists encourages continued land fragmentation.

Due to the presence of diverse sources of livelihood in the wetlands, only few people reported to depend on single source to support their needs. Figure 3 shows various sources of livelihood in the study area.

The sources of livelihood in the study villages varied spatially depending on the location of the village and accessibility to wetland resources. However, crop production and livestock keeping are the main sources of livelihood as the majority of the respondents are engaged in these activities (Figure 3). It was noted that more people in Mpunguti village are engaged in crop production and livestock keeping than those in Katumbasongwe village. This is probably due to the fact that households in Mpunguti village have access to a large arable land, and a high percentage of them are keeping cattle than those in Katumbasongwe village. Non-farm activities include fishing, handcraft, local beer brewing and selling, kiosks making and selling palm oil (Figure 4).

It was also noted that there are more households in Katumbasongwe village involved in handcraft activities than in Mpunguti village (Figure 4). This is due to closeness of Katumba songwe to Lake Nyasa and Nsesi stream which provides macrophytes as well as easy access to market. The macrophytes are utilized by local communities in making local carpets (*'mikeka' in Swahili)* and fish traps (*'migono'*, in Swahili) which contribute to household incomes.

Wetland resource utilization patterns

Major land uses in the studied villages include crop production (mainly rice cultivation), settlement and livestock grazing, especially during dry season. Other uses are fishing, macrophyte harvesting for handcraft activities and water for domestic activities. Rice is mainly cultivated along Lake Nyasa in the wetlands, and upland crops such as maize, cassava, sweet potatoes, banana and cocoa are cultivated on homestead farms in Katumbasongwe. Settlements are concentrated along Songwe River for easy access to water for domestic purposes as well as availability of fertile land along the river. Rice in Mpunguti village is cultivated in Matela (Nyakyusa) (seasonal wetlands) located away from Lake Nyasa as well as Songwe River.

The extent of exploitation of wetland resources greatly depends on the nature of the household strategies and

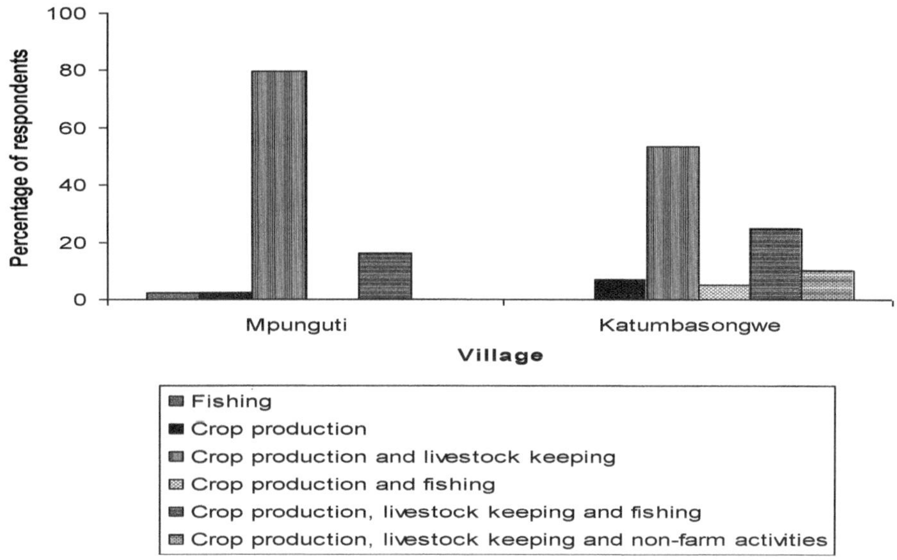

Figure 3. Major sources of livelihood in the study villages.

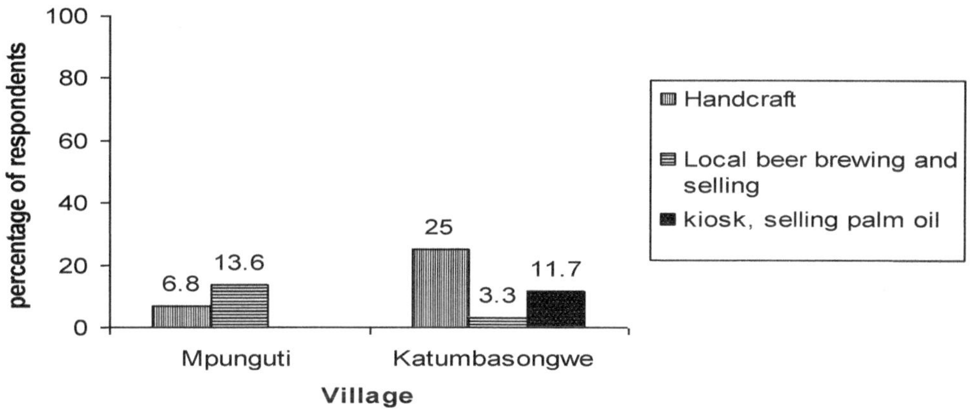

Figure 4. Non-farm sources of livelihood in the study villages.

the ownership of the livelihood assets. Households in the study villages can be characterized into three main socio-economic groups (Table 2).

Table 2 shows that the respective wealth groups are characterised by similar levels of resource endowment across the studied villages. Through focus group discussion it was revealed that the extents of involvement in the wetland utilisation vary considerably from one socio-economic group to another given their differences in the level of resource endowment and flexibility in resource mobilization.

The *well-off* group having access to more assets and high flexibility in mobilisation of resources, owns and cultivates the largest and productive land mainly land for rice production. In contrast to other groups, the *well-off* is very successful in exploiting the existing market opportunities and the ecological potential through commercialisation of rice production. Most of the *well-off* households using their accumulation strategies secure high income from crop sales mainly rice and significant income from business and rent of assets such as ox-plough

Households in the *middle wealth* group represent the majority of households in the study villages (Table 2). Such group is endowed with few resources/ assets and not very flexible in mobilising and allocating the agricultural resources compared to the *well-off* households. The *middle wealth* group households cultivate average farm sizes for rice as compared to the former group. They are not exclusive subsistence farmers, as they produce crops mainly for their own consumption and also engage to a limited extent in the production for the market. Most of them are not very successful as compared to the *well-off* group in terms of

Table 2. Socio-economic groups in Katumbasongwe and Mpunguti Villages.

Group	Social characteristics	Percentage	
		K/songwe	Mpunguti
The well-off (WLO) *'Ntajiri'/'Nkabi'*	Own > 5 cattle, own > 3-5 acres of land for rice production, have house roofed with corrugated iron sheets. Own motor bike and/ or bicycles, have surplus food, own more than one pair of oxen (*Kambaku)* and plough. Can hire a tractor for land cultivation, use fertilizers and hire labourers. Rents out farm land and ox-ploughs to other socio-economic groups and because of financial stability are involved in other non farm business. They have food security throughout the year after harvest	20	22.8
The middle wealth group (MWG) *'Hali ya kawaida'*	Own 1 - 5 cattle, and some own 2-3 pigs. Own 1.5 - 2.5 acres of land, few cultivate rice, majority has a house made of mud bricks and roofed with grass, own a bicycle, sometimes can hire labour when necessary, but most of the time depend on family labour and some possess a pair of oxen and plough (*Kambaku*), they are involved in handcraft have food security for 8 months after harvest	49.2	54.4
The low wealth group (LWG) *'Ndondo'*	Have no cattle but a few chickens, farm size below 1 acre, do not cultivate rice, have a mud house roofed with grass, often work as casual labourers and use hand hoe, have food security for less than 3 months after harvest	30.8	22.8

Table 3. Major crops cultivated in the SRB.

Village	Farm Location	% Respondents	
		Rice	Maize
	Wetland only	90.0	36.7
	Upland only	4.6	31.7
Katumbasongwe	Both	3.3	8.3
	Not cultivating rice or maize	2.1	23.3
	Total	100	100
	Wetland only	90.9	63.6
	Upland only	2.3	13.6
Mpunguti	Both	6.8	15.9
	Not cultivating rice or maize	0.0	6.9
	Total	100	100

optimizing gains from the changing marketing conditions and other livelihood in the studied villages. Diversification to off-farm activities involves selling of local beer, fishing, and handcraft and to the lesser extent working as hired casual labour.

The poor group represents 30.8 and 22.8% of the population in Katumba songwe and Mpunguti villages respectively, Most of the households in the *poor* group are constrained in terms of having access to fewer resources/ assets compared to the former wealth groups (Table 2). The nature of such resources makes them less flexible, which in turn restricts them from pursuing certain strategies. For example, the *poor group* are often unable to exploit existing opportunities to optimize gains from wetland resources. To a large extent households in this group are unable to meet their household consumption needs from farming, thus are increasingly driven to off-farm based livelihood strategies such as selling casual labour and fishing .

Wetland cultivation and its contribution to household food and incomes

Agricultural production is the major economic activity in the two villages contributing to food security and income generation. The results indicate that rice production is undertaken by majority of the respondents in both villages as indicated in Table 3.

Wetland resources in Songwe Basin which includes Majoka, Matela, Mbalasi and Kakumbe wetlands, have vital role in ensuring food security in the study area and Kyela district in general. The study shows that 81.7 and 85.6% of the sample population in Katumbasongwe and

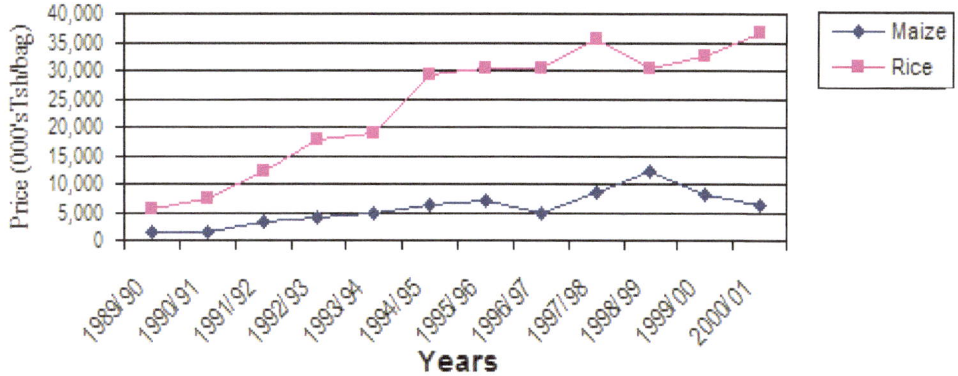

Figure 5. Wholesale Prices of Maize and Rice in Mbeya Town. Source: MAFS (2003).

Mpunguti village respectively depended largely on wetland resources for their livelihoods.

Crop production ranked first among wetland-related activities followed by fishing, livestock keeping and handcrafts activities in both villages. This indicates that the livelihoods of people in the study area mainly rely on cultivation in the wetlands. Similar observations on the role of wetlands in supporting the livelihoods of households were reported from other wetlands in the southern highlands (Majule and Mwalyosi, 2005). However, high demand of wetland resources coupled with increasing human and livestock populations in the study villages is threatening the well-being of the wetland environment. Focus group discussion revealed the disappearance of some wild animals in SRB.

Maize and other crops like beans, groundnuts and sweat potatoes are grown in the wetland on a small scale immediately after harvesting rice. This is aimed to take the advantage of moisture residue available in the soil as the area start experiencing drought conditions. Through focus group discussions in the villages it was revealed that cocoa is mainly grown in the uplands because it is not adapted to water-logged conditions. Such crops is supporting livelihood of only a few people as compared to rice production. Findings from this study also revealed that the wetlands contribute most to both food and cash income from agriculture. Similar results have been revealed from other parts with similar ecological conditions (Kangalawe and Liwenga, 2005; Yanda et al., 2005; Kulindwa et al., 2006).

Results from the study villages show that a portion of rice, maize and other crops such as beans and groundnuts are sold in Kyela market. However, among the crops rice is highly marketable with high demands in urban centres such as Mbeya and Dar es Salaam where they fetch high price compared to crops such as maize. Prices of maize and rice from Mbeya Town, main market for rice were used to reflect changes in the wholesale prices for the two crops in the study area (Figure 5)

The wholesale prices of both crops have increased over time. However, assuming that the marginal cost of production has remained constant for both crops, the wholesale price of rice has increased substantially compared to maize. Such a situation has made rice production more attractive to farmers especially those who can meet production costs.

In the study villages price of crops vary depending on the type of crop and on the season of the year. For instance, during the harvesting season one tin (18 kg) of unshelled rice is sold at Tshs 2,500/= to 6,000/[1]= depending on the supply of the product. Middlemen are the main buyers of the agricultural products in Kyela.

Table 4 provides a summary of estimated annual income from rice and maize produced in the wetland. The prices depend on the market price as suggested by the respondents in the study villages. The table show the well-off socio-economic group, benefited more in terms of their earnings from crops cultivated in the wetlands due to high endownment of assets.. Such group is able to mobilize adequate resources in terms of money, labour, land and agricultural inputs at the right time to optimize crop yield. For instance, the well-off group earns over 500,000 and 1,000,000 Tsh per year in Katumbasongwe and Mpunguti villages respectively (Table 4). On the other hand the well-off socio-economic group in Katumbasongwe earns less than that of Mpunguti (Table 4). The variation in income accrued from rice production among other factors may be attributed to difference in endowment of livelihood assets such as land for rice farm, financial capital to purchase farm inputs and use of improved farm management practices. People in Mpunguti village have access to large farm sizes and are a bit better-off than those in Katumbasongwe village.

The LWG have been earning little from rice production mainly below Tshs.100,000 per household, for various reasons, including possession of very small plots in the wetlands, lack of financial capacity to buy farm inputs such as fertilizers and pesticides as well as inability to hire additional labour to manage their plots. Also because

[1] 1 US $ =Tsh. 1200

Table 4. Income generated from sales of rice and maize main crops in the study area.

Villages	Socio-economic groups	Estimated Income per season (Tsh)
`Katumbasongwe	The LWG	<100,000
	Medium group	100,000 to 500,000
	Well-off	Above 500,000
Mpunguti	The LWG	<100,000
	Medium group	100,000 to 1,000,000
	Well-off	Above 1,000,000

of their poverty they spend much of their time as cheap labour working for the well-off socio-economic group during cropping season, consequently paying little attention to their own farms. This calls for the need to improve micro credit services to empower this group in terms of finance so that they are able to manage their farms better to increase their incomes and ensure food security for their families. The currently provided micro credit through Savings and Credit Cooperative Societies (SACCOs) service according to them is favouring only the well-off socio-economic group who have the ability to pay the debt even before harvesting.

Livestock production and its contribution to livelihoods in the SRB

The study showed that 95 and 88% of the sample population keep livestock in Mpunguti and Katumbasongwe villages respectively, which comprise cattle, pigs and chicken. The number of livestock kept varies depending on the type of livestock and the financial well-being of an individual household. However, it was found out that the largest herd in the study area had only 45 cattle. The livestock are mainly used for food (milk and meat), provision of draught power, sources of revenue through sales of live animals and animal products. About 91.0 and 94.5% of the respondents in Mpunguti and Katumbasongwe village respectively use ox plough (*Kambaku)* during land preparation. It was revealed that few livestock keepers sell their animals, especially cattle, due to the reason that such animals provides main source of animal power for ploughing and other farm activities. It was found out that only 27 and 40% of the livestock keepers in Mpunguti and Katumbasongwe villages respectively have sold cattle in the past two years. Such differences between the two villages could be attributed to various reasons, one being lack of grazing areas in Katumbasongwe village. It was revealed that most of the livestock keeper belongs to the well off group with few households from the medium wealth group. Households in the LWG are mainly involved in keeping small stock such as chicken which do not need high capital investment. Income obtained from the sale of these animals is considerably lower than those obtained from the sale of cattle and pigs. It was observed that keeping livestock such as cattle requires high initial capital to purchase the animals as well as maintaining them, some thing which not many people can afford especially the LWG.

Contribution of fishing to the livelihoods of the community in the SRB

Fishing is mostly done in Songwe River where there are lower fish catches compared to Lake Nyasa. About 47.4% of the households mostly from Katumbasongwe village earn above 100,000 Tshs per year (Figure 6). Fishermen from this village fish in Lake Nyasa as well as Songwe River. Focus group discussion revealed that the fish obtained from Lake Nyasa are normally bigger and in large quantity, thus fetching a good price. This may be one of the reasons why some fishermen from Katumbasongwe village earned more than 500,000 Tsh per year (Figure 6).

Fishing from Lake Nyasa, Songwe River and seasonal ponds has made a significant contribution to the income of the people who engage in fishing activities in the study area. The fishermen have integrated their fishing activities with agriculture such as crop and livestock enterprises. The peak fishing season is between late March and early May each year. This period coincide with heavy rains in the SRB. During such time most of the area is flooded, creating good fish habitat as well as conducive environment for fishing.

According to the fishermen there has been a low fish catch in recent years, attributed among other reasons to the destruction of breeding sites due to agricultural activities especially in areas where Songwe River enters Lake Nyasa. Other reasons mentioned were increased siltation in Lake Nyasa due to soil brought in by Songwe River from the upper catchments. Reduced catch can also be due to lack of good fishing gears since most of the fishermen are using local/ traditional fishing gears.

Contribution of handicraft activities to the community livelihoods

Figure 7 shows that respondents in Katumbasongwe

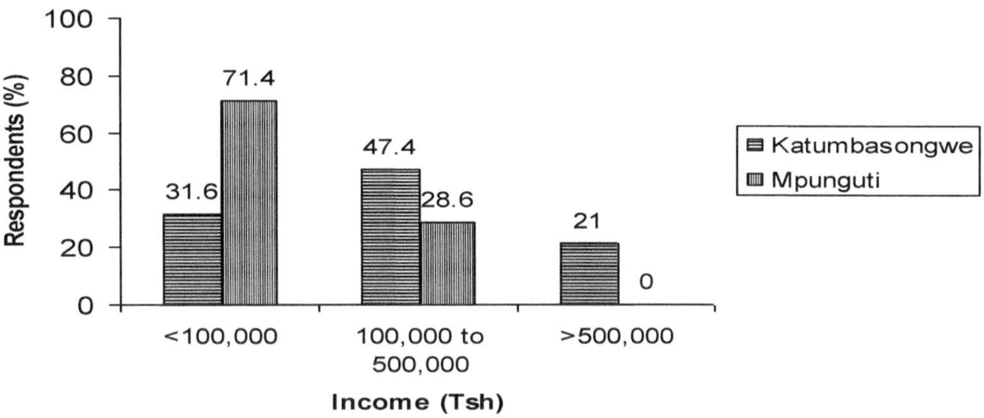

Figure 6. Annual incomes from fish sales in the SRB.

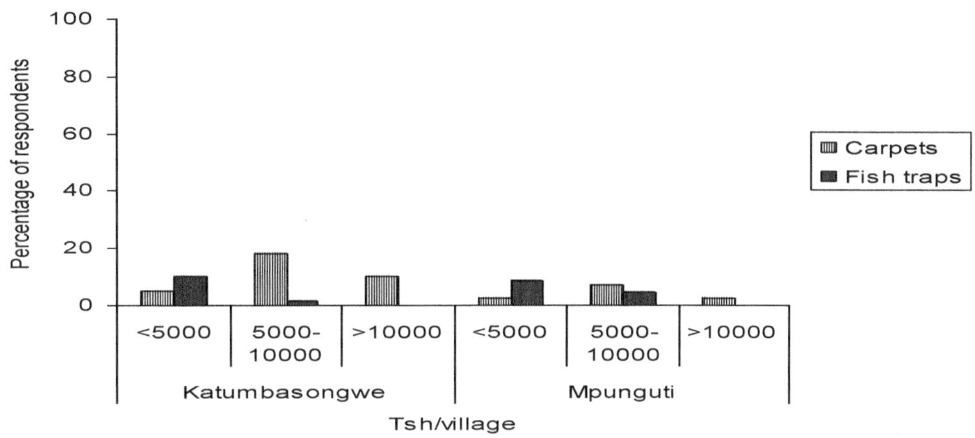

Figure 7. Market Value of Handcraft Products from the Wetlands in the SRB.

village earned more than those in Mpunguti village in relation to handicraft activities. Handicraft activities benefit the people surrounding these wetlands through incomes accrued from sales of hand made products. Carpets made of *Cyperus spp* play a vital role during traditional ceremonies, such as funerals and weddings. Also the *Cyperus spp* is used in making fish traps (*'migono in swahili')* which is commonly used in the study area. The contribution of handicrafts in terms of income depends on the quality, type and size of the final products. Local carpets (*mikeka/kilago in Swahili))* generate more income for the local people than fish traps *'migono'*. Carpets could be sold for more than 10,000 Tsh, as compared with fish traps which were reported to generate no more than 10,000 Tsh per unit.

The survey revealed that the two villages benefited differently from the income generated from handicraft activities. About 18.3% of the sample population in Katumbasongwe reported to sell their macrophyte products (carpets) for between Tshs 5000 and 10000 as compared to 9.2% of Mpunguti village. The study revealed that this difference is attributed to excessive vegetation clearance in Mpuguti village, which has resulted in the loss of macrophytes used for handcraft activities.

Handicraft activities were reported to be carried out mostly by the poor socio-economic groups. This implies that degradation of the wetland resources such as macrophytes may have negatively affects the well-being of this group. Sustainable management of these wetland resources can be seen as a prerequisite for helping the poor community around the wetlands to tackle income poverty.

Conclusion

The wetlands in SRB comprises various natural resources which make a significant contribution to livelihoods including income, food and employment of the local community. Most of the community depends on arable land for agriculture especially rice production, and pasture for livestock grazing. Small-scale fisheries in the wetlands are source of food security, employment, and generate additional income to the rural communities living in such areas. However, the wetlands have experienced spatial and temporal change in land use patterns with most of it being converted into agriculture at the expense of other land use/ cover types. This has been among other factors due to an increase in population pressure as well as market demand particularly for rice. The nature and extent of wetland resource utilization in the SRB vary considerably from one socio-economic group to another depending on the flexibility and access to the livelihood assets. The Well-off households with more access to resources including livestock, high income, ox-ploughs and large productive farm size are intensively involved in commercial rice production and have expanded their agricultural land frontier. The Intermediate wealth group unlike the well-off group they have few resources such as small land holdings, and low incomes. In contrast to the former group, they are less flexible in terms of access to agricultural resources including productive rice farm and labour. The LWG are resource constrained with very limited access to assets and have few livelihood options open to them compared to the former groups. The group is largely involved in small scale fishing, labour selling and weaving activities to supplements the low incomes from agriculture. Given the current resource utilisation patterns, provision of support to the agricultural production systems is important in enhancing agricultural productivity as well as sustainability of the wetlands resources in the SRB.

ACKNOWLEDGMENT

This study was undertaken with financial support from Swiss National Centre of Competence in Research (NCCR) through the Institute of Resource Assessment (IRA), University of Dar es Salaam in Tanzania. The authors appreciate this support. The authors acknowledged support from Staff members of the District Agricultural and Livestock Office in Kyela District, Mbeya Region and the villagers in the study area.

REFERENCES

Boserup E (1965). The conditions of Agricultural Growth. Allen and Unwin.

Davis TJ (Editor) (1994). The Ramsar Convention Manual, *A guide for the Convention on Wetlands of International Importance especially as waterfowl habitat, Ramsar Convention Bureau,* Gland, Switzerland. P. 207.

Ellis F (1998). Household Strategies and Rural Livelihood Diversification. J. Dev. Stud. 35(1):1-38.

Kangalawe RYM, Liwenga ET (2005). Wetlands management in the Kilombero valley Tanzania, pp. 134-166: in Sosovele H, Boesen J, Maganga F (eds.). *Social and Environmental Impact of Irrigation Farming in Tanzania*, Selected cases. Dar es Salaam University Press. Dar es Salaam.

Kulindwa K, Ikiara M, Chande A, Milly N (2006). ‘‘To Conserve or to Convert the Lake Victoria Basin Wetlands”, An Economic Valuation of Mara Wetland. Inter-University Council of East Africa Vicres Project.

Lema M, Majule AE (2009). Impacts of Climate Change, variability and adaptation strategies on Agriculture in semi arid areas of Tanzania: The Case of Manyoni District in Singida Region, Tanzania. Afr. J. Environ. Sci. Technol. 3(7):206-218.

Lillesand TM, Kiefer RW (2000). *Remote Sensing and Image Interpretation,* Fourth Edition. John Wiley & Sons, Inc.

Lyimo JG (2005). Changes in Agricultural Land use and Household Production. A case of Small farm holders in the Usangu Plains, Mbarali District. *Doctoral Thesis Faculty of Science Universirty of Copenhagen.*

MAFS (2003). Ministry of Agriculture and Food Security. Crop production statistics.

Majule AE, Kalonga S (2009). Role of Local Knowledge and Innovations in Conserving Forest Biodiversity in Kilombero Wetlands, Tanzania. ICFAI J. Environ. Econ. VII:1.

Majule AE, Mwalyosi, RBB (2005). Enhancing Agricultural Productivity through Sustainable Irrigation. *A case of Vinyungu Farming System in selected Zones of Southern Highlan, Tanzania*. A chapter in a Book Social and Environmental Impacts of Irrigation farming in Tanzania: Selected Cases: Edited by H. Sosovele, J. Boesen and F. Maganga. Dar es Salaam University Press. ISBN 9976 60 431 9.

Majule AE, Omollo JO (2008). The Performance of maize during acid amelioration with organic residues in soils of Mtwara, Tanzania. Tanzania J. Sci. 34:21-30.

Malthus TC (1960). First Essay on Population. Modern Library, New York.

Mwakaje AG (2009). Wetlands, livelihoods and sustainability in Tanzania. Afr. J. Ecol. 47(Supplement 1):179-184(6).

Netting R (1993). Smallholders, Householders. Farm Families and the Ecology of Intensive, Sustainable Agriculture. Stanford: Stanford University Press.

Netting R, Stone MP (1993). Agricultural Expansion Intensification and Market Participation among the Kofyar, Jos Plateau, Nigeria.

NORPLAN (2003). *Feasibility Study Report For the stabilisation of the course of the Songwe River (Volume 2)*. ‘Songwe River Basin Land Use and Natural Resources Characterization’. Norplan A.S. in Association with COWI, DHI Water and Environment, W&PES.

Palela E (2000). The Impacts of Anthropogenic Factors on Wetlands: The Case of Msimbazi Valley Dar es Salaam”, *Unpublished M.A Thesis (GEM)* University of Dar es Salaam.

Stuart SN (1990). The Threats to Biodiversity and Genetic Resources in Tanzania, pp. 238-239. In: A. Kauzeni, H.F. Bitanyi and M.A.K. Ngoile (Eds). *Proceeding of seminar on National Conservation Strategy for Tanzania*, 12-17 November, 1990. Dodoma. NEMC, SIDA.

United Republic of Tanzania (URT) (1978). Population Census Regional Profile Mbeya.

United Republic of Tanzania (URT) (1988), Population Census Regional Profile Mbeya: Takwimu-Bureau of Statistics President’s office, Planning Commission Dar es Salaam, Tanzania.

Yanda PZ, Majule AE, Mwakaje AG (2005). Wetland utilisation, poverty alleviation and environmental conservation in semi arid areas of Tanzania – *The case of Singida Region, Tanzania.* Proceedings of East Africa “Integrated River Basin Management” Conference Held on 7th-9th 2005 at ICE Hall, Sokoine University of Agriculture Morogoro, Tanzania.

Yanda PZ, Mwalyosi RBB, Shishira EK, Mung’ong’o C, Majule AE (2004). ‘Development of the Songwe river transboundary catchment management project’. *Natural resources baseline studies*. A final Consultant Report Submitted to WWF. Institute of Resource Assessment (IRA), University of Dar es Salaam, Tanzania.

Ensuing economic gains from summer pearlmillet (*Pennisetum glaucum* L.) due to different dates of sowing and land configuration

Deshmukh S. P., Patel J. G. and Patel A. M.

Department of Agronomy, Navsari Agricultural University, Navsari, India.

This study was undertaken at Navsari, India (20.95°N 72.93°E, elevation of 10 m above Sea level) during summer season 2010 to work out effect of different land configuration and dates of sowing on pearlmillet. The results gained confirm that ridges and furrow along with early sowing date during summer season is beneficial to obtain higher net returns from pearlmillet crop. Ridges and furrow sowing of pearlmillet showed superiority in all needed agronomic trait of pearlmillet crop during summer and the increase in grain and stover yield under ridges and furrow was 20.00 and 17.71% over flat bed sowing, respectively. The economic returns by application of ridge and furrow bed sowing gave net returns of Rs.14081 ha^{-1} and B:C ratio of 1.88 while, in flat bed treatment those were Rs.10181.04 ha^{-1} and 1.67, respectively. Sowing pearlmillet during last week of January or first week of February increases the yield as well as net returns from the crop. Pearlmillet sown on 5 February (3.24 t ha^{-1}) gave significantly highest grain yield followed by 25 January (3.04 t ha^{-1}) sowing, while 15 February sowing gave significantly lowest yield. Highest net returns of Rs.15541 ha^{-1} with B:C ratio of 2.03 were obtained with 5 February sowing treatment, followed by 25 January sowing treatment with Rs.13971 ha^{-1} and 1.93 B:C ratio respectively. Whereas, the lowest net returns and B:C ratio was acquire from 15 February, that is, Rs. 8341 ha^{-1} and 1.55, respectively.

Key words: Summer pearlmillet, land configuration, ridges and furrow, dates of sowing.

INTRODUCTION

India is one of the major producers of millets in world. The predominant millets grown in India are sorghum (*jowar*), pearl millet (*bajra*), finger millet (*ragi*), kodo millet (*kodo*), proso millet (*cheena*), little millet (*kutki*), foxtail millet (*kangni*) and barnyard millet (*sawa*); of which barnyard millet, little millet and kodo millet are endemically domesticated in Indian subcontinent. Millets are warm weather grasses belonging to C_4 group of plants and considered as physiologically efficient. Their cultivation in India extends from sea-level up to 2,000 m above mean sea-level and often grown in diverse soils, climates and harsh environments. Millets have been important food and feed crops producing more reliable harvests than many other crops.

Pearlmillet is commonly known in India as *Bajri* or *Bajra.* Pearlmillet (*Pennisetum glaucum* L.) is also known as 'bullrush millet', originated in tropical western Africa, where the greatest number of both wild ancestors and cultivated forms occur. It belongs to family Gramineae (Poaceae). The cultivated species are *P. glaucum* L. (2n=14) used for grain and fodder and *Pennisetum purpureum* L. (2n=28) used for green and dry fodder. It is highly cross pollinated crop due to protogynous nature of the flowers (more than 85% outcrossing) diploid annual (2n=2x=14) with a large genome size (2450 Mbp). Average composition of the edible portion of seed is

12% moisture, 10-12% protein, 3-5% fat, 60-70% carbohydrates, 1.5-3% fibre and 1.5- 2% ash. Its nutritional value is somewhat superior to maize, rice, sorghum and wheat. Stalks are also used for thatching, as fuel and made into mats for winnowing.

Waterlogging in heavy black soil and salt injury due to saline irrigation water are the important factors for low productivity of pearlmillet in south Gujarat mainly during summer season. Under these conditions land configuration can play a vital role to overcome these problems by providing easy and uniform germination as well as good growth and development of plants. Land configuration increases water use efficiency as reported by Chiroma et al. (2008) and also increases availability of nutrients to crops. It is particularly useful in areas having saline irrigation water because it helps to avoid direct contact of young plants with saline irrigation water. The superiority of ridges and furrow system could be ascribed to proper drainage of excess water coupled with adequate aeration at the time of irrigation or heavy rainfall. Parihar et al. (2009) reported that ridges and furrow sowing method improved grain as well as stover yield of pearlmillet and succeeding mustard over the flat bed method of sowing.

Sowing time is the most important non-monetary input influencing crop yield. Sowing at optimum time improves the productivity by providing suitable environment at all the growth stages. Upadhyay et al. (2001) have reported higher grain yield of summer pearlmillet when sown on 15 march and found reduction in grain yield with delay in sowing. Identifying suitable time of sowing for pearlmillet during summer is important to have proper growth and development of plants, save the crop from early monsoon showers and timely vacate the field for succeeding *kharif* crop.

But until now very fewer literatures is available on cultivation of summer pearlmillet and keeping this in view, the present investigation was undertaken.

MATERIALS AND METHODS

The field experiment was conducted at College Farm, N. M. College of Agriculture, Navsari Agricultural University, Navsari during summer season of 2010. Geographically, Navsari is located at 20.95°N latitude, 72.93°E longitude and at height of 9 m above the mean sea level.

Normally, the summer season commences from the middle of February and ends by the middle of June. The temperature starts rising from February and reaches the maximum in the months of April and May, which are the hottest months of season. The weekly mean maximum and minimum temperature varied from 29.8 to 38°C and 13.2 to 28.8°C, respectively during the course of investigation. The maximum and minimum relative humidity ranged from 68 to 91% and 26 to 73%, respectively and daily sunshine hours from 7.5 to 10.8 were available during the crop period. There was no rainfall during the crop season. Thus, the weather conditions were normal and congenial for satisfactory growth of pearlmillet crop.

The soils of South Gujarat are locally known as 'Black Cotton Soil'. The soil of Navsari campus falls under the great group *Ustochrepts* and has been placed under Jalalpur series. These soils are dominated by montmorillonite clay, which cracks heavily after drying. The soil of the experimental site was dark grayish brown type having medium to poor drainage and good water holding capacity.

Data on soil analysis as analyze using standard methods from Jackson (1967) revealed that soil of experimental field was clay in texture, medium in available nitrogen and phosphorus and high in available potassium. The soil was slightly alkaline in reaction with normal electrical conductivity.

Pearlmillet crop was sown with 4 kg ha^{-1} seed rate at 45 × 10 cm spacing with drill method in North-South direction as per treatments in the experimental field. The fertilizer dose used throughout experiment was 120-60-00 NPK kg ha^{-1}. The first irrigation was given just after sowing for germination and subsequently irrigations were applied accordingly during crop period as required. Two hand weeding were done to check the weed population. No serious disease and pest were observed during the crop growth period hence no use of insecticide was needed. The data collection for growth and yield attributes was done from randomly selected five plants and then by calculating the average value for each plot. After maturity of crop, harvesting was followed by removing the earheads from plant and subsequently the plants were harvested by sickle from each plot. The yield observations were calculated from whole net plot plant population and then converted to hectare basis.

The treatment combinations comprise of two levels of land configurations *viz*., flat bed (L_1) and ridges and furrow (L_2) and three dates of sowing, that is, 25 January (D_1), 5 February (D_2) and 15 February (D_3). The data recorded on various characters during the period of investigations were statistically analysed by appropriate procedure to the factorial randomize block design. The significance of differences was tested by the 'F' test at 5% level (Panse and Sukhatme, 1967). The critical differences were calculated when the differences among treatments were found significant by 'F' test. Summary tables of treatment means have been prepared and presented with standard error of mean (S.Em ±) and coefficient of variation (C.V. %).

RESULTS AND DISCUSSION

Effect of weather on pearlmillet crop in summer season

The results on growth and yield attributes, yield, nutrient content and uptake as well as water use efficiency data were presented in the previous chapter indicated profound effect of environment factors. Weather condition plays an important role on the growth and yield of pearlmillet in summer season. It is evident from the meteorological data that the weather condition prevailed during the entire crop period was favorable and congenial for the normal growth and development of pearlmillet crop. No severe incidences of diseases and pests were observed during the entire crop growth period. Thus, it is expected that the variation observed in the experimental results were mainly due to the treatment effects only.

Effect of soil

The soil properties *viz.,* texture, pH, EC and available nutrients were estimated before sowing. The soil of experimental field was found clayey in texture, slightly

Table 1. Plant height (cm) leaf are index, dry matter (g) and number of tillers of summer pearlmillet at 20, 40, 60 DAS and at harvest as influenced by land configurations and dates of sowing.

Treatment	Plant height (cm)				Leaf area index				Dry matter (g)				Number of effective tillers
	20 DAS	40 DAS	60 DAS	At harvest	20 DAS	40 DAS	60 DAS	At harvest	20 DAS	40 DAS	60 DAS	At harvest	60 DAS
Land configurations (L)													
L_1 = Flat bed	23.51	51.83	156.73	162.70	1.37	3.47	4.99	1.11	6.10	25.27	40.18	45.24	2.83
L_2 = Ridges and furrow	23.84	53.86	166.66	176.28	1.38	3.61	5.36	1.16	6.27	26.59	42.35	47.62	3.19
S.Em.±	0.30	0.56	1.60	1.97	0.017	0.04	0.08	0.01	0.07	0.35	0.51	0.54	0.05
C.D. at 5%	NS	1.61	4.60	5.66	NS	0.13	0.23	0.04	NS	1.01	1.46	1.55	0.14
Dates of sowing (D)													
D_1 = 25 Jan.	24.11	54.66	167.32	175.09	1.40	3.62	5.52	1.20	6.27	26.67	42.11	47.36	3.09
D_2 = 5 Feb.	23.73	53.20	162.65	168.49	1.38	3.57	5.08	1.12	6.20	26.05	41.74	46.99	3.06
D_3 = 15 Feb.	23.19	50.66	155.11	164.89	1.33	3.43	4.91	1.08	6.08	25.06	39.96	44.94	2.89
S.Em.±	0.35	0.69	1.96	2.41	0.021	0.05	0.10	0.02	0.09	0.43	0.63	0.66	0.06
C.D. at 5%	NS	1.97	5.63	6.93	NS	0.15	0.28	0.05	NS	1.24	1.79	1.90	0.17

alkaline (pH 7.8), low in available nitrogen (176 kg/ha), with medium availability of phosphorus (32 kg/ha) and fairly rich in potash (350 kg/ha) as analyze using standard methods from Jackson (1967), which is suitable for proper growth and development of pearlmillet crop.

Effect of land configuration on growth and growth attributes

Different techniques of land configuration showed remarkable influence on crop growth. Significant differences in plant height, leaf area index, dry matter accumulation and number of effective tillers per plant (Table 1) at different growth stages of crop was observed due to the effect of various methods of land configuration.

The periodical plant height and leaf area index recorded at various growth stages were significantly higher except 20 DAS under ridges and furrow (L_2). This might be due to maintenance of proper air moisture regimes under ridges and furrow sowing which might have improved the drainage resulting in good supply of required moisture, available nutrients, soil aeration, soil environment and better growth and development. The results were in conformity with those reported by Ugale et al. (1995) for plant height in *kharif* pearlmillet and by Patel et al. (2008) in sorghum.

Better plant height and leaf area index under ridges and furrow (L_2) reflected into significant increase in dry matter accumulation except in 20 DAS) and number of effective tillers per plant. Kiran et al. (2008) observed similar results in sorghum in terms of dry matter accumulation.

Different land configuration treatments did not produce any significant effect on days to 50% flowering and physiological maturity (Table 2).

Effect of dates of sowing on growth and growth attributes

Growth components of pearlmillet *viz.,* plant height, leaf area index, dry matter accumulation, effective number of tillers, days to 50% flowering and days to maturity varied significantly due to time of sowing (Tables 1 and 2). Only the plant height, leaf area index and dry matter accumulation measured at 20 DAS was not significant due to changes in sowing time. Summer pearlmillet sown at normal time on 25 January registered measurable increase in growth components as compared to late sown crop. Thus, results were in favour for sowing of summer pearlmillet on 25 January (D_1) far to on 5 February (D_2) than late sowing, that is, on 15 February (D_3). This is probably due to early sown crop may enjoy favourable climatic conditions in term of

Table 2. Influence of land configurations and dates of sowing on physiology and yield of summer pearlmillet.

Treatment	50% flowering (days)	Physiological maturity (days)	Girth of earhead (cm)	Length of earhead (cm)	Test weight (g)	Grain yield (t ha^{-1})	Stover yield (t ha^{-1})
Land configurations (L)							
L_1 = Flat bed	49.85	81.89	9.00	19.22	11.27	2.65	6.72
L_2 = Ridges and furrow	50.41	82.52	9.64	20.42	11.80	3.18	7.91
S.Em.±	0.50	0.80	0.15	0.21	0.15	0.08	0.18
C.D. at 5%	NS	NS	0.42	0.59	0.42	0.23	0.53
Dates of sowing (D)							
D_1 = 25 Jan.	51.17	84.06	9.13	20.06	11.88	3.04	7.78
D_2 = 5 Feb.	50.61	82.06	9.98	20.59	11.72	3.24	7.95
D_3 = 15 Feb.	48.61	80.50	8.85	18.81	11.01	2.46	6.21
S.Em.±	0.62	0.98	0.18	0.25	0.18	0.10	0.23
C.D. at 5%	1.77	2.83	0.51	0.73	0.52	0.28	0.65

temperature and other climatic parameters during various crop growth stages, which reflected into better growth. Similar results of summer pearlmillet growth were observed by Andhale et al. (2007b), Upadhyay et al. (2001), Patel and Patel (2002) and Patel et al. (2004).

Days required to 50% flowering and physiological maturity was also remarkably influenced under varying sowing time. Late sown crop advanced the flowering and maturity over early sowing in summer pearlmillet and 15 February sowing took minimum days to 50% flowering and maturity. It might be due to the availability of required photoperiod to early sown pearlmillet crop at reproductive stage. Similar findings were also reported by Andhale et al. (2007b) in pearlmillet and by Pawade (2010) in maize.

Effect of land configuration on yield attributes and yield

The yield attributes *viz.*, length and girth of earhead and test weight were significantly influenced by land configuration treatments (Table 2).

Significantly higher length and girth of earhead and test weight was recorded under ridges and furrow (L_2) sowing. This might be due to better growth of plant in terms of dry matter accumulation under ridges and furrow sowing which might have adequately supplied more photosynthates for development of sink. The present findings were in accordance with those of Ugale et al. (1995) in pearlmillet and Patel et al. (2008) in sorghum for length and girth of earhead and with Kumar (2008) in maize with respect to test weight.

The better performance of pearlmillet crop observed in terms plant height, dry matter accumulation, and length and girth of earhead and test weight obtained under ridges and furrow (L_2) sowing treatment which in turn converted into the maximum grain and stover yields. An increase in grain and stover yield under ridges and furrow was 20.00 and 17.71% over flat bed sowing, respectively. This might be due to the cumulative effect exerted from better improvement in drainage, soil environment, aeration, root development, optimum moisture-air equilibrium throughout the crop growth besides supply of available nutrients to the crop resulting in better growth and development ultimately reflected in better grain and stover yields. These findings were corroborated the results of Patel et al. (2008) and Kiran et al. (2008) in *rabi* sorghum. Parihar et al. (2009) also found resembling results in pearlmillet for grain and stover yield with sowing on ridges and furrow treatment over flat bed.

Effect of dates of sowing on yield attributes and yield

Various yield attributes *viz.*, length and girth of earhead and test weight were significantly influenced under varying sowing time (Table 2). Crop sown either on 25 January (D_1) or 5 February (D_2) recorded higher values for almost all the above yield characters than late sown crop, that is, on 15 February (D_3).

Better growth of plant in terms of plant height, leaf area and dry matter accumulation under 25 January (D_1) or 5 February (D_2) sowing reflected into better development of yield attributes under early sown crop. Moreover, congenial climatic conditions during early sowing also play vital role in development of yield attributes. These findings are substantiated with those reported by Patel and Patel (2002) and Patel et al. (2004). Thus, the overall better growth and higher values of most of the yield attributes under 5 February sowing resulted into maximum grain yield of 3.24 t ha^{-1} (Table 2), however, it remained statistically at par with 25 January sowing with grain yield (3.04 t ha^{-1}). Late sowing on 15 February recorded lowest grain yield of 2.46 t ha^{-1}. Crop sown on 5 February increased the grain yield by 6.57 and 31.70%, respectively over 25 January and 15 February sowing.

The reason for higher yield in D_1 or D_2 might also be due to ideal maximum temperatures around 29 to 32°C during vegetative and flowering periods resulting in better translocation to reproductive structures, seed set and seed development. When the crop is sown on 15 February, the flowering period coincides with higher mean maximum temperature of around 37 to 38°C, which adversely affected the seed set and translocation of nutrients resulting in poor grain yields. The yield attributing characters such as length and girth of earhead and 1000 grain weight were also significantly higher with the crop sown on D_1 or D_2 dates. These result land support to those reported by Shinde et al. (2003).

Similarly, better development of various growth parameters *viz.,* plant height, dry matter accumulation and leaf area under early sowing reflected into significant variation in stover yield recorded under different sowing times. Though, 5 February sowing produced highest stover yield (7.95 t ha^{-1}) it remained statistically at par with 25 January (7.78 t ha^{-1}). Whereas lowest stover yield (6.21 t ha^{-1}) was recorded under late sowing on 15 February. The crop sown on 5 February increased the stover yield in pearlmillet by 2.19 and 28.02%, respectively over 25 January and 15 February sowings. Similar findings were also reported by Patel and Patel (2002) at S.K. Nagar (Gujarat), Patel et al. (2004) at S. K. Nagar (Gujarat) and Deshmukh et al. (2009).

Effect of land configuration on nutrient content and uptake

Significantly the highest N, P and K content and uptake by grain and stover were recorded under ridges and furrow (L_2) over flat bed (L_1) sowing, except P content in stover. Higher N content (1.746 and 0.712%) and N uptake (55.650 and 56.172 kg/ha), P content (0.350 and 0.092%) and P uptake (11.165 and 7.280 kg/ha) and K content (0.449 and 0.990%) and K uptake (14.289 and 78.244 kg/ha) by grain and stover of pearlmillet, respectively were recorded with the treatment ridges and furrow (Table 3).

This might be attributed to better root growth due to better aeration and good drainage might have also increase microbial activity with optimum moisture and nutrient availability for its growth. The findings were in accordance with those earlier reported by Patel et al. (2008), Parihar et al. (2009) and Dhimmar and Damame (2010).

The total uptake is a function of total biological yield and the nutrient concentration in the tissues. The data showed profound increase in grain and stover yield of summer pearlmillet that ultimately increased the uptake of N, P and K by pearlmillet.

Effect of dates of sowing on nutrient content and uptake

At harvesting, treatment-wise, grain and stover samples were collected and analyzed for N, P and K content. Using content and yield values, uptake of these nutrients was calculated separately for grain and stover. The content of N, P and K in grain was affected significantly due to changes in sowing time. However, P and K content of stover were non-significant, but N content showed significant results with change in sowing time.

Apart from difference in nutrients content, all nutrients *viz.*, N, P and K uptakes were found significant in both grain and stover due to different dates of sowing treatments (Table 3). In all these cases (grain and stover content and uptake), treatment D_2 (5 February sowing) showed superiority over rest of the sowing date treatments *viz.*, D_1 and D_3. This increase in content and uptake by grain and stover might be due to increased yield of grain and stover under treatment D_2. Similar results were also observed by Andhale et al. (2007a).

Effect of land configuration on economics

It is obvious from the data (Table 4) reported in that ridges and furrow treatment irrespective of land configuration markedly increased the returns over flat bed treatment giving returns of Rs.14081 ha^{-1} and B:C ratio of 1.88. Whereas, the net returns and B:C ratio obtained from flat bed treatment (L_1) were Rs.10181.04 ha^{-1} and 1.67, respectively. This might be due to higher yields of crop gained from ridges and furrow treatment (L_2). It clearly brings out the fact that adoption of ridges and furrow techniques of land configuration was more paying than flat bed techniques. Results of Kiran et al. (2008) and Parihar et al. (2009) add to the compliance of obtained results.

Effect of dates of sowing on economics

From the economics (Table 4) point of view, maximum

Table 3. NPK content and uptake of grain and stover of pearlmillet crop as influence by land configurations and dates of sowing.

Treatments	Grain		Stover		Grain		Stover		Grain		Stover	
	N content (%)	N uptake (kg/ha)	N content (%)	N uptake (kg/ha)	P content (%)	P uptake (kg/ha)	P content (%)	P uptake (kg/ha)	K content (%)	K uptake (kg/ha)	K content (%)	K uptake (kg/ha)
Land configurations (L)												
L_1 = Flat bed	1.713	45.462	0.700	47.063	0.343	9.089	0.090	6.073	0.839	22.220	0.776	52.131
L_2 = Ridges and furrow	1.746	55.650	0.712	56.172	0.350	11.165	0.092	7.280	0.850	27.036	0.790	62.434
S.Em.±	0.011	1.436	0.004	1.318	0.002	0.273	0.0008	0.187	0.003	0.669	0.005	1.407
C.D. at 5%	0.032	4.130	0.010	3.791	0.006	0.785	NS	0.537	0.008	1.919	0.014	4.036
Dates of sowing (D)												
D_1 = 25 Jan.	1.739	53.072	0.710	55.391	0.349	10.650	0.091	7.110	0.844	25.721	0.786	61.198
D_2 = 5 Feb.	1.750	56.778	0.713	56.324	0.350	11.368	0.092	7.356	0.851	27.563	0.783	62.224
D_3 = 15 Feb.	1.700	41.819	0.694	43.137	0.340	8.362	0.090	5.564	0.837	20.599	0.780	48.425
S.Em.±	0.014	1.759	0.004	1.615	0.003	0.334	0.001	0.229	0.003	0.820	0.006	1.723
C.D. at 5%	0.039	5.059	0.013	4.643	0.007	0.961	NS	0.658	NS	1.24	1.79	1.90

Table 4. Economics of summer pearlmillet as influenced by land configurations and dates of sowing.

Treatments	Net returns (Rs.)	B:C ratio
Land configurations *(L)*		
L_1 = Flat bed	10181.04	1.67
L_2 = Ridges and furrow	14081.00	1.88
Dates of sowing (D)		
D_1 = 25 January	13971.00	1.93
D_2 = 5 February	15541.00	2.03
D_3 = 15 February	8341.00	1.55

78.244 kg/ha) by grain and stover of pearlmillet, net returns of Rs.15541 ha^{-1} with B:C ratio of 2.03 was obtained with 5 February sowing treatment (D_2), followed by 25 January sowing treatment (D_1) with Rs.13971 ha^{-1} of net returns and 1.93 B:C ratio. Whereas, the net returns and B:C ratio obtained from 15 February (D_3) were Rs. 8341 ha^{-1} and 1.55, respectively. This might be due to higher yields of pearlmillet crop gained from D_2 and D_1 treatments. The results resembled with Sukhadia and Dhoble (1992).

Conclusion

From the results of the experimentation, it can be concluded that for getting higher profitable production of summer pearlmillet should be sown on 5 February on ridges and furrow. In other words, land configuration and dates of sowing ensue in higher economic returns from pearlmillet crop in summer season.

REFERENCES

Andhale RP, Shinde SH, Sinare BT, Tambe AD (2007b). Effect of sowing dates and fertilizer levels on phenology and heat unit accumulation in pearlmillet (*Pennisetum glaucum* L.) hybrids. J. Maharashtra Agric. Univ. 32(3):401-402.

Andhale RP, Shinde SH, Sinare BT, Tambe AD (2007a). Effect of sowing dates and fertilizer levels on uptake of N, P and K in pearlmillet (*Pennisetum glaucum* L.) hybrids in summer season. J. Maharashtra Agric. Univ. 32(3):399-400.

Chiroma AM, Alhassan AB, Khan B (2008). Yield and water use efficiency of millet as affected by land configuration treatments. J. Sustain. Agric. 32(2):321-333.

Deshmukh LS, Jadhav AS, Raskar SK (2009). Effect of sowing dates on grain and fodder yield of pearl millet (*Pennisetum glaucum*) genotypes in summer season. Karnataka J. Agric. Sci. 22(1):186-187.

Jackson ML (1967). "Soil Chemical Analysis". Prentice Hall of India, New Delhi. pp. 183-192.

Kiran JA, Lingaraju BS, Ananda N (2008). Effect of in situ moisture conservation practices and nitrogen levels on growth, yield and economics of rabi sorghum under rainfed condition. Crop Res. 35(1 and 2):13-16.

Kumar A (2008). Growth, yield and water use efficiency of different maize (*Zea mays*)-based cropping systems under varying planting methods and irrigation levels. Indian J. Agric. Sci. 78(3):244-247.

Panse VG, Sukhatme PV (1967). "Statistical Methods for Agricultural workers" Indian Council of Agricultural Research, New Delhi, pp. 187-197.

Parihar CM, Rana KS, Parihar MD (2009). Crop productivity, quality and nutrient uptake of pearlmillet (*Pennisetum glaucum*) and Indian mustard (*Brassica juncea*) cropping system as influenced by land configuration and direct and residual effect of nutrient management. Indian J. Agric. Sci. 79(11):927-930.

Patel AM, Patel DR, Patel GA, Thakor DM (2004). Optimization of sowing and fertilizer requirement of barley (Hordeum vulgare) under irrigated condition. Indian J. Agron. 49(3):171-173.

Patel BJ, Patel IS (2002). Response of summer pearlmillet (Pennisetum glaucum) to different dates, methods of sowing and nitrogen levels under North Gujarat agro-climatic conditions. Crop Res. 24(3):476-480.

Patel HH, Patel SD, Patel TU (2008). Effect of land configuration and nutrient management in rabi sorghum (*Sorghum bicolor*) under South Gujarat conditions. National symposium on "New Paradigms in Agronomic Research", Nov., 19-21, 2008, Navsari, Gujarat. pp. 251-252).

Pawade PD (2010). Response of winter maize (*Zea mays* L.) to sowing time and inter row spacing under South Gujarat condition. M.Sc. (Agric.) thesis submitted to N.A.U., Navsari.

Shinde GG, Aglave BN, Deshmukh AS (2003). Ideal sowing dates and genotypes for summer irrigated sorghum in Maharashtra, India. Int. Sorghum Millet Newslett. 44:83-86.

Sukhadia NM, Dhoble MV (1992). Productivity and water-use efficiency of rainy-season crops under different dates of sowing. Indian J. Agron. 37(4):669-675.

Ugale SD, Wani AG, Patil BD, Bhingrde MT (1995). Performance of different sowing methods of pearlmillet on lighter type of soils under rainfed conditions. J. Maharashtra Agric. Univ. 20(2):311-312.

Upadhyay PN, Dixit AG, Patel JR, Chavda JR (2001). Response of summer pearlmillet to time and method of planting, age of seedling and phosphorus grown on loamy sand soils of Gujarat. Indian J. Agron. 46(1):126-130.

ONVAREF: A decision support system for onion varietal reference

Mir S. A.[1], Mir I. A.[2], Quadri S. M. K.[3], Ahmad N.[4] and Angchuk S.[5]

[1]KVK Pombay, Kulgam, Kashmir- 192233, India.
[2]ARIS, Cell, SKUAST Kashmir, India.
[3]Post Graduate Department of Computer Science,University of Kashmir, Hazratbal Campus-190006, India.
[4]Central Institute of Temperate Horticulture, (ICAR), Old Airport, Srinagar, India.
[5]KVK Kadakh (Leh), India.

This paper presents the development of a decision support system (DSS) called ONVAREF for screening of new onion varieties being released. The system is a windows oriented, user-friendly, database driven, having excellent graphical user interface (GUI), designed using 4GL, which works on searching and matching principle to spawn large database comprising of dozens of characteristics suitable for judging originality of a variety under screening. It is also capable of assessing closeness of existing varieties with the one being screened. The accuracy of decision making process provided by ONVAREF has been found 93.33% as compared to 53.33% by manual screening. It is 123% time efficient as compared to paper based assessment manual from which the computer programme was evolved.

Key words: Decision support system, modeling, database driven decision support system (DSS), database management system (DBMS), decision making.

INTRODUCTION

Screening of newly developed onion varieties amidst ever-increasing duplicacy of research effort is a complex process that demands evaluation of many characteristics having different parameters. There is a substantial amount of empirical evidence that human initiative judgment and decision making can be far from optimal and it deteriorates further with added complexity and stress (Adams et al., 2000). In this situation, the precision of human judgment and decision making is very important. Design and development of tools and techniques aiding agricultural decision process is gaining popularity, particularly with the advancement in electronically processing of data (Power, 2002). Decision support systems can help to reduce uncertainty and improve decision-making process by providing access to data through procedures and analytical reasoning (Uehera and Tsuji, 1993). For agriculture production and marketing of sustainable development, decision support system (DSS)'s have allowed rapid assessment of several production systems around the world to facilitate decision making at farm and policy levels.

The concept of decision support system research has evolved from technical studies of organization decision making and technical work on interactive computer system, mainly carried at Massachusetts's Institute of Technology (MIT) in 1960s (Keen et al., 1978). It became recognized as self sustained research area widely in the 1970s and gained high reputation in the 1980s. Data warehousing and Online Analytical Processing (OLAP) further broadened the realm of DSSs in 1990s. Arrival of new millennium introduced web based analytical applications, giving DSS an important role in decision making process.

The applications of DSS seem natural considering the widespread use of decision making agents in and out of the field. With the advent of powerful computational power, efficient database management system (DBMS) with OLAP, computer programs aimed to support decision process has improved a lot. These systems have already been applied in different areas of agriculture (Jenkins and Williams, 1998), such as Crop-Protection, Agronomy, Plant Pathology, Pomology, Soil Science, Agricultural Engineering, Forestry, Environmental Science, Agro meteorology among others, and have the potential to be an important tool in the decision making process for primary producers and their advisors (Ritchie, 1995).

Use of DSS in agriculture for providing information and recommendation on efficient utilization of fertilizer (Chai et. al., 1994), reducing herbicide use (Coble, 1994), plant protection (Yialouris et al., 1996), variety specific information (Detlefsen et al., 2004), management of environment risks (Didier, 2005), integrated nutrient management (Mosseddaq, 2005), forest management (Rui and Jose, 2005), crop disease control (Tagir et al, 2005), agricultural practices and extension (Suarez et. al., 2005) and sustainable land use planning (Klik, 2006) have been designed and implemented successfully mainly for improving economic returns, changing farming practices and minimizing environmental risks.

Division of Vegetable Science, Sher-e-Kashmir, University of Agricultural Sciences and Technology of Kashmir (SKUAST Kashmir) acted as nodal centre for screening of onion varieties. The process was manual based and results were always error prone, resulting in inefficient decision making. Realizing the need of error free and quality decision in screening process viz-a-viz potential and successes of agricultural decision support systems, ONVAREF was designed and developed for quality decision making to discourage duplicity of research effort. The system is database driven having efficient graphical user interface, with options to generate reports and proper validation checks. It has been build based on the manual which consists of tens of evaluation characteristics having hundreds of parameters sufficient to distinguish one onion variety with the other. The system has incorporated 19 existing varieties of onion, along with their evaluation characteristics, which were existing at the time of system development. The system has a provision to include newly release varieties along with their physio-chemical characteristics.

Decision support systems (DSS)

DSS are interactive, computer-based systems and subsystems intended to help decision makers to use communication technologies, data, documents, knowledge and/or models to complete decision process tasks, giving emphasis to structural, semi-structured and unstructured decisions (Keating and McCown, 2001; Yialouris et al., 1997). They can improve the quality and effectiveness of decision making by processing a lot of data and providing alternative solutions. The term DSS has been used and defined in various ways depending upon the author's point of view (Power, 2002; Druzdzel and Flynn, 1999). Finlay (1994) defined it as "a computer-based system that aids the processing or decision making". While Turban (1995) defined it specifically as "an interactive, flexible, and adaptable computer based information system, especially developed for supporting the solution of a non-structured management problem for improved decision making". There are also several definitions given by different authors, which fall between these two extremes (Little, 1970; Keen and Scott, 1978; Moore and Chang, 1980; Sprague and Clarlson, 1982). In order to avoid exclusion of any of the existing types of DSSs, we define them roughly as interactive computer based systems that aid in making a quality decision.

Overview of DSS

Researchers and technologists have built and investigated DSSs for about 48 years. This concept has evolved from technical studies of organizational decision making and technical work on interactive computer systems mainly carried at Massachusetts's Institute of Technology (MIT) in 1960s (Keen and Scott Morton, 1978). The first DSS for agriculture cited in the literature is *Televis*, which was developed by Norway in 1957 for plant protection. Another such system named as *Guntz-Divoux* was developed by France in 1963. First computer-aided DSS for production scheduling was investigated, which was running on IBM 7049 (Ferguson and Jones, 1969). Michael S. Scott Morton's dissertation in 1967 is regarded as major historical turning point wherein building, implementing and testing an interactive and model-oriented DSS. During 1971 Scott Morton demonstrated "how computer and analytical models could help managers to make key planning decisions". DSS became recognized as self-sustained research area in 1970s, when business journals started publishing articles on management decision systems, strategic planning systems and decision support systems (Sprague and Watson, 1979). DSS gained high reputation during 1980s when its application development started in Universities and organizations, which expanded their scope. During 1982, Ralph Sprague and Eric Carlson's book "Building Effective Decision Support System" further explained the framework of database, model base, dialog generation and software management. It provides a practical and understandable overview of how organization could or should build DSS.

Table 1. Existing onion varieties incorporated in the ONVAREF.

S/N	Onion variety
1	Panjab Naroya
2	B4
3	Phule Suvarna
4	Arka Niketan
5	Pusa Madhavi
6	Pillpatti Local
7	Agrifound Light Red
8	Gujarat White Onion
9	Hisar-2
10	VL-3
11	Arka Pragati
12	Udaipur-102
13	Yello Globe
14	PRO-6
15	Pusa White Round
16	N-2-4-1
17	Pusa red
18	Arka Pitambar
19	Agrifound Rose

During 1990, Bill Inmon, father of data warehouse, and Ralph Kimball, the doctor of DSS, actively promoted DSS development using relational database technologies.
Beginning in approximately 1995, the WWW and global Internet provided a technology expansion platform for further extending the capabilities and deployment of computerized decision support. In 2000, Application Service Providers (ASPs) began hosting the application software and technical infrastructure for decision support capabilities. Emergence of enterprise knowledge portals, knowledge management, business intelligence and communication technologies integrated DSS to an integrated environment (Bhargava and Power, 2001). Most software developers have regarded WWW as a serious platform for implementing all types of DSSs. Manipulation of quantitative models, accessing and analyzing of large databases and building of group DSSs are major applications of DSS (Ecom, 2002; Arnott and Pervan, 2005). Recently, intelligent decision support systems have added new capabilities to these systems (Dhar and Stein, 1997).

MATERIALS AND METHODS

Development and evaluation of Agricultural DSS has widely been documented in literature (Renner et al., 1999). However design, development and evaluation of ONVAREF follows as related research (Batchelor et al., 2004a; Hochman et al., 1994; Cox, 1996; Jakku,and Thorburn 2010; Adelman, 1992). Data design comprises of 30 parameterized characteristics already identified by the domain experts pertaining to 19 already existing onion varieties (Tables 1 and 2).

In order to study the accuracy of DSS with respect to manual procedure, 100 trial cases each involving Domain experts, who used to screen onion varieties using manual procedure and those who got trained to use DSS instead of manual procedure were conducted. The maximum scores that an evaluator could obtain was 109 (109 option corresponding to 30 characteristics) and the minimum score that a respondent could obtain was 0. The accuracy index was calculated based on the following formula

$$\text{Accuracy index} = \frac{\text{No. of correct options taken into consideration}}{\text{Total options}} * 100$$

Data analysis has been done using R software.

System design

The technical structure of ONVAREF-DSS is designed in a way that allows interactive operations of the decision maker without any specific knowledge. The DSS output includes recommendations on originality of variety, its closeness with other existing varieties, cultivation and management practices and all attributes that makes evaluation of onion variety possible.

The system has been developed using Microsoft Visual Basic as front end tool (Graphical User Interface (GUI), with embedded SQL statements) and Microsoft Access as backend to handle structure of database and data. The chosen languages Visual Basics and Structured Query Language (SQL) provides more than three times and nine times, respectively, the functionality of a LOC than other conventional languages. The comparison characteristics of different languages used as front end and back end have been given in literature (Jones, 1998).

All the database operations have been implemented through SQL calls embedded in Microsoft Visual Basics. For data entry and reports, interactive Microsoft Visual Basics forms and reports have been used. We have also used the Microsoft Jet Database Engine

Table 2. Characteristics of Onion varieties incorporated in ONVAREF.

S/N	Onion varieties
1	Foliage attitude
2	Bulb width of neck
3	Time of harvesting
4	No. of leaves
5	Bulb colour of skin
6	Leaf maximum diameter
7	Foliage length
8	Bulb thickness of rings
9	Leaf waxiness
10	Leaf glossiness
11	Bulb colour of flesh
12	Leaf colour
13	Foliage / leaf cranking
14	Bulb diameter of root disc
15	Leaf foliage fall
16	Pseudo stem diameter
17	Pseudo stem length
18	Bulb diameter
19	Bulb cross section
20	Bulb bolting tendency
21	Bulb height
22	Bulb shape
23	Bulb thickness of skin
24	Bulb skin adherence
25	Bulb firmness of flesh
26	Bulb position of root disc
27	Bulb pred. no. of axes
28	Bulb dry matter content
29	Splitting note bulb lets
30	Seasonal adaptability

for connectivity between form end and back end. The connection is dynamic with automatic configuration at the time of installation.

Knowledge representation

Knowledge representation has been expressed as a series of IF/Then statements, which are mostly followed in development of decision support systems (Gonzalez-Andujar, 1993). The rule is made up of a list of IF conditions(s) and a list of THEN conclusions(s) or statement(s) about the appropriate solution of the problem. If the computer determines that certain IF condition is true, it executes THEN statement(s) and draws conclusion(s) in the form of statement(s). Knowledge base of ONVAREF is implemented as a set of tables designed used in a proper schema.

System architecture

The ONVAREF is database driven, singly user system. The front end side consists of a GUI, rule-based reasoning and interface engine. The back end handles storage using the capabilities of a DBMS. The DBMS performs the ordinary DB operations such as insert, select, delete and update. GUI accesses the DBMS using embedded SQL statements to alter or get results. The Modular structure and interface description of ONVAREF has been shown in Figures 1 to 4 and Entity_Relationship Diagram (ER- Diagram) is shown in Figure 2. The historical evolution of the DBMS are described in Date (1995).

RESULTS AND DISCUSSION

100 trial cases each involving Domain experts who used to screen onion varieties using manual procedure and those who have closely watched the development of the system were involved to run many trial cases. The results of the trial cases are presented in Table 3 which testifies the accuracy of the system. Error-free decisions made by DSS were 93.33% as compared to 53.33% while 10% errors were reported in 4 cases (6.67%) as compared to 21 cases (35%). Also no case was reported in which more than 10% errors could be seen using DSS as compared to 7 cases (11.67%) using manual procedure.

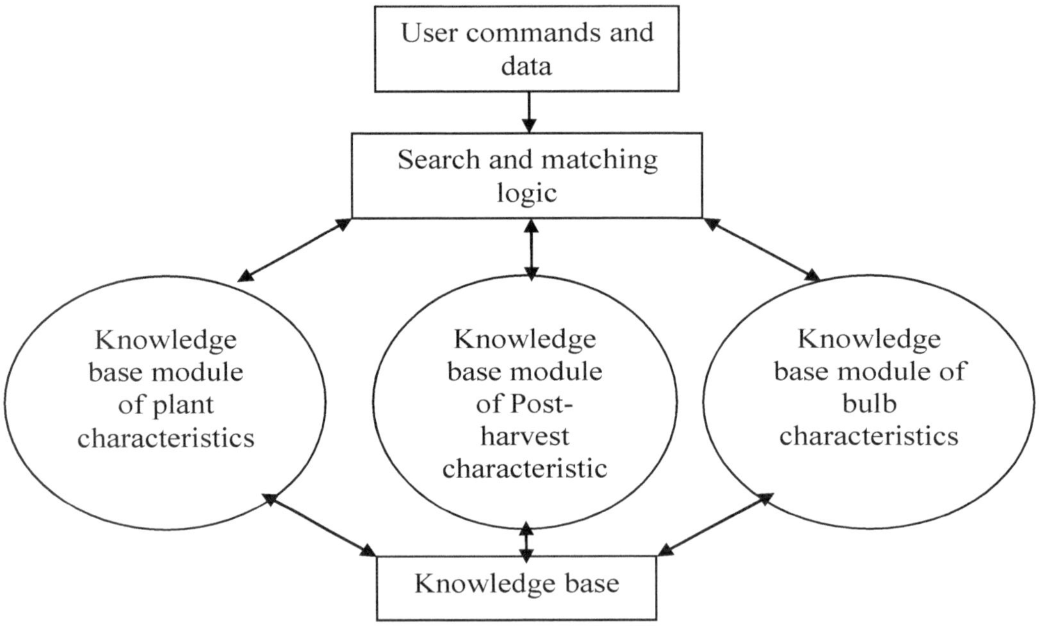

Figure 1. Modular structure and interface description of ONVAREF.

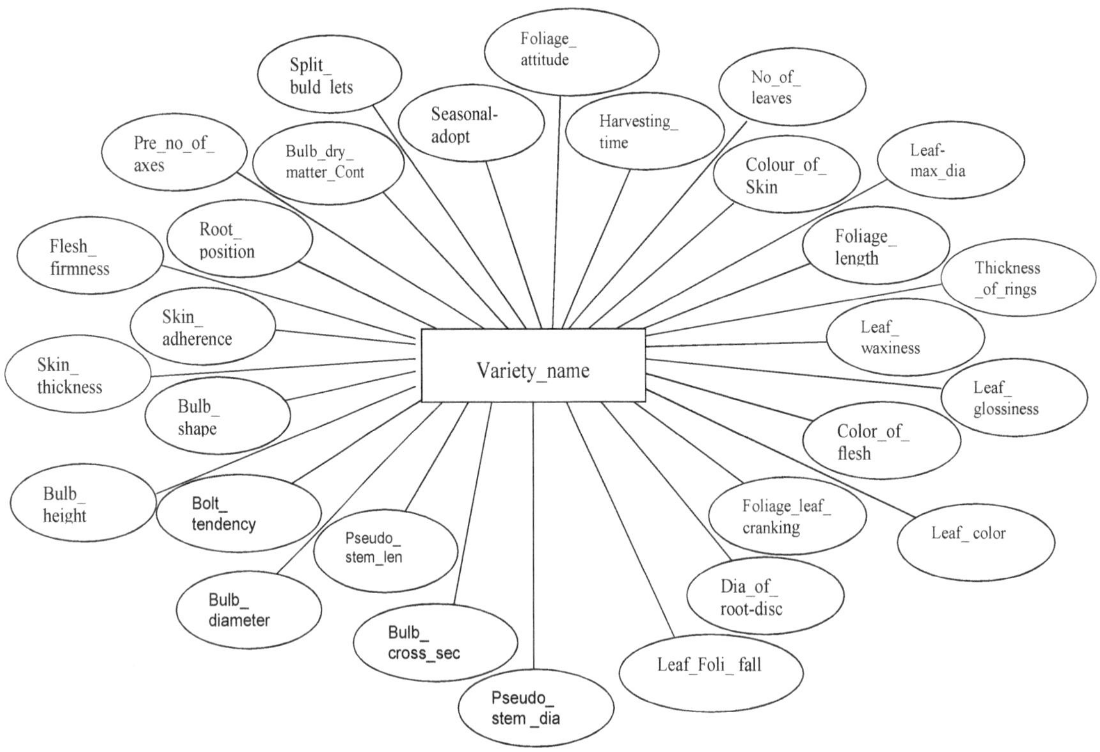

Figure 2. Entity-relationship diagram.

It also clearly indicates that system is considerably reducing workload and saving time (138%) related to variety evaluation / screening process, as and when made accessible. The perception of the users about using ONVAREF is given in Table 4. It is evident that more than 10% users have rated usefulness and

Table 3. Distribution of trial-cases results about accuracy of ONVAREF Vs manual procedure.

Parameter	Using ONVAREF (N=60)		Using manual procedure (N=60)	
Mean score	108.83		103.08	
Standard deviation	0.69		9.04	
Range	105 – 109		72 – 109	
Average time taken for screening of one variety	09.7 min		21.67 min	
Decision	**No.**	**Percent**	**No.**	**Percent**
No errors	56	93.33	32	53.33
Up to 10% errors	4	6.67	21	35.00
More than 10% errors	0	0	07	11.67
Total	60	100	60	100

Table 4. Distribution of end-users based on their perception about utility of ONVAREF.

Particular	(N = 60)		
	Excellent	**Good**	**Fair**
Usability	9 (15%)	48 (80%)	3 (5%)
Portability	8 (13.30%)	51 (85.00%)	1 (1.70%)
Extendability	5 (8.30%)	55 (91.70%)	0 (0.00%)
Usefulness	12 (20.00%)	42 (70.00%)	6 (10.00%)
Operateability	3 (5.00%)	57 (95.00%)	0 (0.00%)
Adoptability	13 (21.70%)	35 (58.30%)	12 (20.00%)

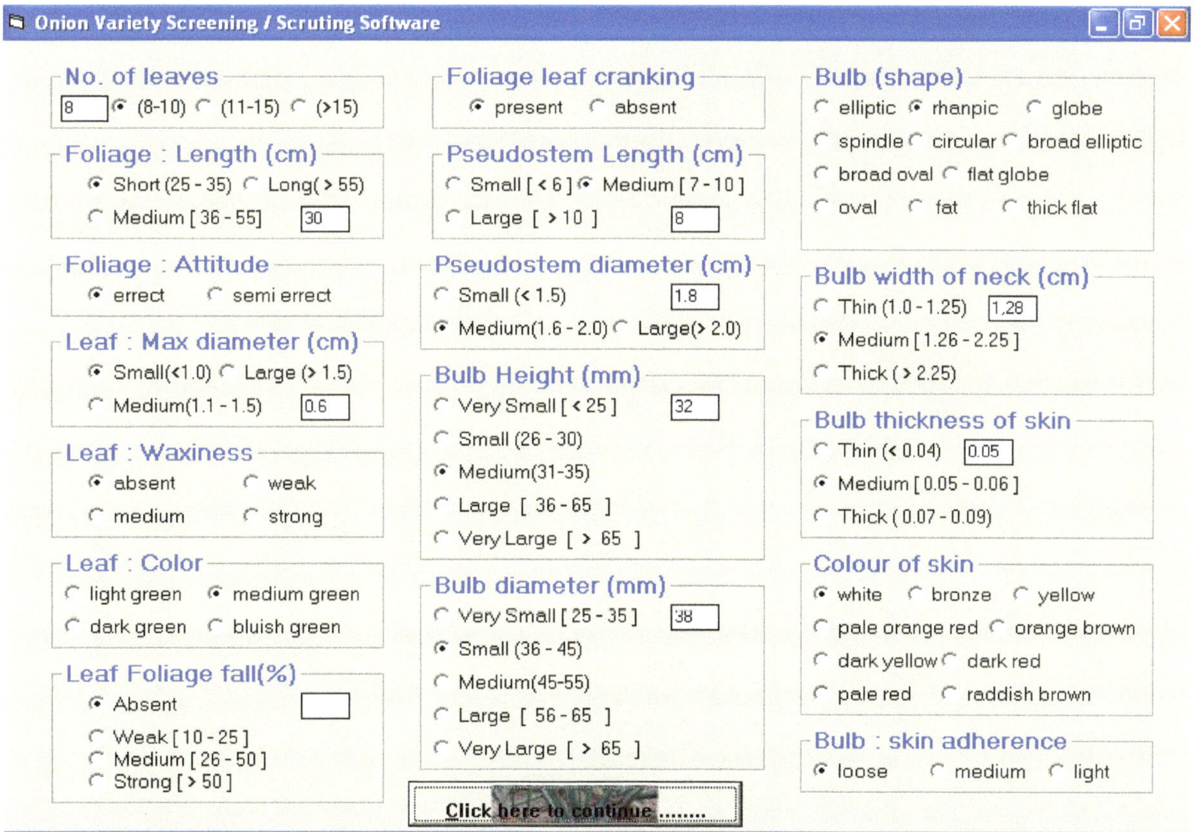

Figure 3. User interface of ONVAREF -1.

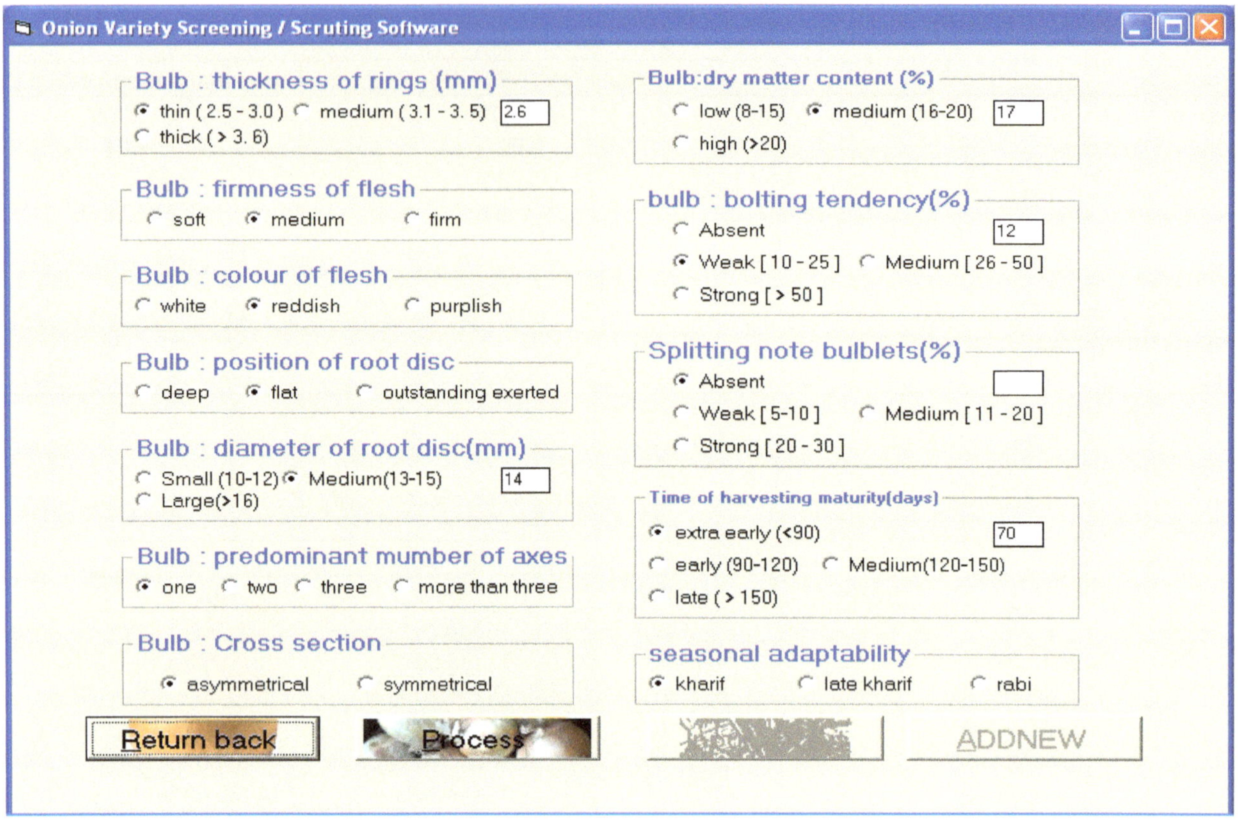

Figure 4. User interface of ONVAREF-2.

adoptability as excellent while the rest have been rated from 3-9%. Also more than 50% users have rated usability, portability, extendability and operateability as good while more than 40% users have rated usability and usefulness as good. Moreover, nearly 12% users have rated adoptability as fair. The results are in agreement with the recommendations given by various researchers regarding design, development, evaluation and implementation of agricultural DSS viz, user friendly, portable, useful, extendable, with maximum adoptability score, which most of the agricultural based DSS lacks. (Hochman et al., 19940; Renner et al., 1999 and Cox PG, 1996)

Conclusion

ONVAREF has been successfully implemented for screening process of newly developed onion varieties. It has proved to be accurate, time saving, as well as ascertaining closeness of particular variety under screening with the existing one which has paved the way to put the variety under more screening process before releasing recommendation, which was not possible with the manual procedure being used. In short, design and development of ONVAREF is an effort towards quick, timely and reliable decision support for evaluation of Onion varieties mainly to discourage duplicacy of research effort.

Abbreviations: DSS, Decision support system; **OLAP,** online analytical processing; **DBMS,** database management system; **ASPs,** application service providers; **SQL,** structured query language; **GUI,** graphical user interface.

REFERENCES

Adams ML, Cook SE, Corner R (2000). Managing Uncertainty in Site-Specific Management: What is the Best Model? Prec. Agric. 2:39-54.

Adelman L (1992). Evaluating Decision Support Systems and Expert Systems. New York Wiley.

Arnott D, Pervan G (2005). A critical analysis of decision support systems research. J. Inform. Technol. 20(2):67-87.

Batchelor WD, Paz JO, Thorp KR (2004a). Development and evaluation of a decision support system for precision farming. Proceedings of the 7th International Conference on Precision Agriculture. July 25-26, 2004, Minneapolis, MN. ASA, CSSA, SSSA Inc., Madison, WI, USA.

Bhargava H, Power DJ (2001). Decision support systems and web technologies: a status report. In: Americas Conference on Information Systems, 3-5 August 2001, Boston, MA.

Chai KL, Costello TA, Wells BR, Norman RJ (1994). Expert system for fertilization management of rice. App-eng-agric. 10:849-855.

Coble HD (1994). Computer decision aids for reducing herbicide use. Proc-annu-meet-Northeast-weed-sci-Soc. College Prak, Md.: The

Society. 48:155-159.
Cox PG (1996). Some issues in the design of agricultural decision support systems. Agric. Syst. 52(2-3):355-381.
Date CJ (1995). An Introduction to Database Systems, 6th Edition, Eddison-Wesley, 1995.
Detlefsen NK, Jensen AL (2004). A stochastic model for crop selection. Agric. Syst. 81(1):55-72.
Dhar V, Stein R (1997). Intelligent Decision Support Methods: The Science of Knowledge. Prentice-Hall, Upper Saddle River, NJ.
Didier B, Christophe LP, Souleymane D, Geraldine A (2005). Perspectives of modelling the farmer's seed system for in situ conservation of sorghum varieties in Mali. Paper presented at EFITA/WCCA, 2005, 25-28 July 2005, Vila Real, Portugal.
Druzdzel MJ, Flynn RR (1999). Decision Support Systems. Encyclopedia of Library and Information science. A. Kent, Marcel Dekker, Inc.
Ecom SB (2002). Decision Support Systems Research (1970-1999). Edwin Mellen Press, Lewiston, NY.
Ferguson RL, Jones CH (1969). A Computer Aided Decision System. Manage. Sci. 15(10):550-562.
Finlay PN (1994). Introducing Decision Support Systems. Blackwell Publishers, Oxford, UK; Cambridge, MA.
Gonzalez-Andujar JL (1993). Cereal aphids expert system (CAES): Identification and decision making. Comput. Electr. Agric. 8:293-300.
Hochman Z, Pearson CJ, Litchfield RW (1994). User's attitude and roles in the development and evaluating of Knowledge Base Decision Support System for agricultural advisors. Agric. Syst. 44(2):217-235.
Jakku E, Thorburn PJ (2010). A conceptual framework for guiding the participatory development of agricultural Decision Support Systems 103(9):675-682.
Jenkins TG, Williams CB (1998). DECI-Decision evaluator for the cattle industry. In: Proc. 6th World Cong. Genet. Appl. Livest. Prod. Armidale, Austr. 27:461-462.
Jones C (1998). Estiamting Software Costs, McGraw-Hill, 1998.
Keating BA, McCown RL (2001). Advances in farming systems analyses and intervention, Agric. Syst, 70:555-579.
Keen PGW, Scott Morton MS (1978). Decision Support Systems: An Organizational Perspective. Reading, MA: Addison-Wesley.
Keen PGW (1980). Decision support systems: a research perspective. In: Fick, G.1258 and Sprague, R. H. (Eds.), Decision Support Systems: Issues and Challenges. Pergamon Press, Oxford; New York.
Klik A (2006). Agricultural Decision Support System for Soil and Water Conservation Measures. Computer in Agriculture and Natural Resources, 4th World congress conference, proceedings of the 24-26 July 2006 (Orlando, Florida USA) Publication Date 24 July 2006 701P0606.
Little JDC (1970). Models and Managers: The Concept of a Decision Calculus. Manage. Sci, 16(8).
Moore JH, Chang MG (1980). Design of Decision Support Systems. Data base, 12(1&2).
Mosseddaq F, Dnidane S, Lahlou M (2005). Integrated Wheat N Nutrition Management in Morocco: A Decision Support Model Paper presented at EFITA/WCCA, 2005, 25-28 July 2005, Vila Real, Portugal.
Power DJ (2002). Decision Support Systems: Concepts and Resources for Managers. Greenwood/ Quorum, Westport, CT.
Power DJ (1997). What is a DSS?. DSstar, The On-Line Executive J. Data-Intensive Decision Support, 1(3).
Renner KA, Swintin SM, Kells JJ (1999). Adaptation and evaluation of the WEEDSIM weed management model for Michigan. Weed Sci. 47:338-348.
Ritchie JT (1995). International consortium for agriculture Systems Application (ICASA): Establishment and Purpose. Agric. Syst. 49:329-335.
Rui PR, Jose GB (2005). MetaForest- a web-based decision system to support forest management involving multiple-ownership. Paper presented at EFITA/WCCA, 2005, 25-28 July 2005, Vila Real, Portugal.
Sprague Jr. RH, Watson HJ (1979). Bit by Bit: Toward Decision Support Systems. California Manage. Rev. 22(1):60-68.
Sprague Jr, RH, Carlson ED (1982). Building Effective Decision Support Systems. Prentice-Hall, Inc., Englewood Cliffs, NJ.
Suarez de Cepeda, Recio M, Rubio, F (2005). Decision Support System for Farm Mechanization. Paper presented at EFITA/WCCA, 2005, 25-28 July 2005, Vila Real, Portugal.
Tagir Z, Ibragimov I, Ibragimova T, Sergey SS (2005). Decision Support Systems for Cereal Crop Disease Control. Paper presented at EFITA/WCCA, 2005, 25-28 July 2005, Vila Real, Portugal.
Turban E (1995). Decision support and expert systems: management support systems. Prentice Hall, Englewood cliffs, NJ.
Uehera G, Tsuji GY (1993). The IBSNAT project. In: Systems approach to agricultural development. Penning de Vries, F.W.T., Teng, P.S. (eds.) Kluwer Academic Publishers, The Netherlands. pp. 505-514.
Yialouris C, Kollias V, Lorentzos N, Kalivas D, Sideridis A (1997). An Integrated Expert GIS for Soil Suitability and Evaluation. J. Geogr. Inform. Dec. Anal. 1:90-100.
Yialouris CP, Sideridis AB (1996). An expert system for tomato disease. Comput. Electr. Agric. 14:61-76.

15

Agriculture as a tool for development in Angola

Jose de Assuncao Sambo Tomas

Department of Agricultural Economics, College of Management, Zhejiang University, Hangzhou, China.

After a long period of war in Angola, agriculture can contribute significantly to achieving higher sustainable economic growth and greater rural development in all 18 provinces. The Minister of Agriculture in Angola manages Ministry of Agriculture and Development Rural (MADR) including the Forest Sector, which is designated as the Institute of Development Forest (IDF). The most important is the recuperation of infrastructure, trade liberalization and other economic policy reforms to promote agriculture and increased rural welfare in Angola. It should be given more dynamics based on regional economic cooperation, or regional convergence. The increase in agricultural trade in Angola through open economy by the region, with appreciable-too-large infrastructure, agricultural sectors and with revealed international and regional–inter-regional comparative advantage in the agricultural production and yields, agricultural land, agricultural inputs, dairy products, cereal grains, and agricultural raw materials is the most appropriate avenue for improving agricultural growth and its contribution to rural development.

Key words: Revealed comparative advantage, convergence, agriculture, rural development.

INTRODUCTION

Agriculture in Angola has a tremendous potential. Angola is a potentially rich agricultural country. In this study, the typically revealed comparative advantage system was used as a real evidence for the rapid growth of the Angola economy, which is based on ideology of open economic and administrated by the Government of Movimento Popular de Libertação de Angola (MPLA). Based on this system, the growth relationships with China and Angola as well as the benefit of this relationship for the development of agriculture in Angola were compared. From 1975, one of the rich developing countries in Africa (Angola) became a very poor economy (1975 - 2000). Fortunate relationship with Republic Communist of China gave a new breath for Angola from 2002 and it gave a new way to the rapid growth of the economy. The open economy policies was based on the regional economic cooperation, or the regional convergence. Increasing agricultural infrastructure and the trade politics in Angola depends on the government role in trade liberalization and other economic policy reforms to promote agriculture and increased rural welfare in Angola.

ECONOMIC DEVELOPMENT, AGRICULTURE AND ECONOMY POLICIES

Perspectives of economic development by some traditional partners of Angola include Brazil and Portugal which are the traditional partners of Angola in well alimentary and industrial products. Russia is a traditional partner in military science, politics and exploration of diamonds and the United State of America is a traditional partner of Angola in the oil and gas exploration but, in this transition period, the real economic partner for Angola is absolutely the communist country of China. The peace period in Angola came in 2002. Today, agriculture now contributes significantly to the economy of Angola.

The economy of Angola hit higher levels of growth rate exploration of oil in the provinces of the Cabinda (75% of oil "petroleum"), Zaire (24%) and Luanda (1%), and the exploration of diamantes in the provinces of Lundas (80%), Bie (18) and Luanda (2%). Fuelled by the high record of international oil prices and the robust growth in both the oil and the diamante sector; Angola has experienced exceptionally high growth rates in recent years. In 2006, real gross domestic product (GDP) reached 18.6% following the already impressive 20.6% in 2005. The International Monetary Fund (IMF) projected that the GDP growth remains high at 23.4% in 2007 and 26.6% in 2008. Meanwhile, inflation has fallen from over 300% in 1999 to 12% in 2006, and surging oil revenues have led to large fiscal and external current account surpluses (Indira and Alex, 2008; IMF Washington, D.C., 2007).

With the end of the war, rapid post-conflict reconstruction has become the government's priority. The People's Republic of China (China) has in particular, played an important role in assisting these efforts. Chinese financial and technical assistance has kick-started over 100 projects in the areas of energy, water, health, education, agriculture, telecommunications, fisheries, and public works. On the occasion of Chinese prime minister; Wen Jibao's visit to Angola in June 2006, Angolan president; Eduardo dos Santos described bilateral relationship as being "mutually advantageous"; partnerships that were "pragmatic" and had no "political pre-conditions" (Angonoticiais, 2006). Mellor (1995) said that, agriculture can contribute more in any economic growth, and economic development is a pertinent and insightful theme developed among others. More specifically, this theme contends that, sustained industrial development and economic growth in low-to-middle-income countries requires policies for economic development that yield incentives for efficient expansion of not only industry but also agriculture and the rural sector, despite the expected long-term decline in the relative importance of agriculture in growing economies (Johnson, 1995). Increase in agricultural output stimulates the demand for industrial inputs such as fertilizer and farm equipment ("backward linkage") as well as expanding the supply of agricultural goods used as inputs to non-agricultural production ("forward linkage") (DeRosa, 1988). Agricultural crop and livestock production are generally characterized as "weak" backward linkage and "medium strong" forward linkage, respectively (Hirschman, 1958).

However, agricultural growth also raises the real income of farm households and hence it is demanded not only for food, forest products and other agricultural products or agriculture structure, but also, for industrial goods and services and infrastructures. Consumption linkage effects are critical to the influence of agricultural growth on the overall growth performance of the rural and national economy. Ranis et al. (1989) said that, rural industry growth in turn provides "additional impetus for further increases in agricultural productivity, leading to a mutually supportive cycle of agricultural and industrial growth". The economic growth based on the oil and diamante is the way for rapid development strategies that recognizes the importance of ensuring agricultural and rural development along with industrialization, offer the promise of reducing poverty directly and, more generally, the disparity of access in rural areas relative to urban areas to social services, such as education and health-care. Development strategies should be expected to increase access to many privately provided services, such as banking and retailing of consumer goods, in rural areas. The policies towards agriculture and the rural sector, regional convergence with greater equality of access to economic opportunities, physical infrastructure, and social service facilities in rural areas, should be expected to result in greater agricultural and rural incomes, increasing the effective demand for manufactured producer and consumer goods (Appendix). Greater dynamism in agriculture led by expanded trade is required to achieve higher sustained economic growth and greater rural development in Angola.

AGRICULTURAL PRODUCTION, TRADE AND COMPARATIVE ADVANTAGE

Agriculture in Angola has a tremendous potential. Angola is a potentially rich agricultural country, with fertile soils, a favourable climate, and about 57.4 million ha of agricultural land, including more than 5 to 8 million ha of arable land and about 8 to 14% is currently being used. Before independence from Portugal in 1975, Angola had a flourishing tradition of family-based farming and was self-sufficient in all major food crops except wheat. The country exports coffee and maize, as well as crops such as sisal, bananas, tobacco and cassava. By the end of the colonial period, a variety of crops and livestock were produced in Angola. In the north, cassava, coffee, and cotton were grown; in the central highlands, maize was cultivated and in the south, where rainfall is lowest, cattle herding was prevalent. In addition, there were large plantations run by Portuguese that produced palm oil, sugarcane, bananas, and sisal. Commercial farmers grew these crops; primarily Portuguese, and also peasant farmers, who sold some of their surplus to local Portuguese traders in exchange for supplies. The commercial farmers were dominant in marketing these crops, and enjoyed substantial support from the colonial government in the form of technical assistance, irrigation facilities, and financial credit. They produced the great majority of the crops that were marketed in the cities or exported. By the 1990s, Angola was producing less than 1% of the volume of coffee it had produced in the early 1970s, while production of cotton, tobacco and sugar cane had ceased almost entirely. Poor global market prices and lack of investment have severely limited the sector after independence.

REVEALED COMPARATIVE ADVANTAGE IN ANGOLA

The commodity patterns of Angola and China or International and Inter-regional trade can be employed to examine the "revealed" comparative advantage (RCA) based on the regional economic cooperation or the regional convergence countries. The RCA indicator employed here is one commonly specified in applied analysis of international trade relations (Balassa, 1965, 1979). It relates the importance of each member of convergence country in policies of development. The economy of Angola can be improved by building infrastructure and as a supplier of agricultural products, and other products to the world market (and alternatively to markets of this group or another countries) relative to all competing exporting countries. In symbols, the indicator is computed as follows:

International trade

$$RCAj(i,w) = [Xj(i,w)/TX(i,w)]/[Xj(w,w)/TX(w,w)] \quad (1)$$

Inter-regional or regional trade

$$RCAj(i,A) = [Xj(i,A)/TX(i,A)]/[Xj(w,A)/TX(w.A) \quad (2)$$

Where, $Xj(i,..)$ represents exports of commodity, j by country i to the world (w) and Angola (A); $Xj(w,...)$ represents exports of commodity j by the world (that is exports of the commodity by all competing producers) to the same two markets, and $TX(i,..)$ and $TX(w,..)$ represent total exports of all commodities by country i and the world, respectively. Broadly speaking, RCA values greater than unity indicates comparative advantage and RCA values less than unity indicates comparative disadvantage.

RCA values reflect not only fundamental factors such as natural resource endowments but also trade distorting factors, such as foreign and domestic protection and also foreign and domestic subsidies to production and trade. In case of this study, the RCA values computed for Angola trade in agriculture reflect trade-distorting farm policies in Angola and the China financial investment in Angolans economic game. In another hand, two indexes were used, which are Balassa Index and Vollrath Index.

Balassa Index measures the revealed comparative export advantages (RXA) as follows:

$$RXA = (X_{ij}/X_{it})/(X_{nj}/X_{nt}) = (X_{ij}/X_{nj})/(X_{it}/X_{nt}) \quad (3)$$

Where, X represents exports of commodity, i is a country, j is a commodity (or industry), t is a set of commodities (or industry) and n is a set of countries. If $RXA > 1$ it shows that the country i has a comparative advantage in commodity j. If $RXA < 1$, it shows the disadvantages. Vollrath index (Vollrath, TL) also uses the same system to explain the relative import advantage (RTA) as follows:

$$RMA = (M_{ij}/M_{it})/(M_{nj}/M_{nt}) = (M_{ij}/M_{nj})/(M_{it}/M_{nt}) \quad (4)$$

$$RTA = RMA - RXA \quad (5)$$

Where, *M* represents import and the rest is the same as Ballasa Index. If *RTA* is positive, it shows the relative trade advantage and vice-versa.

INTERNATIONAL AND REGIONAL TRADE

The exceptional comparative advantage in International and regional trade deeply shows that the Chinese investment in Angola was apparent in the computed RCA statistics in the Financial Cooperation (Tables 1 and 2). The bulk of Chinese financial assistance in Angola was reserved for key public investment projects in infrastructure, telecommunications, and agro-businesses under the Angolan government's national reconstruction program. The China Construction Bank (CCB) and China's EximBank provided the first funding for infrastructural development in 2002. The Angolan Ministry of Finance had little input in these arrangements since CCB and Exim-Bank funding was provided directly to Chinese firms. Financial relationship between China and Angola grew in the late 2003, when the Angolan Ministry of Finance and the Chinese Ministry of Trade formally signed the "framework agreement" for new economic and commercial cooperation. On March 21, 2004; the first $2-billion financing package for public investment projects was approved. The loan was payable over 12 years at a deeply convectional interest rate plus a spread of 1.5%, with a grace period of up to three years. Also, EximBank of China have been invested $545,009,187.18 in Angola, for complementary action projects about "those money that was spent for the acquisitions of new technology, equipments and instruments" (Angolan Ministry of Finance, 2008). In agriculture, $249 million permitted the acquisition of new agricultural machinery as well as the rehabilitation of irrigation systems in the localities of Luena, Caxito, Gandjelas, and Waco-Kungo.

Table 1. Projects Financed by EximBank of China (Phase I).

Sector	Number of contract	Total value (US$)
Health	9	206,100,425.42
Education	8	217,158,670.63
Social Communication	1	66,905,200.00
Transport	1	13,840,468.00
Energy and Water	8	243,845,110.58
Agriculture	3	149,753,214.00
Public Works	1	211,684,100.65
Total	**31**	**1,109,287,188.28**

Source: Angolan Ministry of Finance (2007) review 2009.

Table 2. Projects Financed by EximBank of China (Phase II).

Sector	Number of contract	Total value (US$)
Health	1	43,805,500
Education	3	229,642,314
Post and Telecommunications "social communication"	4	276,307,189
Energy and Water	3	144,902,615
Agriculture	1	54,006,958
Fisheries	3	266,847,509
Public works	2	89,490,000
Total	17	1,105,002,085

Source: Angolan Ministry of Finance (2007) review 2009.

The comparative disadvantage of the Angola provinces in agriculture stands in contrast to their strong comparative advantage in crude petroleum. The one important exception is fruits and vegetables. Individual Angolan also exhibits substantial comparative advantage in selected agricultural categories and products, relative in colonial period, period of unique party and multiparty period, evidence in international competitiveness in livestock, meats, and dairy products. The Inter-regional RCA statistics gauge the competitiveness of the Angola countries as exporters to one another. As might be expected, the competitiveness of Angola as suppliers of not only crude petroleum but also light manufacturers was reduced, reflecting the similarity of mineral fuel deposits and labor resources in Angola. Of greater interest here is the evidence that the Angola countries have appreciable inter-regional comparative advantage in agriculture, led in particular by strong comparative advantage in livestock, meats, and dairy products, as well as fruits and vegetables. This may reflect not only some sharp differences in agricultural resources and production capabilities between Angola and China but also natural barriers to trade with countries outside, including transportation and handling costs, and general preferences of consumers for regional varieties of agricultural products. In summary, under current economic and other circumstances, the comparative advantage of Angola is based on two periods, before dependence and from dependence of Angola, which differs depending on whether their exports are destined for world markets or markets in neighboring countries (Tables 3 to 5).

BILATERAL TRADE IS A RESULT OF INTERNATIONAL AND REGIONAL TRADE

Political and economic game between two friendly commercial trades has been growing. Figure 1 shows the sharp increase in Sino-Angolan trade since the mid-1995s. From 1995, bilateral trade between the two countries ranged on average from $150 million to $700 million. In 2000, trade exceeded $1.8 billion, and by the end of 2005, it increased four-fold to $6.9 billion. Within a year, it nearly doubled to $12 billion, making Angola, China's largest trading partner in Africa. Crude oil represents over 95% of all Angolan exports, and it is also China's main Angolan import. Over the last six years, China has been the second largest importer of oil from Angola besides the United States, representing roughly 9.3 to 30% of Angola's total oil exports. Despite the fact that, the United State lead in the imports of Angolan oil,

Table 3. Agricultural production in Angola.

				Cereals in tons								
	Total	Plant forest and wood	Meat and dairy product	Wheat	Rice	Beans and peanuts	Sugar	Coffee	Cotton and Sisal	Fruits and vegetables	Roots and tubers
				… to 1974 colonial period government system							
Portuguese	269,509	16 ha X	233,218	X	16.000	79.000	32.000	21.000	50.000	517.000	9.837.000
				1975 to 2001 Period of Party Unique government system							
Angolans		0.6 ha X	90.012	X	3.000	56.000	5.000	6.000	7.000	69.000	3.759.000
				2002 to 2009 Multiparty government system							
Angolans		2ha X	99.000	X	X	59.000	X	7.000	X	70.000	4.998.000

Source: JAST, 2009±.

Table 4. Exports and Imports of Angola by period.

Selected agricultural categories and products												
All produce	Manufactured produce	Mineral fuels	Agricultural produce	Food, live animals	Livestock meats, dairy	Rice	Sugar	Coffee	Cotton and Sisal	Fruits and vegetables	Agricultural raw material	Wood
Colonial period-govern by Portuguese												
10,951(100)	9,342(85)	X(x)	1,473(14)	833(8)	18 (0)	147 (1)	115 (1)	120 (1)	86 (1)	174 (2)	86 (1)	265 (3)
Dependence period from 1975-govern by Party of MPLA												
502(100)	379(75)	49(10)	65(13)	55(11)	10(2)	0 (0)	1 (0)	3 (1)	0 (0)	10 (2)	1 (0)	478 (4)

Source: JAST, 2009±.

Table 5. Revealed comparative advantage in international and regional trade.

Selected agricultural categories and products												
All produce	Manufactured produce	Mineral fuels	Agricultural produce	Food, live animals	Livestock meats, dairy	Rice	Sugar	Coffee	Cotton and Sisal	Fruits and vegetables	Agricultural raw material	Wood
Colonial period-govern by Portuguese												
	0.54	X	1.76	2.27	1.36	8.14	0.89	10	5,0	9.88	1.04	7.00
Dependence period from 1975-govern by Party of MPLA												
	1.08	5.98	1.00	0.95	0.05	0.00	0.16	X	X	1.98	0.21	6.00

Source: JAST, 2009±.

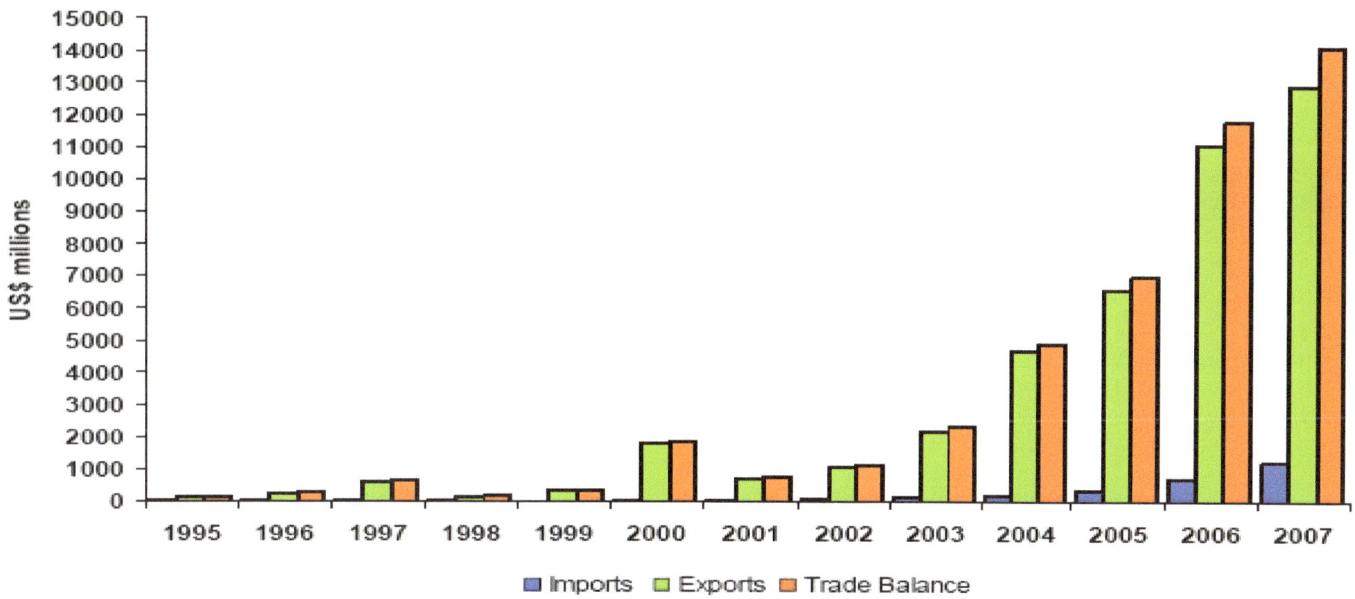

Figure 1. China- Angola Trade Vplume (1995 – 2006).

Angolan oil exports to China since 2002 have increased seven-fold, compared to only 3.5 times that of the United States.

Angola's oil exports to China reached $3.9 billion in 2004, making it a major supplier and ranking it third after Saudi Arabia and Iran. By 2005, Angola had overtaken Iran with exports totaling 17.46 million tons, 45.5% of China's African oil imports. During the first six months of 2006, Angola temporarily leapfrogged Saudi Arabia as the largest supplier of crude oil to China, with 23.45 million tons of crude shipped from the African nation that year. Angolan oil imports now represent over 18% of China's total oil imports, and this proportion is increasing. In recent years, Chinese imports to Angola have also seen a significant increase. In 2004, China became Angola's fourth-largest trading partner at $194 million, up from being its seventh-largest trading partner the year before. In 2006, China kept its position despite the fact that Chinese exports to Angola quadrupled, with steel and iron bars, batteries, cement, and automobiles as the principal imports.

In early 2007, China surpassed Brazil and South Africa as the second-largest trading partner besides Portugal. Imports reached $368 million, an increase of 106% from the same period of the year before. Despite the rise in the value of both imports and exports in the period analyzed, Angola has consistently run a large trade surplus with China, owing to the rapidly rising rate of Chinese oil importation. With increase in the infrastructural projects and greater competitiveness of Chinese exports compared to European exports in Angola, it is expected that in the next few years, the penetration of Chinese products in Angola will rise significantly, equalizing the level of Portuguese imports.

CHINESE INTEREST IN ANGOLA BY CONVERGENCE INTERNATIONAL REGIONAL TRADE

China has great interest in oil and other industries of Angola. Throughout the 1990s, bilateral trade between the two countries ranged on the average from $150 million to $700 million. In 2000, trade exceeded $1.8 billion, and by the end of 2005, it increased four-fold to $6.9 billion. Within a year, it nearly doubled to $12 billion, making Angola China's largest trading partner in Africa. Following the opening of China's first credit line to Angola in March 2004, China Petrochemical Corporation, better known as Sinopec Group, acquired its first stake in Angola's oil industry, 50% of the BP-operated block 18. Sonangol Sinopec International (SSI)-a joint venture between Angolan National Oil Company, Sonangol and Sinopec was created to explore the stake on the block. Sinopec Group holds a 55% stake in the joint venture and Sonangol the remaining 45%. Although BP's former license partner Shell had agreed to sell its stake to India's state-owned Oil and Natural Gas Corporation (ONGC), the Chinese on their first involvement in the Angolan oil industry sidelined ONGC with an offer that media sources estimated at $725 million (ONGC, 2006). The company reportedly spent an additional $1.5 billion for the development and exploration of the block (International Herald Tribune, 2006).

Additionally, the two parties also signed a memorandum to jointly study the plans for the exploration of the shallow offshore blocks 3/05 and 3/05A (previously known as block 3/80) that was withdrawn from the French Group Total in late 2004. Later that year, the two parties agreed that China Sonangol International Holding (CSIH), a subsidiary of Sonangol E. P., jointly owned by Beiya

International Development Limited, would explore the 25% stake. In March 2006, the Sino-Angolan joint venture SSI acquired three new Angolan offshore oil blocks with proven reserves of 3.2 billion barrels. It offered $750 million for 20% of ENI-operated block 15 and made a record bid of $1.1 billion as signature bonus payment for each of the relinquished offshore of block 17/06 (27%) and 18/06 (40%). In addition to the bids for the rights to prospect for oil, the joint venture earmarked $200 million for social projects. Both companies also agreed to jointly study plans for a new $3 billion oil refinery in Lobito (Sonaref) with a capacity of 240,000 barrels per day (Angola Press News Agency, 2006).

The Chinese have also shown an interest in Angola's diamond. The Angolan Council of Ministers accepted a joint venture agreement between Angola's state-owned diamond company, National Enterprise of the Angola Diamantes, Exploration and Production (Endiama, EP), and CIF. It also approved Endiama's participation in the creation of Endiama China International Holding Limited (Endiama China) to prospect, produce, and market diamonds, including diamond cutting and production of jewelry in Hong Kong but the Chinese interest in Angola have not ended.

CONCLUSIONS

The government of Angola has developed strategies to revitalize agricultural sector. The relationship and convergence with Chinese government gives new dynamic to Angolan's government. It will increase agricultural output in many sectors. Therefore, Angola faces three key socioeconomic challenges which include eradicating poverty, ensuring food security, and protecting the environment. To address these challenges, Angola will have to generate growth in the agricultural sector. During the years from 2002 to 2009, the GDP share of the industrial, agricultural and service sectors had changed dramatically while industrial sector became the second biggest sector after service sector. That is a completely reverse picture of the sector shares when compared to the Colonial period governed by Portuguese (1974), Party Unique period governed MPLA (1975 – 2001) and Multiparty period governed by MPLA (2002 – 2009). It showed that during this period, with the help of the Chinese Government, it speeded to about 2 billions for the recuperation of infrastructure.

Chinese financing offers better conditions than commercial loans, lower interest rates, and longer repayment time. Non-Chinese credit lines that Angola secured in 2004 demanded higher guarantees of oil, with no grace period and with high interest rates. Since Angola started pumping oil in the early 1970s, the United States has been the main importer followed by China. Although the United States remain the number one importer of Angolan oil, its share has significantly decreased in contrast to those of China and South Africa, which have more than doubled. Although, there are signs of growing Chinese FDI to Angola, such investment remains small when compared to Western FDI. Outside the extractive industry, the Portuguese still dominate, with over $200 million invested in 2006 alone, representing 57% of total FDI to Angola. Angola welcomes any investment in the non-oil sectors and the recent trade mission by Japanese, British, Turkish, Indian, and Pakistani businessmen clearly demonstrates this. Likewise, despite China's efforts to enter the oil sector, production is still dominated by Western companies such as Chevron, Total, BP, and ExxonMobil. With Angola's increased dependence on China's energy security, one can expect China to be visible in the 2008 bidding round for 10 new blocks. The Chinese Government and Chinese private sector have been interested in full cooperation with Angolans Government in Agricultural Science especially in the forest industry. It is no secret that the reason is Maiombe forest, which gives more medicinal plants. According to the needs for rehabilitation or recuperation of all infrastructure and structure in all 18 provinces and development of agriculture and other industries in Angola, government needs more that 2 billion of the Chinese money (they need more or about 18 billion US$), but real success of the government in finance and economic activity is clearly politics in national and international game transparency.

ABBREVIATIONS

GDP, Gross domestic product; **IMF,** International Monetary Fund; **RCA,** revealed comparative advantage; **CCB,** China Construction Bank; ONGC**,** Oil and Natural Gas Corporation.

REFERENCES

Angola Press News Agency, Jose PM (2006). Sonangol rubrica acordo para desenvolvimento da refinaria do Lobito, [J] Luanda, Jornal de Angola.

Angonotícias (2005). Comissão Permanente do Conselho de Ministros reuniu sob orientação do Presidente da República, [J] Luanda Diary informative.

Balassa B (1965). Trade Liberalisation and 'Revealed' Comparative Advantage", The Manchester School. P. 33.

DeRosa DA 1988. Asian preferences and the gains from MFN tariff reductions. [B] World Econ. 11:377-396

Hirschman AO (1958). *The strategy of economic development.* New Haven [J] Yale University Press.

Indira C, Alex V (2008). ANGOLA and CHINA - A Pragmatic Partnership, (Working Paper Presented at a CSIS Conference), "Prospects for Improving U.S.-China-Africa Cooperation", [J] Chatham House, London.

International Herald Tribune (2006). China Buys More Angolan Crude than Saudi, [J] Diary of Portugal.

Johnson DG (1995). The limited but essential role of government in agriculture and rural life. In *Agricultural competitiveness: market forces and policy choice*, eds., G.H. Peters and D.D. Hedley. Hants, England: Dartmouth Publishing Company for the International Association of Agricultural Economists [J] England press.

Mellor JW (1995). Agriculture on the road to industrialization. Baltimore:

Food Policy Research Institute. [J] Johns Hopkins University Press for the International.
Ranis G, Stewart F, Reyes EA (1989). Linkages in development: A Philippine case study. [J] Working Paper Series No. 89-02. Manila: Philippine Institute for Development Studies.
Sinopec Beats ONGC (2006). Bets Angola Block, [J] *Financial Express*, as cited by China Institute, University of Alberta.
The Economist Intelligence Unit (2004-2005). Angola at a glance, [J] Angolans Country Report 2005.

APPENDIX

Agricultural policy

The Government of the Republic of Angola has shown that it is committed to promoting the revitalization of the agricultural sector. The Ministry of Agriculture and Rural Development had developed specie programmer "The Economist Intelligence Unit (2004-2005)" which states that the government's program in the agricultural sector gives special importance to:

(1) Rehabilitation and/or construction of basic infrastructure to support production.
(2) Agriculture expansion and rural extension.
(3) Seed propagation, production and multiplication.
(4) Conducting basic and applied research on agriculture and livestock.
(5) Encouragement of agriculture, livestock and forestry production.
(6) Technical and professional training.

Specific goals related to crop production set forth for the 2004/2005 programs are:

(1) To increase food production by restoring the ability to supply inputs to farmers and products to private agriculture and livestock-linked businessmen.
(2) To integrate agriculture-oriented refugees and demobilized people, into the agriculture, livestock and forestry activity.
(3) To promote sustainable rural trade by emphasizing the commercialization of cereals, roots and tubers.
(4) To boost agriculture expansion and rural extension programs.
(5) To rehabilitate and reactivate production-supportive infrastructure such as hydro-electric projects, by-roads and other rural engineering infrastructure
(6) To boost the rehabilitation of coffee, sugar and sisal production and promote its commercialization.
(7) To provide incentives and stimulate the creation of middle- and large-sized agriculture and livestock farms.
(8) To ensure the sustainable management of natural resources.

Determinants of farmers' choice of innovative risk-reduction interventions to wastewater-irrigated agriculture

Ezekiel N. Ndunda* and Eric D. Mungatana

Centre for Environmental Economics and Policy in Africa (CEEPA), Department of Agricultural Economics, Extension and Rural Development, Faculty of Natural and Agricultural Sciences, University of Pretoria, Pretoria 0002, South Africa.

This paper identifies the innovative methods used by urban farmers to reduce the health and environmental risks linked to wastewater-irrigated agriculture in Nairobi, Kenya. A study involving 317 urban and peri-urban farmers was conducted and innovative methods identified for risk-reduction in wastewater irrigation. According to the results, the farmers' choice of adaptation measures in wastewater irrigation was: No intervention (49.8%), crops restriction (21.1%), protective clothing (12.6%), safer application (8.8%), and irrigation cessation (7.6%). The estimated model had a robust explanatory ability since the likelihood ratio statistics were statistically significant (χ^2 =222.13; p=0.000). The marginal analysis results show that the following factors significantly (p=0.005) influence the farmers' choice of low-risk measures in wastewater irrigation: Household size, farming experience, membership to farmers group, access to credit, access to certified seed, access to media, crop income, awareness to World Health Organization irrigation guidelines, and awareness to wastewater hazards. Therefore, it was concluded that education support and creation of awareness about health risks in wastewater irrigation are important for enhanced adoption of risk-reduction technologies among the farmers. There is a need to design policies and programs that support farmers in safe wastewater irrigation, while raising their awareness on the health hazards attributed to untreated wastewater reuse.

Key words: Low-risk measures, marginal effects, multinomial logit, urban farmers, wastewater irrigation.

INTRODUCTION

Studies show that about 20 million hectares of land in developing countries is irrigated with wastewater and at least 10% of the world's population consumes foods produced by irrigation with wastewater (Hamilton et al., 2007; Jiménez and Asano, 2008; Scott et al., 2004; WHO, 2006). However, many developing countries are confronted with apparent limitations in implementing conventional wastewater treatment systems. This has exposed many poor urban and peri-urban farmers in developing countries to health risks due to exposure to polluted wastewater. Therefore, the utilization of risk-reduction options is a low-cost critical risk-reduction measure in wastewater-irrigated agriculture (Keraita et al., 2008). Non-conventional methods commonly used in control of health risks include: crops restriction, safer application techniques, cessation of irrigation before harvesting and using protective clothing (Drechsel et al., 2008; Keraita et al., 2007).

Agriculture is the mainstay of Kenyan economy and growth of the sector is vital for the overall social and

*Corresponding author. E-mail: ezndunda@yahoo.com.

economic development of the country. The sector contributes 24% directly and 27% indirectly to the national GDP (AEO, 2012; GOK, 2009). Millions of rural and urban farmers in the country rely on the agriculture sector for their livelihoods. The urban and peri-urban farming sub-sector is a source of food security, employment creation, and poverty alleviation to the urban population in Kenya (Addo, 2010; GOK, 2010a). However, the agriculture sector is challenged by water-scarcity, which is currently 548 cubic metres per capita per year (NCAPD, 2010; NEMA, 2011). This is much lower than the Falkenmark water stress index (FWSI) that sets the threshold of severe water deficit at 1000 cubic metres per capita per year (Falkenmark et al., 1989). Projections indicate that water endowment in the country will shrink to 250 cubic metres per capita per year by 2025, which is much lower than the bordering countries (GoK, 2010b; NEMA, 2003; World Bank, 2010). The scarcity of freshwater resources has led many urban and peri-urban farmers to rely on untreated or partially treated wastewater for irrigation agriculture. Consequently, knowledge of the innovative low-risk interventions and factors affecting farmers' choice of the risk-reduction interventions is important for informing policy in order to reduce risks facing many urban farmers in Kenya.

Studies show that wastewater is increasingly being embraced as a feasible substitute to freshwater sources for irrigation, especially as the water scarcity increases and more reliable and economic technologies are developed to treat urban wastewater (Buechler and Devi, 2006; Drechsel et al., 2006; Ensink et al., 2003; Qadir et al., 2010; Rutkowski et al., 2007; Srinivasan and Reddy, 2009; van der Hoek, 2004). Since many studies on wastewater reuse concentrate on quality analysis and risk-reduction measures in irrigated agriculture, there is still a knowledge gap on the factors affecting the choice of the suggested low-risk intervention. This poses a serious challenge since farmers' response to wastewater-related health risks and also their choice of risk-reduction interventions is influenced by various socio-economic and institutional factors. The knowledge about these factors can support policy intervention measures aimed at minimizing the risks to public health and environment.

This paper analyses the factors affecting the choice of innovative risk-reduction interventions to reduce the health risks attributed to wastewater-irrigated agriculture in peri urban and urban Kenya. The case study was conducted in Kibera informal settlements, which is the largest slum in sub-Saharan Africa but lacks sewerage infrastructure. Most of the raw sewage from this informal settlement is discharged into Motoine-Ngong river without treatment hence threatening the livelihoods of many urban and peri-urban farmers in Nairobi. The case study includes urban and peri-urban farmers in the Motoine-Ngong river basin to guide policy-makers on the approach to promote utilization of low-risk methods in wastewater irrigation.

Econometric model

This study employed a multinomial logit model (MNL) to identify the factors that influence the farmers' choice of risk-reduction interventions in reuse of wastewater for agriculture in the Motoine-Ngong river basin in Nairobi. The random utility model may be used to motivate this unordered-choice model such that, for the ith farmer that is faced with J choices of risk-reduction options the utility of choice j is:

$$U_{ij} = \beta_{ij} X_{ij} + \varepsilon_{ij} \tag{1}$$

Therefore, when the farmer makes choice j of risk-reduction intervention, it is usually assumed that U_{ij} is the highest utility among the J utilities. The MNL for this study was based on the probability that a risk-reduction choice j is made as follows:

$$\Pr\left(U_{ij} > U_{in}\right) = \quad for \quad all \quad other \quad n \neq j \tag{2}$$

The MNL specification is:

$$\Pr ob\left(y_i = n/x_i\right) = \frac{\exp(X_i \beta_n)}{1 + \sum_{j=1}^{J} \exp(X_i \beta_j)} \quad Where \ j = 0,1,2,\ldots,J;\ n = 1,2,\ldots,J \tag{3}$$

The parameter y_i represents the alternative risk-reduction interventions to wastewater irrigation, X_i denotes a vector of all the explanatory variables of the ith observations, and β_i is a vector of all coefficients in the jth regression. However, the coefficients obtained from the estimation of equation (3) are challenging to deduce. Therefore, marginal effects of the factors on the probabilities are obtained through differentiation of Equation (4):

$$\delta_j = \frac{\partial \Pr(y_i = n)}{\partial X_i} = \Pr(y_i = n)\left[\beta_j - \sum_{j=0}^{J} \Pr(y_i = n)\beta_n\right] = \Pr(y_i = n)\left(\beta_j - \bar{\beta}\right) \tag{4}$$

The risk-reduction interventions considered in this study include: Crops restriction, protective clothing, safer application techniques and irrigation cessation. The independence of irrelevant alternatives (IIA) assumption was considered in order for the MNL estimates to be consistent.

RESEARCH METHODOLOGY

The location of this study is in the Motoine-Ngong river basin of Nairobi in Kenya. The total area of the river basin from the source to the confluence with Nairobi river is approximately 127 km^2. Motoine-Ngong river passes through the sprawling Kibera slum, which has an average population density of 6000 persons per hectare. Due to poor environmental sanitation and lack of sewerage infrastructure in Kibera slum, the informal settlement is a major contributor to pollution of the Motoine-Ngong River (UNEP, 2003). It is estimated that about 280 tonnes of municipal solid waste is generated in the slum per day. Also, the Biochemical Oxygen

Demand (BOD_5) from solid waste in Kibera slum is approximately 6,650 kg per day. The generated urban waste, which includes human waste dumped into channels, drains into the river. Many urban and peri-urban farmers rely on the wastewater either directly or indirectly for irrigation agriculture. This study was based on a cross-sectional household survey data collected from urban and peri-urban farmers using wastewater for irrigation agriculture in the Motoine-Ngong river basin. Whereas the selected sample size was 325 respondents (Equation 5), 8 questionnaires were rejected due to incomplete information and hence 317 were used in the analysis.

The sample size was obtained using the following formula (Kothari, 2004):

$$n = \frac{z^2 * p(1-p)}{e^2} \tag{5}$$

$$= \frac{2.576^2 * 0.98(1-0.98)}{0.02^2} = 325$$

Parameter n represents the sample size, z is the confidence level at 99% (standard value of 2.576), p denotes the estimated extent of wastewater irrigation in this study area (98%), and e refers to the margin of error at 2%.

A structured questionnaire was administered to urban and peri-urban farmers between December 2011 and February 2012. This study purposively selected Kibera slum due to high population of urban and peri-urban farmers who rely on wastewater- irrigated agriculture for livelihoods and also the lack of sewerage infrastructure. A representative sample of the farmers was randomly selected for interview in this research. In order to analyse the determinants of farmers' choice of innovative risk-reduction interventions to wastewater-irrigated agriculture, the dependent variables are crops restriction, protective clothing, safer application techniques and irrigation cessation. The considered independent variables are: household size, farming experience, age of the household head, education level of household head, extension on crop and livestock, membership to farmers group, support from non-governmental organizations (NGO), access to credit, access to certified seed, access to media, log of crop income, awareness to WHO irrigation guidelines, and awareness to wastewater hazards. These variables were selected based on literature and availability of survey data.

RESULTS AND DISCUSSION

Descriptive statistics

The descriptive results show that households have an average size of 4.61 members in Kibera slum (Table 1), which compares well with the current mean household size estimation of 5.0 persons per household in the slum (Umande Trust, 2012). Household size may not necessarily have a positive relationship with adoption behaviour since large families may be forced to divert some labour into non-farm activities to increase income (Yirga, 2007). Respondents interviewed had an average farming experience of 5.2 years and hence able to make informed decisions in crop and animal husbandry. Several studies show that farming experience promotes the adoption of improved technologies (Maddison, 2006; Nhemachena and Hassan, 2007). The results show that household heads have an average age of 40.22 years. Since the age of household head may be related to farming experience, its relationship with adoption behaviour may be positive or negative.

The education of household head in this study area was 7.94 years. More years of education may be linked to an increased access to information and hence technology adoption. The study by Maddison (2006) shows a positive relationship between the education level of the household head and adoption behaviour. Results of this study show that the log of crop income is Kshs.7.06. The increase in crop income may provide an incentive for household head to adopt improved technologies. The results show that the mean access to agricultural extension services is 26.5% for the sample of selected farmers. A study Yirga (2007) shows a positive relationship between access to extension services and adoption behaviour of farmers.

This study shows that about 39.4% of urban and peri-urban farmers have membership in farmers' groups. These farmers' groups provide an important platform for exchange of important information for urban and peri-urban agriculture. The farmers who had some support from the non-governmental organizations were 26.5%. This support may provide essential resources that can promote the adoption behaviour of urban and peri-urban farmers. Descriptive results show that 35.3% of farmers have access to credit facilities. The increased access to credit facilities may enable farmers to purchase essential farm inputs such as irrigation facilities and hence ease the resource constraint. Therefore, increased access to credit facilities has a positive relationship with the adoption behaviour of farmers (Pattanayak et al., 2003). Access to certified seeds in this study area was about 48.6%. This may have a positive relationship with adoption behaviour of farmers (Buah et al., 2011). The results show that the access to media in the sample studied was 44.8%. This study hypothesizes that increased media access has a positive impact on technology adoption among the urban and peri-urban farmers. According to this study, only 23.1% of the sample of farmers was aware of the wastewater irrigation guidelines by the World Health Organization (WHO). The guidelines' awareness is considered to have a positive relationship with the adoption behaviour. Also, awareness to wastewater hazards was 52.7% in this study sample. This study hypothesizes that increased awareness to wastewater hazards is positively related to adoption behaviour.

Table 2 presents the adaptation strategies of urban and peri-urban farmers in wastewater irrigation to reduce health and environmental risks. About 49.8% of the interviewed farmers had not adopted any innovative risk-reduction interventions in wastewater-irrigated agriculture.

The urban and peri-urban farmers who practiced crop restrictions to reduce wastewater-related risks were approximately 21.1%. About 12.6% of the urban and peri-urban farmers relied on protective clothing for reduction

Table 1. Description of explanatory variables.

Independent variable	Mean	S.D.	Description
Household size	4.612	1.744	Continuous
Farming experience (years)	5.195	6.329	Continuous
Age of the household head (years)	40.215	11.223	Continuous
Education level of household head (years)	7.935	2.601	Continuous
Log of crop income (kshs.)	7.063	1.032	Continuous
Percentage			
Access to agricultural extension services		26.5	Dummy, 1 if visited and 0 otherwise
Membership to farmers group		39.4	Dummy, 1 if a member and 0 otherwise
Supported by NGO		26.5	Dummy, 1 if supported and 0 otherwise
Access to credit		35.3	Dummy, 1 if has access and 0 otherwise
Access to certified seed		48.6	Dummy, 1 if has access and 0 otherwise
Access to media		44.8	Dummy, 1 if has access and 0 otherwise
WHO irrigation guidelines' awareness		23.1	Dummy, 1 if aware and 0 otherwise
Awareness to wastewater hazards		52.7	Dummy, 1 if aware and 0 otherwise

S.D. is standard deviation

Table 2. Farmers' choice of adaptation measures in wastewater irrigation.

Variable	Percent of respondents
No intervention	49.8
Crops restriction	21.1
Protective clothing	12.6
Safer application	8.8
Irrigation cessation	7.6
Total number of respondents	317

of health risks in wastewater irrigation. There were on average 8.8% of the farmers in this study sample who had adopted safer application techniques in wastewater irrigation.

The cessation of irrigation before harvesting was adopted by about 7.6% of the interviewed farmers involved in wastewater reuse in agriculture. These innovative risk-reduction measures employed by urban and peri-urban farmers in Kenya were similar to other findings in the wastewater irrigation literature in developing countries (Keraita et al., 2007, Keraita, 2008; Knudsen et al., 2008; Marenya and Barrett, 2007; Obuobie et al., 2006; Weldesilassie et al., 2011).

Econometric analysis

The results of multinomial logit (MNL) model estimated for this study are presented in Table 3. In this model, the base category was no intervention variable while the other dependent variables were crop restrictions, protective clothing, safer application and irrigation cessation.

Under the independence of irrelevant alternatives (IIA) assumption, it is expected that there would not be any systematic change in the coefficients if one of the outcomes from the model is excluded. This study used the Hausman test (Hausman and McFadden, 1984) to confirm the IIA assumption in the model. The Hausman test failed to reject the null hypothesis on the IIA assumption at 95% confident level. This suggests that the MNL model is appropriate to identify the determinants of farmers' choice of innovative risk-reduction interventions to wastewater-irrigated agriculture in Kenya. The likelihood ratio statistics for this study were statistically significant (χ^2=222.13; p=0.000), which implies that the model has a robust explanatory ability. Since the MNL model estimates provide only the direction of the impacts of explanatory variables on response variable, a further analysis to obtain marginal effects was conducted (Table 4). The marginal effects provide the expected change in probability of a particular innovative risk-reduction intervention selected by farmers with respect to a unit change in explanatory variable.

Household size

The results show that the adoption of innovative risk-reduction interventions in wastewater irrigation significantly declines with an increase in household size. A unit increase in household size results in 3.6% (p=0.045) decline in the probability of using crop restrictions and 2.1% (p=0.026) decrease in the probability of using protective clothing. Also, the results show that a unit increase in household size decreases the probability of using safer wastewater application by 1.1% (p=0.048) and irrigation cessation by 0.3% (p=0.097). A

Table 3. Parameter estimates of the multinomial logistic low-risk wastewater irrigation model.

Explanatory variable	Crops restriction		Protective clothing		Safer application		Irrigation cessation	
	Coefficient	P level	Coefficient	P level	Coefficient	P level	Coefficient	P level
Household size	-0.305**	0.020	-0.300**	0.040	-0.252	0.156	-0.120	0.483
Farming experience	0.081**	0.040	0.080*	0.060	-0.053**	0.017	-0.098	0.300
Age of the household head	-0.033	0.179	-0.036	0.197	-0.004	0.898	0.007**	0.033
Education level of household head	0.054**	0.047	0.022*	0.080	0.201**	0.024**	0.012	0.310
Access to agricultural extension services	-0.350***	0.010	0.251	0.570	-0.511	0.361	-0.654	0.276
Membership to farmers group	1.184***	0.003	1.432***	0.001	1.488***	0.003	2.085***	0.000
Supported by NGO	0.473**	0.025	1.155**	0.011	1.064**	0.039	1.242**	0.021
Access to credit	2.886***	0.000	2.310***	0.000	2.077***	0.000	1.488***	0.007
Access to certified seed	1.026**	0.012	0.800*	0.082	0.111	0.833	1.205**	0.045
Access to media	1.272***	0.001	0.292	0.495	1.137**	0.020	0.574	0.256
Log of crop income	0.378**	0.041	0.292	0.167	0.638***	0.007	-0.234	0.350
Awareness to WHO irrigation guidelines	0.879*	0.053	1.597***	0.001	0.521	0.375	-0.592	0.489
Awareness to wastewater hazards	0.714*	0.063	1.168***	0.009	0.997**	0.049	0.673**	0.039
Constant	-5.366***	0.001	-4.966***	0.008	-9.457***	0.000	-3.458	0.123
Diagnostics								
Base category					No intervention			
LR chi-square					222.13***			
Log likelihood					-315.774			
Pseudo - R^2					0.2602			
Number of observations					317			

*, ** and ***, Significant at 1, 5 and 10% level respectively.

similar study by Adeoti (2009) found that household size has a negative impact on adoption of irrigation technology in Ghana. Similar results were obtained from .Therefore, it can be inferred that the bigger the household size the lower the chance of adopting risk-reduction measures in wastewater-irrigated agriculture.

Farming experience

Farming experience of the household head significantly influences the choice of risk-reduction measures in wastewater-irrigated agriculture. The results of this study show that a unit increase in farming experience raises the probability of using crops restriction to reduce health risks by 1.3% (p=0.020) and that of using protective clothing by 1.8% (p=0.001). Similarly, the probabilities of using safer application methods and practising irrigation cessation in wastewater reuse increase by 2.6% (p=0.032) and 1.2% (p=0.089) respectively with a unit increase in farming experience. The results from a related study in Nigeria shows that farming experience has positive effect on adoption of improved agricultural technologies (Agwu et al., 2008). Thus, it can be deduced that the greater the farming experience the more likely is the household head likely to adopt risk-reduction measures.

Age of the household head

The age of household head has a negative and non-significant impact on the adoption of risk-reduction measures in wastewater irrigation. A unit increase in the age of household head reduces the adoption of crop restriction method by 3.4%, use of protective clothing by 1.3%, employment of safer application techniques by 1.1 %, and irrigation cessation by 2.6%. A study conducted in Western Kenya to identify the determinants of adopting Imazapyr-resistant maize technologies shows that age is positively related to technology adoption (Mignouna et al., 2011). This shows that the older the household head the less the likelihood of adopting risk-reduction measure in wastewater irrigation.

Education level of household head

As expected, the increase in education of the household head has a significant and positive impact on the adoption of the considered risk-reduction measures in wastewater irrigation. A unit increase in education level of household head increases the probability on using crop restriction by 14.4% (p=0.019) and wearing protective clothing by 15.1% (p=0.000). The probability of using safer

Table 4. Marginal effects from the multinomial logistic low-risk wastewater irrigation model.

Explanatory variable	Crops restriction		Protective clothing		Safer application		Irrigation cessation		No intervention	
	Coefficient	P level	Coefficient	P level	Coefficient	P level	Coefficient	P level	Coefficient	P level
Household size	-0.036**	0.045	-0.021**	0.026	-0.011**	0.048	-0.003*	0.097	-0.067***	0.009
Farming experience	0.013**	0.020	0.018***	0.001	0.026**	0.032	0.012*	0.089	-0.019	0.341
Age of the household head	-0.024	0.210	-0.013	0.258	-0.011	0.816	-0.026	0.553	-0.016	0.236
Education level of household head	0.144**	0.019	0.151***	0.000	0.101**	0.027	0.074**	0.046	-0.019	0.211
Access to agricultural extension services	-0.046	0.401	0.018	0.334	-0.031	0.359	-0.027	0.267	0.057	0.499
Membership to farmers' group	0.093*	0.090	0.051**	0.045	0.068*	0.086	0.090**	0.014	-0.342***	0.000
Supported by NGO	0.034**	0.046	0.101**	0.022	0.159**	0.019	0.053**	0.015	-0.217***	0.009
Access to credit	0.355***	0.000	0.119*	0.079	0.048***	0.003	0.027**	0.041	-0.540***	0.000
Access to certified seed	0.127**	0.026	0.050***	0.004	0.022**	0.043	0.046*	0.059	-0.201**	0.011
Access to media	0.172***	0.000	-0.017	0.672	0.060	0.105	0.008	0.757	-0.223***	0.002
Log of crop income	0.045*	0.086	0.017	0.399	0.041**	0.016	-0.022*	0.087	-0.082**	0.030
Awareness to WHO irrigation guidelines	0.083***	0.000	0.186***	0.007	0.023***	0.004	-0.046**	0.019	-0.226***	0.009
Awareness to wastewater hazards	0.060**	0.023	0.092**	0.028	0.049**	0.014	0.015***	0.002	-0.215***	0.004

*, ** and ***, Significant at 1, 5 and 10% level respectively.

application of wastewater increases by 10.1% (p=0.027) and that of irrigation cessation increase by 7.4% (p=0.046) with a unit increase in education of household head. A study on smallholder agricultural productivity in sub-Saharan Africa shows that low level of education limits technology adoption (Muzari et al., 2012). Therefore, higher education of household head promotes the adoption of innovative risk-reduction measures in wastewater irrigation.

Access to agricultural extension services

Access to agricultural extension services has a non-significant and negative impact on crop restriction, safer application and irrigation cessation. However, the access to extension services has a positive but non-significant effect on protective clothing. The unit increase in the access to agricultural extension services reduces the probability of using crop restriction by 4.6%, the probability of employing safer application by 3.1% and the probability of using irrigation cessation by 2.7%. In contrast, a unit increase in access to agricultural extension services increases the probability of wearing protective clothing by 1.8%. A similar study in Nigeria by Ajayi and Okunlola (2005) shows that agricultural extension services have a positive impact on adoption of root crops technologies. Therefore, it can be inferred that higher access to agricultural extension services may inhibit adoption of risk-reduction technologies in the reuse of untreated wastewater for urban agriculture.

Membership to farmers' group

The results show that membership to farmer's group has a significantly positive effect in adoption of risk-reduction interventions in wastewater irrigation. A unit increase in membership to farmers' group increases the probability of adopting crop restriction by 9.3% (p=0.090), the probability of wearing preventive clothing by 5.1% (p=0.045), the probability of using safer application by 6.8% (p=0.086) and the probability of adopting irrigation cessation by 9.0% (p=0.014). A similar study conducted in Cameroon shows that membership to farmers' group has a positive effect on the adoption intensity of improved yam seed technology (Nchinda et al., 2010). This indicates that higher membership in farmers' groups is an important factor in promotion of risk-reduction measures for urban wastewater users.

Supported by NGO

The farm households which had received support

from NGOs were more likely to adopt innovative risk-reduction interventions to minimize health risks linked to wastewater irrigation. An increase in NGOs support by one unit significantly raises the probability of adopting crops restriction by 3.4% (p=0.046), probability of wearing protective clothing by 10.1% (p=0.022), probability of employing safer application methods by 15.9% (p=0.019), the probability of using irrigation cessation method by 5.3% (p=0.015). Another study to determine the diffusion technology in Benin shows that Support by NGO positively influences the adoption of improved technology for rice parboiling (Dandedjrohoun et al., 2009). This implies that the significant involvement of NGOs in urban and peri-urban agriculture in Nairobi played a vital role in enhancing the adoption of risk-reduction technologies in wastewater irrigation. Thus, the higher the NGOs support for wastewater users, the greater the adoption of innovative risk-reduction intervention in wastewater-irrigated agriculture.

Access to credit

Access to credit for the farmers has a positive and significant effect on the possibility of adopting crops restriction, protective clothing, safer application and irrigation cessation in wastewater irrigation. A unit growth in access to credit facilities increases the probability of adopting crops restriction by 35.5% (p=0.000) and also raises the probability of using protective clothing by 11.9% (p=0.079). Similarly, a unit increase in access to credit facilities raises the probability of using safer application techniques by 4.8% (p=0.003) and the probability of adopting irrigation cessation by 2.7% (p=0.041). The study by Mohamed and Temu (2008) in Zanzibar shows that the access to credit has a positive effect on the adoption of agricultural technologies. This can be used to infer that the greater the access to credit facilities to wastewater users the more likely is the adoption of risk-reduction measures.

Access to certified seed

Access to certified seed is positively related to the adoption of risk-reduction interventions in wastewater irrigation. The results show that a unit increase in access to certified seeds increases significantly the probability of adopting crops restrictions, protective clothing, safer applications and irrigation cessation by 12.7% (p=0.026), 5.0% (p=0.004), 2.2% (p=0.043), and 4.6% (p=0.059) respectively. Results of a study in Nigeria on sustainable rice productivity and rural farmers' welfare, the access to certified seed positively impacts on the adoption of improved agricultural technology (Awotide et al., 2012). Therefore, it may be deduced that a higher access to certified seed motivates the urban farmers to adopt risk-reduction measures to lower the hazards attributed to wastewater reuse.

Access to media

Access to media has a positive impact on the choice of risk-reduction measures in wastewater irrigation. The results show that the crop income has a positive and significant impact on adoption of innovative risk-reduction measures in wastewater irrigation. A unit increase in the log of crop income increases the use of crops restriction method by 4.5% (p=0.000) and wearing of protective clothing by 1.7%. A study on rice farming technologies in China by Chi (2008) shows that access to media enhances the adoption of technology among rice farmers. This implies that a well organised media access for urban and peri-urban farmers can be used to disseminate important information on wastewater reuse. This can in turn lead to increased adoption of low-risk technologies in wastewater irrigation hence minimizing health hazards.

Log of crop income

As expected, the results show that a unit increase in the log of crop income significantly raises the probability of adopting crop restriction by 4.5% (p=0.086) and safer application techniques by 4.1% (p=0.016). Also, the probability of using protective clothing increased by 1.7% once the log of crop income was raised by a unit. However, a unit increase in the log of crop income results in significant decrease in the probability of using irrigation cessation method by 2.2% (p=0.087). These results is consistent with the findings from a study in Ethiopia which shows that farm income is a key determinant in farmers' decisions to adopt agricultural technologies (Asfaw et al., 2011). Also, another study on sustainable soil conservation technologies in Iran shows that farm income positively effects adoption of sustainable soil conservation practices (Rezvanfar et al., 2009). Hence, it may be inferred that higher crop income encourages adoption of innovative measures to reduce health risks from wastewater reuse.

Awareness to WHO irrigation guidelines

Awareness to WHO irrigation guidelines has a positive and significant impact on the adoption of innovative risk-reduction measures in wastewater irrigation. A unit increase in awareness to WHO irrigation guidelines raises the probability of using crops restriction by 8.3% (p=0.000) and the probability of wearing protective clothing by 18.6% (p=0.007). Similarly, the probability of adopting safer wastewater irrigation technologies increases by 2.3% (p=0.004) as a result of a unit increase in awareness to the irrigation guidelines. However, the probability of using irrigation cessation in order to reduce health risks in wastewater irrigation declines by 4.6% (p=0.019) with a unit increase in awareness to WHO irrigation guidelines. The review of literature focused on agricultural adoption in United States of America shows that environmental awareness has a positive effect on the adoption of best management practices (Prokopy et al., 2008). This shows that the greater the awareness to WHO irrigation guidelines, the greater is the possibility of adoption of risk-reducing innovative technologies.

Awareness to wastewater hazards

Awareness to wastewater hazards has a significantly positive impact on the adoption of innovative risk-reduction interventions in wastewater irrigation. The results show that a unit increase in awareness to wastewater hazards increases the probability of using crops restriction, wearing protective clothing, adoption of safer application, and employment of irrigation cessation by 6.0% (p=0.023), 9.2% (p=0.028), 4.9% (p=0.014), and 1.5% (p=0.002) respectively. A similar study conducted in Ethiopia shows that (Weldesilassie et al., 2011) lack of health risk awareness is a key limitation for individuals' decision to work on irrigation farms. These results may be used to infer that greater awareness to wastewater hazards enhances the adoption of risk-reduction measures in wastewater-irrigated agriculture.

CONCLUSIONS AND POLICY IMPLICATIONS

This study provides an analysis of the factors that affectthe choice of innovative risk-reduction interventions to reduce the risks related to wastewater reuse in agriculture. The urban and peri-urban farmers indicated that they had employed the following risk-reduction innovations: crops restriction, protective clothing, safer application, and irrigation cessation. The MNL model was used in this study to investigate the socio-economic and institutional factors that condition the choice of the risk-reduction measures in wastewater irrigation. In order to ensure efficient estimations under the IIA assumption, the Hausman test was conducted on the MNL model. Marginal effects from the fitted model were used to measure the expected change in probability of the choices made by farmers with respect to unit change in explanatory variables.

The marginal analysis results show that education level and farm income significantly affect the adoption of innovative risk-reduction interventions in wastewater reuse. Therefore, there is need for policy makers to enhance support of education systems and supply of relevant inputs to promote urban agriculture in an effort to reduce the hazards of urban wastewater irrigation. The results also reveal that access to credit facilities, access to certified seed and access to media significantly influence adoption of innovative measures to reduce the health risks due to wastewater irrigation. This paper recommends provision of affordable credit schemes, supply of certified seeds and production of inexpensive media for urban and peri-urban farmers in order to promote the adoption of risk-reduction interventions in wastewater irrigation. The membership to farmers' group and support by NGO also significantly affect the choice of risk-reduction measures in wastewater reuse for agriculture. Thus, future policies that encourage informal community linkages and also incentivise funding of urban agriculture projects by NGOs can greatly enhance the adoption of innovative risk-reduction measures in wastewater irrigation. Lastly, there is a need to design policies and programs that support farmers in safe wastewater irrigation, while raising their awareness on the health hazards attributed to untreated wastewater reuse. This is likely to reduce health risks to the farmers and consumers of wastewater-produced vegetables and also improve livelihoods of many urban people.

ACKNOWLEDGEMENTS

The authors are grateful to two anonymous reviewers for the insightful comments. This research was financially supported by the Organisation for Social Science Research in Eastern and Southern Africa (OSSREA).

REFERENCES

Addo KA (2010). Urban and peri-urban agriculture in developing countries studied using remote sensing and in situ methods. Remote Sensing 2:497-513

Adeoti AI (2009). Factors influencing irrigation technology adoption and its impact on household poverty in Ghana. J. Agric. Rural Dev. Trop. Subtropics 109(1):51-63.

AEO (African Economic Outlook) (2012). Kenya. [Online] Available from:http://www.africaneconomicoutlook.org/fileadmin/uploads/aeo/Country_Notes/2011/Full/Kenya.pdf [Downloaded: 2012-05-25].

Agwu AE, Ekwueme JN, Anyanwu AC (2008). Adoption of improved agricultural technologies disseminated via radio farmer programme by farmers in Enugu State, Nigeria. Afr. J. Biotechnol. 7(9):1277-1286.

Ajayi MT, Okunlola JO (2005). Impact of agricultural extension services on adoption of root crops technologies in Ondo State, Nigeria. Afr. J. Agric. Exten. 34(2):181-187.

Asfaw S, Shiferaw B, Simtowe F, Mekbib Gebretsadik Haile MG (2011). Agricultural technology adoption, seed access constraints and commercialization in Ethiopia. J. Develop. Agric. Econ. 3(9):436-477.

Awotide BA, Diagne A, Omonona BT (2012). Impact of improved agricultural technology adoption on sustainable rice productivity and rural farmers' welfare in Nigeria: a local average treatment effect (LATE) technique. A paper presented at the African Economic Conference Kigali, Rwanda, October 30 – November.

Buah SSJ, Nutsugah SK, Kanton RAL, Atokple IDK, Dogbe W, Karikari AS, Wiredu AN, Amankwa A, Osei C, Ajayi O, Ndiaye K (2011). Enhancing farmers' access to technology for increased rice productivity in Ghana. Afr. J. Agric. Res. 6(19):4455-4466.

Buechler S, Devi G (2006). Adaptations of wastewater-irrigated farming systems: a case of Hyderabad, India. In: van Veenhuizen, R. (ed.)

Cities Farming for the Future: Urban Agriculture for Green and Productive Cities. RUAF Foundation, IDRC and IIRR.
Chi TTN (2008). Factors affecting technology adoption among rice farmers in the Mekong Delta through the lens of the local authorial Managers: an analysis of qualitative data. Omonrice 16:107-112.
Dandedjrohoun L, Diagne A, Biaou G, N'cho S, Midingoyi, SK (2009). Determinants of diffusion and adoption of improved technology for rice parboiling in Benin. Rev. Agric. Environ. Stud. 93(2):171-191.
Drechsel P, Graefe S, Sonou M, Cofie OO (2006). Informal irrigation in urban West Africa: an overview. IWMI Research Report 102. Colombo, Sri Lanka. Available from: http://www.iwmi.cgiar.org/publications/iwmi_research_reports/pdf/pub102/rr102.pdf [Downloaded: 2012-11-30].
Drechsel P, Keraita B, Amoah P, Abaidoo R, Raschid-Sally L, Bahri A (2008). Reducing health risks from wastewater use in urban and peri-urban sub-Saharan Africa: applying the 2006 WHO Guidelines. Water Sci. Technol. 57(9):1461-1466.
Ensink JHJ, van der Hoek W, Matsuno Y, Munir S, Aslam MR (2003). The use of untreated wastewater in peri-urban agriculture in Pakistan: risks and opportunities. Colombo: International Water Management Institute.
Falkenmark M, Lundqvist J, Widstrand C (1989). Macro-scale water scarcity requires micro-scale approaches: Aspects of vulnerability in semi-arid development. Nat. Resour. Forum 13(4):258-267.
GOK (Government of Kenya) (2009). Ministry of agriculture strategic plan 2008–2012. Nairobi: Government Printer.
GOK (Government of Kenya) (2010a). Draft national urban and peri-urban agriculture and livestock policy. Nairobi: Government Printer.
GOK (Government of Kenya) (2010b). National water harvesting and storage management policy. Available from: http://xa.yimg.com/kq/groups/15900173/66228397/name/National_Water_Storage_Policy.pdf [Downloaded: 2011-09-24].
Hamilton AJ, Stagnitti F, Xiong X, Kreidil SL, Benke KK, Maher P (2007). Wastewater irrigation: the state of play. Vadose. Zone J. 6:823-840.
Hausman JA, McFadden D (1984). Specification tests for the multinomial logit model. Econometrica 52:1219-1240.
Jiménez B, Asano T (2008). Water reclamation and reuse around the world. In: Jiménez, B. & T. Asano (eds.) Water reuse: an international survey of current practice, Issues and Needs. London: IWA Publishing.
Keraita B (2008). Low-cost measures for reducing health risks in wastewater irrigated urban vegetable farming in Ghana. Unpublished doctoral thesis. Copenhagen: University of Copenhagen.
Keraita B, Konradsen F, Drechsel P, Abaidoo RC (2007). Effect of low-cost irrigation methods on microbial contamination of lettuce. Trop. Med. Int. Health 12(2):15-22.
Keraita B, Drechsel P, Konradsen F (2008). Perceptions of farmers on health risks and risk reduction measures in wastewater-irrigated urban vegetable farming in Ghana, J. Risk Res. 11(8):1047-1061.
Kothari CR (2004). Research methodology: methods and techniques. New Delhi: New Age International.
Knudsen LG, Phuc PD, Hiep NT, Samuelson H, Jensen PK, Dalsgaard A, Raschid-Sally L, Konradsen F (2008). The fear of awful smell: risk perceptions among farmers in Vietnam using wastewater and human excreta in agriculture, Southeast Asian J. Trop. Med. Public Health 39(2):341-352.
Maddison D (2006). The perception of and adaptation to climate change in Africa. CEEPA. Discussion paper No. 10. Centre for environmental economics and policy in Africa. Pretoria: University of Pretoria.
Marenya PP, Barrett CB (2007). Household-level determinants of adoption of improved natural resources management practices among smallholder farmers in western Kenya. Food Policy, 32:515-536.
Mignouna DB, Manyong VM, Rusike J,. Mutabazi KDS, Senkondo EM (2011). Determinants of adopting Imazapyr-resistant maize technologies and its impact on household income in Western Kenya. Agric. Bio. Forum, 14(3):158-163.
Mohamed KS, Temu AE (2008). Access to credit and its effect on the adoption of agricultural technologies: the case of Zanzibar. Afri. Rev. Money Finance. Bank. pp. 45-89.
Muzari W, Gatsi W, Muvhunzi S (2012). The impacts of technology adoption on smallholder agricultural productivity in sub-Saharan Africa: a review. J. Sustain. Dev. 5(8):69-77.
NCAPD (National Coordinating Agency for Population and Development) (2010). Population dynamics and climate change: implications for the realization of the MDGs and the goals of vision 2030. Nairobi: National Coordinating Agency for Population and Development.
Nchinda VP, Ambe TE, Nathalie H, Leke W, Che MA, Nkwate SP, Ngassam SB, Njualem DK (2010). Factors influencing the adoption intensity of improved yam (Dioscorea spp.) seed technology in the western highlands and high guinea savannah zones of Cameroon. J. Appl. Biosci. 36:2389- 2402.
NEMA (National Environmental Management Authority) (2003). State of the environment report, Kenya. Nairobi, Government Printer.
NEMA (National Environmental Management Authority) (2011). Kenya: state of the environment and outlook 2010. Supporting the delivery of vision 2030. Summary for decision makers. Malta: Progress Press.
Nhemachena C, Hassan R (2007). Micro-level analysis of farmers' adaptation to climate change in South Africa. IFPRI Discussion paper No.00714. Washington, D.C.: International Food Policy Research Institute.
Obuobie E, Keraita B, Danso G, Amoah P, Cofie O, Raschid-Sally L, Drechsel P (2006). Irrigated urban vegetable production in Ghana: characteristics, benefits and risks. Accra: IWMI–RUAF–CPWF. Available from: http://www.ruaf.org/sites/default/files/contents_0.pdf [Downloaded: 2011-09-25].
Pattanayak SK, Mercer DE, Sills E, Jui-Chen Y (2003). Taking stock of agroforestry adoption studies. Agroforest. Syst. 57(3):173-186.
Prokopy LS, Floress K, Klotthor-Weinkauf D, Baumgart-Getz A (2008). Determinants of agricultural best management practice adoption: Evidence from the literature. J. Soil. Water Conser. 63(5):300-311
Qadir M, Wichelns D, Raschid-Sally L, McCornick PG, Drechsel P, Bahri A, Minhas PS (2010). The challenges of wastewater irrigation in developing countries. Agric. Water Manage. 97:561–568.
Rezvanfar A, Samiee A, Faham E (2009). Analysis of factors affecting adoption of sustainable soil conservation practices among wheat growers. World Applied Sciences Journal, 6(5):644-651.
Rutkowski T, Raschid-Sally L, Buechler S (2007). Wastewater irrigation in the developing world: two case studies from the Kathmandu Valley in Nepal. Agric. Water Manag. 88:83-91.
Scott CA, Faruqui NI, Raschid-Sally L (2004). Wastewater use in irrigated agriculture: Management challenges in developing countries. In: Scott CA, Faruqui NI, Raschid-Sally, L (eds.) Wastewater use in irrigated agriculture: confronting the livelihood and environmental realities. Wallingford: CABI Publishing. pp. 1-10.
Srinivasan JT, Reddy VR (2009). Impact of irrigation water quality on human health: A case study in India. Ecol. Econ. 68:2800-2807.
Umande Trust (2012). Kibera, Nairobi. http://www.umande.org/index.php/where-we-work/77-kibera-nairobi [Downloaded: 2012-05-28].
UNEP (United Nations Environment Programme) (2003). Nairobi river basin phase II: pollution monitoring report. [Online] Available from: http://www.unep.org/roa/Nairobi_River_Basin/Downloads/Phaseii_publications/reports/PollutionMonitoringReportPhase2.pdf [Downloaded: 2012-05-28].
van der Hoek W (2004). A framework for a global assessment of the extent of wastewater irrigation: the need for a common wastewater typology. In: Scott CA, Farunqui NI, Raschid-Sally L (eds.) Wastewater use in irrigated agriculture, confronting the livelihood and environmental realities. Trowbridge: International Development Research Centre.
Weldesilassie AB, Boelee E, Drechsel P, Stephan DS (2011). Wastewater use in crop production in peri-urban areas of Addis Ababa: impacts on health in farm households. Environ. Dev. Econ. 16:25-49.
World Bank (2010). World development indicators: 2010. Washington, D.C.: The World Bank. [Online] Available from: http://data.worldbank.org/sites/default/files/wdi-final.pdf [Downloaded: 2011-09-24].
WHO (World Health Organization) (2006). Guidelines for the safe use of wastewater, excreta and greywater: wastewater use in agriculture,Volume 1. Geneva: World Health Organisation.

Yirga CT (2007). The dynamics of soil degradation and incentives for optimal management in central highlands of Ethiopia. Unpublished doctoral thesis. Pretoria: University of Pretoria.

17

Preliminary studies on polyphenol oxidase activity in plantain (*Musa paradisiaca)* cultivars

Adeosun Olubunmi

Department of Fisheries Technology, Oyo State College of Agriculture, Igbo-ora, Oyo State, Nigeria.
E-mail: moriyike2006@yahoo.com.

Polyphenol oxidase (PPO) has been shown to be responsible for browning reactions and discolouration in many fruit and vegetable products. Crude polyphenol oxidase (PPO) of ripe and unripe plantain (Musa paradisiaca cultivars – Agbagba, Cardaba and plantain hybrid) was isolated and some of the characteristics of the PPO extracts investigated. The activity of the enzyme was evaluated using spectrophotomeric method. Plantain PPO (ripe and unripe) catalyzes oxidation of both various substrates with catechol being the most readily oxidized substrate. The optimum pH of the enzyme was between pH 6 and 7 for the three cultivars. Thermal inactivation data showed that activity of the enzyme was lost after heating for 15 and 4 min at 50 and 80°C, respectively. Browning reactions of plantain PPO were inhibited by sodium metabisulphite (0.1%), ascorbic acid (0.2%), malic acid (1.0%) and the least sodium chloride (>1.0%) which is similar to banana PPO. Plantain polyphenol oxidase was active towards catechol but not with cresol.

Key words: plantain (*Musa paradisiaca*) cultivars, polyphenol oxidase, catechol, enzyme activity.

INTRODUCTION

Plantain (*Musa paradisiaca*) is an important crop in Nigeria as in all humid tropical zones of Africa, Asia, Central and South America (Swennen, 1990). It is grown extensively in the southern states of Nigeria. Plantain is one of the most important sources of food energy throughout the African lowland humid forest zone. Plantain has a unique flavour and aroma and it contains high amount of potassium (K^+) and calcium (Ca^{2+}). Traditionally, plantain pulp is cooked when green and pounded to give "*fufu*"-like dish which can be eaten with vegetable soup or sundried and milled into flour for *'amala'* while the ripe peeled plantain can be eaten directly or boiled or sliced and fried in vegetable oil. Industrially, plantain is an important raw material in the manufacture of plantain flour. Plantain flour is obtained by drying green fruit and can be used as diluents of bread flour. The flour from ripe fruit has high sugar content which makes it a useful component of infant diet and the carbohydrate have been found to be easily digested than that of cereal (Ladele et al., 1984). In some areas, the peel of plantain fruits are used in soap making and as animal feed when dried. In Nigeria, plantain is grown extensively in the southern states namely Anambra, Akwa-Ibom, Bendel, Cross-river, Imo, Lagos, Ogun, Oyo and Rivers. PPO is widely distributed in the plant kingdom and has been shown to be among the first enzymes to be studied according to Schoenbein (1856) as cited by Whitaker (1972). Being widely distributed in nature it has been studied in fruits such as banana (*Musa cavendishii*) (Fatemh et al., 2008), potato (*Colocasia antiquorum* var *esculenta*) (Lee and Park, 2007), grape (Munoz et al., 2004), lettuce (*Luctuca sativa* var. *capitata* L.) (Gawlik-Dziki et al., 2007) and pepper (*Capsicum annuum* L.) (Arnnork et al., 2010). Polyphenol oxidase has been shown to be responsible for browning reactions and discolouration in many fruits and vegetables (Marshall et al., 2000; Yoruk and Marshall, 2003). Enzymatic browning occurs in many fruits and vegetables such as when the tissue is bruised, cut or peeled. The injured tissue gradually darkens on exposure to air due to the oxidation of phenolic compounds to brown pigments. Therefore the objectives of this study are to isolate and determine some properties of enzymes of plantain (*M. paradisiaca*) cultivars and to evaluate methods of controlling activity of the enzyme.

MATERIALS AND METHODS

Procurement of fruit samples

Plantain fruits (*M. paradisiaca*) cultivars (Agbagba, Cardaba and Hybrid) were obtained from the International Institute of Tropical Agriculture (IITA), Ibadan. Unblemished fruit samples were obtained at their pre-climacteric stages and were transported to the laboratory in open top plastic containers within 1 h of harvest.

Sample preparation

The mature green fruits were cleaned with moist cotton wool to remove extraneous matter and fruit latex which may cause discolouration. For extraction of enzyme from unripe green stage (stage 1), the harvested fruits were used as such whilst the remaining fruits were kept in open plastic containers at ambient conditions (28 ± 1°C) and RH 68 ± 2%) until table ripe corresponding to stage 6 of the USA banana ripening chart (Adel, 2005).

Preparation of crude PPO extracts

Preparation of plantain pulps for PPO extraction was accomplished at two physiological stages namely: Mature green (colour index 1) and table ripe (colour index 6) (Adel, 2005). 2 g of fresh fruit pulp was homogenized with 18 ml of 1% polyethylene glycol buffered at pH 7 with 0.1M potassium phosphate (Palmer, 1963). The slurry was centrifuged in a Beckman centrifuge (model J-21B) at 4000rpm for 20 min at 8°C. The resultant supernatant was taken as the crude enzyme extract.

Assay of plantain PPO

The enzyme activity was determined spectrophotometrically by measuring the initial rate of the increase in absorbance at 470nm using CECIL UV spectrophotometer (model CE 202 series 2) as described by Palmer (1963). The sample cuvette contained 1 ml 0.033 M potassium phosphate (pH7), 1 ml 0.005 M substrate solution and quantity of enzyme to give a total reaction volume of 3 ml (Palmer, 1963). The reference cuvette contained the reaction mixture without the enzyme which was replaced with distilled water. Initial experiment was carried out using buffered homogenate (2 g of pulp in 20 ml 0.1 M potassium phosphate pH 7) (Palmer, 1963). One unit of PPO activity was defined as potency of enzyme that increased in absorbance of 0.01 to 0.05 per minute.

Determination of pH optima of PPO activity

The optima pH of plantain PPO was determined by measuring the activity as described in enzyme assay using (1) 0.1 M phosphate buffers pH 6 to 8, (2) 0.1 M citrate buffers pH 3.8 to 6, (3) Standard pH buffer pH 9.2. The cuvette contained the substrate (1 ml), buffer solutions (1 ml) at different pH and enzyme solution (1 ml) to give a reaction mixture of 3 ml (Palmer, 1963).

Determination of optima temperature of PPO activity

The thermal denaturation of the enzyme was determined using the method of Park et al. (1980). 5ml portion of enzyme solution were sealed in test tubes and heated in water bath at 50, 60, 70 and 80°C, respectively (Park et al., 1980). At various heating intervals 1ml of samples were withdrawn and immediately cooled by immersion in ice-water. The PPO activity was measured using spectrophotometer at 470 nm as described in the enzyme assay (Palmer, 1963).

Chemical inhibition of PPO activity

Investigation on the enzyme inhibition was determined by measuring activity using 1ml 0.005 M substrate solution, 1 ml of various concentrations of sodium metabisulphite, ascorbic acid, Malic acid, citric acid and sodium chloride from 0.05 to 5.0%, 1 ml 0.033 M phosphate buffer (pH 7) and 1 ml enzyme solution (Palmer, 1963).

RESULTS AND DISCUSSION

Substrate specificity

All of the three plantain cultivars responded actively towards o-diphenols but not for the monophenols (Table 1). The activity of PPO obtained from unripe plantains (stage 1) was higher than the activity of PPO obtained from ripe plantains. Cathecol was the most readily oxidized by the enzymes. This finding is in line with Sanjeev and Mishra (2011) work that banana PPO is active towards o-diphenolic compounds.

Effect of substrate concentration on plantain PPO

Preliminary study on the effect of substrate concentrations showed that the activity of unripe plantain PPO from different cultivars increased with increase in concentration. Polyphenol oxidase from Agbagba gave maximum activity between 5 and 40 mM after which the activity decreased. Cardaba and Hybrid plantains' PPO have maximum activities at 40 mM and decreased as substrate concentration increased. The activity of PPO extracted from ripe plantain of different cultivars also increased as substrate concentration increases. Ripe plantain PPO (Agbagba) has maximum activity at concentration of 50 mM while ripe Cardaba and ripe Hybrid PPO have maximum activity at 40 and 10 mM respectively. This corroborated the report of Yang et al. (2000) that banana PPO acts best at substrate concentration of 2.8 mM using dopamine as substrate. The difference in substrate concentration may be due to genetic difference between plantain and banana.

Effect of pH on plantain PPO activity

Maximum pH activity occurred at pH 6.4 and 7.0. The result showed decrease in enzyme activity at alkaline and acidic pH values for different cultivars. It is reported

Table 1. Substrate specificity of plantain PPO.

Substrate 40 Nm	Unripe plantain fruits			Ripe plantain fruits		
	Agbagba	Cardaba	Hybrid	Agbagba	Cardaba	Hybrid
Catechol	0.650	0.520	0.440	0.470	0.370	0.360
Pyrogallol	0.360	0.280	0.130	0.135	0.110	0.085
Quinic acid	0.460	0.250	0.120	0.125	0.090	0.085
M-cresol	0.000	0.000	0.000	0.000	0.000	0.000

Table 2. Effect of chemical concentration on plantain PPO.

Inhibitors	Concentrations (%)				
	0.05	0.1	0.2	0.5	1.0
Sodium metabisulphite	Brown colour after 5 min and becomes darker with time	No reaction for 24 h	No reaction for 24 h	No reaction for 24 h	No reaction for 24 h
Ascorbic acid	Colour developed after 2 min	Colour developed after 10 min	No reaction	No reaction	No reaction
Malic acid	Brown colour after 2 min	Brown colour after 5 min	Brown colour after 5 min	Brown colour after 5 min	No reaction
Sodium chloride	Brown colour in 2 min	Brown colour in 2 min	Brown colour in 5 min	Brown colour in 5 min	Brown colour after 20 min

that banana PPO is optimally active at pH = 7.0 (Sanjeev and Mishra, 2011) which is similar to results obtained for plantain PPO.

Effect of heat on plantain PPO activity

The activity of the PPO from all the cultivars decreased with increase in temperature. This is because the enzymes are proteinous and are denatured by heat. The enzyme lost its activity in 15 and 4 min when held at 50 and 80°C respectively.

Effect of chemical treatment on plantain PPO activity

Table 2 shows the effect of chemical inhibitors on the activity of plantain PPO using catechol as substrate. Polyphenol oxidase activity was inhibited at 0.1 and 0.2% solution of sodium metabisulphite and ascorbic acid respectively. Malic acid inhibited the PPO at 1% solution while sodium chloride inhibited the PPO at concentration greater than 1.0%.

Enzymes extracted with buffered polyethylene glycol from the pulp of unripe and ripe plantain cultivars are by substrate specificity a polyphenol oxidase which is similar to enzymes of banana (Palmer, 1963; Sanjeev and Mishra, 2011). The results reported here show that plantain PPO (ripe and unripe) catalyzed oxidation of various substrates with catechol being the most readily oxidized substrate. Browning reactions of plantain PPO were inhibited by sodium metabisulphite(0.1%), ascorbic acid(0.2%), Malic acid (1.0%), and the least sodium chloride (> 1.0%) which is similar to banana (Palmer,1963; Sanjeev and Mishra, 2011).

Conclusion

Polyphenol oxidase has been found to occur in the pulp of plantain fruits. The enzyme catalyzed the oxidation of various phenolic compounds and this study indicated that the cathecol is most readily oxidized. Plantain PPO extracted using buffered polyethylene glycol was relatively stable and had optimum value was pH 6.8 with cathecol as the substrate.

REFERENCES

Adel AK (2005). Banana ripening Chart. Produce Facts: Department of Plant Science, University of California, Davis, CA 95616. postharvest.ucdavis.edu/Produce/ProduceFacts/Fruit/fd http://postharvest.ucdavis.edu assessed on 12/08/2011

Arnnork P, Ruangviriyachai C, Mahachai R, Techawongsteien S, Chanthai S (2010). Optimization and determination of polyphenol oxidase and peroxidase in hot pepper (*Capsicum annuum* L) pericarb. Int. Food Res. J. 17:385-392

Fatemh SN, Kamahbeen H, Masoumeh KB (2008). The Banana Pulp Polyphenol Oxidase is a Tyrosinase. J. Biol. Sci. 2008 ISSN 1727-3048. pp. 1-8.

Gawlik-Dziki U; Zloteck U, Swieca M (2007). Characterization of polyphenol oxidase from butter lettuce (*Luctuca sativa* var. *capitata L).* Food Chem. 107:129-135

Ladele OA; Makanju OO, Olaofe O (1984). chemical constituents of Plantain. Nig. J. Nutr. Sci. 5(1):35-38

Lee MK, Park I (2007). Studies on inhibition of enzymatic browning in some foods by Du-Zhong (*Eucommia uimides* Oliver) leaf extract. Food Chem. 114:154-163.

Marshall MR, Kim J, Wei C (2000). Enzymatic browning in fruits, vegetables and seafoods. FAO report at

http://www.fao.org/ag/ags/agsi/ENZYMEFINAL/ assessed on 12/08/2011
Munoz O, Sepulveda, M, Schwaerz M (2004). Effects of enzyme treatment on anthocyanin pigments from grapes skin from Chilean wine. Food Chem. 87:487-490
Palmer JK (1963). Banana Polyphenol oxidase: Preparation and Properties. Plant Physiol. 38:508-513.
Park YK, Sato,HH, Almeida TD, Moretti RH (1980). Polyphenol oxidase of Mango (*Mangifera indica* var. Haden). J. Food Sci. 45:1619-1621
Sanjeev KD, Sarad KM (2011). Purification and Biochemical Characterization of Ionically Unbound Polyphenol Oxidase from *Musa paradisiaca* Leaf. Preparat. Biochem Biotechnol. 41 (2):187-200.
Swennen R (1990). Plantain cultivation under West African conditions. A reference monval of International Institute of Tropical Agric. (IITA) Ibadan, Nigeria.
Whitaker JR (1972). Polyphenol oxidase. In: Principle of Enzymology for the food Sciences. Marcel Dekker. Inc. New York. pp. 571-582.
Yang CP, Fujita S, Ashrafuzzaman, M; Nakamura, N, Hayashi N (2000). Purification and characterization of polyphenol oxidase from banana (*Musa sapientum*) pulp. J. Agric. Food Chem. 48(7):2732-2735
Yoruk R, MR Marshall (2003). Physicochemical properties and function of polyphenol oxidase: A review. J. Food Biochem.27 (5):361-422

The effects of agricultural subsidies applied in Turkey on sunflower cultivation area and the producers' income

Arif Semerci

Department of Agricultural Economics, Faculty of Agriculture, Mustafa Kemal University, Tayfur Sökmen Kampüsü, Antakya, Hatay, Turkey

Sunflower takes an important place in terms of raw materials provided for the oil industry and of added value, as well as bring-in income for producers in Turkey's economy. Sunflower is an important source of income by standing the main product in alternation of the wheat in the sense of manufacturers in Thrace Region which constitutes the European part of Turkey. In this study, the effect of premium support application on sunflower cultivation area and production amount in Thrace where approximately 65% of sunflower production in Turkey is met and also the effect of support payments on producer income are investigated. In this study, it is concluded that premium supports are inefficient in increasing sunflower cultivation area and production whereas it is concluded that it has an important role on producer income and on the determination of the market price. In fact, from a survey that was carried out, it emerges that the area-based support (diesel oil, fertilizer etc. support) together with premium support increase the revenue of sunflower up to 27.74% per area unit and the gross profit in proportion to 98.87%. This study indicates that in order to increase the sunflower production in Turkey the amounts of sunflower premium support, diesel oil and fertilizer should be rearranged in compliance with current market conditions. Moreover, as well as for other agricultural products, the use of certified seed should also be included into agricultural support programmes, production on well irrigated areas should be widened and the use of oil crops varieties with high oil ratio should be extended.

Key words: Oil crops, sunflower, agricultural subsidy, agricultural income, gross profit.

INTRODUCTION

Alike food sector, agriculture sector has a significant role in national economics because of its contribution to employment and income, its role in meeting the needs of industrial raw materials and its direct or indirect benefits on exports. According to 2009 data, in Turkey the agricultural sector account for 51 billion US$ of gross domestic product and sunflower is one of the leading agricultural products for manufacturers and oil industry especially in Thrace, at the European side of the Turkey (MARA, 2011a, b).Turkey which accounts for 2.46% of sunflower world cultivated area and 3.26% of production is among the top ten countries for production of sunflower according to production data from the year 2009 (Anonymous, 2011a).

Sunflower which accounts for 3.60% of field crops cultivated areas in Turkey, also meets approximately 55% of vegetable oil production (TURKSTAT, 2011; Ozcelik and Fidan, 2003). Besides, 4.99% of 2,328,731 farmers in the country are working on sunflower production (MARA, 2011a). Despite the high agricultural production potential, the self-sufficiency rate of sunflower in our country is only 46.60% (Anonymous, 2011b). In 2008, 3 billion US$ of vegetable oil and oil seeds was imported in order to meet the current deficit (Anonymous, 2009).

Because of the low elasticity of demand, exposure to climate conditions, intense economic instability (price and income) of agricultural sector, government interventions are carried on cultivation areas and the sector is diversely supported across the globe (Gunaydin, 2006). Turkey is one of the countries supporting the agricultural sector in line with the various agricultural policies implemented in accordance with the developments in the world (Zemheri, 2010). Use of certified seeds, premium support system, diesel oil and fertilizer compose are the main factors of agricultural subsidy studies in Turkey especially since the early 2000's in order to ensure an increase in the production of products which are in short supply. However, it is not exactly known the extent these practises reflect to manufacturer agricultural income and how the agricultural production is affected.

By the study carried out in Thrace where approximately 65% of sunflower for oil industry is produced in Turkey, the effects of premium supports on sunflower cultivation areas and amount of production with the aim of increasing oleaginous seeds production was investigated. In addition to this, in this study, it is also put forth the contribution of support payments to manufacturer prosperity for increasing sunflower production. In some studies conducted in various countries, support of agricultural production and the effects of these supports on agricultural production, farmer's welfare and the trade of agricultural products were analyzed (Hennessy, 1998; Mayrand et al., 2003; Anderson et al., 2006; Keeny, 2009). Some studies reside in Turkey related to agricultural support effects on sunflower production, reflection to producer income and foreign trade regime (Gaytancioglu, 1999; Koc, 2005; İlkdogan, 2008).

According to a study based on a 'Theoretical Microeconomic Model' whose data were obtained from 1,000 agricultural enterprices in Norway, it is concluded that agricultural subsidies have significantly positive effects on input use and the level of output (Henningsen et al., 2009). As a result of research conducted in Czech Republic, despite the country's agriculture having a competitive market, in comparison to some of European countries it emerges that agricultural subsidies have low share in producer income. In this study, it is determined that the lower percentages of support level weaken the competitiveness of agricultural area, and the regeneration and modernization of agriculture. As a result, it is concluded that the subsidies which are applied according to type of production might affect the production diversity (Střeleček et al., 2009). A Premium support system is an alternative support model that can be applied by assuring stability in the sense of prices and the producer in agriculture of Turkey, depending on the research work of Sahinoz et al. (2007). In a study published by the Vegetable Oils and Fats Industrialists Association, sunflower production is supposed to be decreasing at an average annual rate of 6% in the period 2010 to 2013 whether premium supports which is being paid to oil seed producers in Turkey will not be paid (Koc, 2005). In another study which investigates the effects of premium support systems in terms of agricultural subsidies in Turkey, it is educed that the subsidy got by farmers has significant effects on increasing the sunflower cultivated area and premium supports paid for sunflower are efficient in protecting the producers against to market prices (Erdal and Erdal, 2008).

Recently, agricultural production support policies applied in Turkey have actually changed a lot are deeply changing. A new model of agricultural production and support model (Turkey Agricultural Basin Production and Support Model) has just begun to be implemented in Turkey. The aims of the new model are; determination of the agricultural basins, preparation of a clear agricultural inventory, provision of opportunities for production planning, and making projections of demand for the future (Sahin et al., 2010).

The amount of resources allocated to support agricultural production varies depending on the development levels of countries. Direct payments are the largest item with 74% of agriculture subsidies (approximately €55 billion) in the European Union's budget in 2008. Direct payments are allocated with a share of 99% to agricultural markets, 0.5% to fisheries sector and 0.5% to plant and animal health. The second largest item which has a 23.5% share with €12.9 billion is rural development. The amount of direct subsidies that USA farmers have received in the years from 1996 and 2001 is 70 billion US$. In the period from 2002 to 2008, the farmers are estimated to have benefited from direct state assistance of 100 billion US$ and 80 billion US$ is planned to be distributed before 2012 (Acar and Bulut, 2009). The amount of resources allocated to agricultural subsidies in 2010 in Turkey is given in the Table 1.Premium support payments in Turkey in 2010 account for 36.95% of agricultural support payments and 24.57% of total agricultural subsidies. The share of total agricultural support payments stood at 0.82% of Gross Domestic Product. The support payments for development of oil seed production in 2008 in Turkey are given based on the products in Table 2. In terms of premium supports the price sunflower seed is ranked second in oil seed support payments in Turkey.

MATERIALS AND METHODS

The primary data used in this study were obtained by question-

Table 1. Funds Allocated to Agricultural Support in Turkey (2010).

Criteria	US$
Premium supports	1.340.745.000
Agricultural support payments	3.627.382.000
Implementation of low interest rate credit	487.536.000
Total agricultural support payments	5.455.663.000
Total gross domestic product	665.891.300.000

Kaynak: MARA, General Directorate of Agricultural Production and Improvement, 2009.

Table 2. Premium support payments on oil seed plants and olive.

Products	Number of enterprices	Production area (ha)	Amount due to supports (ton)	Sum of supports (US$)	Product rate in sum of support (%)
Safflower	292	1.399	1.733	248.552	0.04
Corn for grain	65.929	383.650	3.557.650	46.460.339	8.18
Canola	1.610	6.868	16.704	2.403.263	0.42
Cotton Unseed	69.208	441.581	1.905.437	422.260.035	74.35
Soybean	950	5.199	16.180	2.647.695	0.47
Sunflower	89.984	420.505	705.216	92.108.143	16.22
Olive	5.126	27.610	13.845	1.810.883	0.32
Total	227.973	1.286.812	6.216.765	567.938.910	100.00

Resource: MARA, General Directorate of Agricultural Production and Improvement, 2009

naires from specified 571 agricultural enterprises in 53 locations on the European side of provinces of Edirne, Kırklareli, Tekirdağ, İstanbul and Çanakkale with Stratified Random Sampling Method in production period of 2009. In determination of stratified random sampling method, two different techniques are commonly applied by agricultural economic studies. These are: Neyman Method and Proportional Distribution Method (Erkan and Cicek, 1996). Because of the incompliance of sunflower cultivation areas in locations and enterprises to normal distribution, Neyman Method is used in this research. The equality of Neyman Method is given below (Yamane, 1967).

$$n = \frac{\Sigma\ (N_h S_h)^2}{N^2 D^2 + \Sigma N_h (S_h)^2}$$

In the formula; n = sample size, N_h = number of units in layer h (frequency of farm size);S_h = standard deviation of layer h; N = total number of units; D = d/z; d = a certain percentage of deviation from average(1%-5%-10%), and z = degrees of freedom in t-distribution scale (N-1) and expresses "t value" belongs to a certain confidence limit (90%-95%-99%).

Within the scope of the research, the effects of premium supports applied in Turkey on the cultivation areas and production amount of sunflower for the next production period were investigated. To that end, taking account of two periods as premium supported and unsupported, the relationship between cultivation areas and production amount was investigated with "correlation analysis" by using SPSS programme (Green et al., 2000). With this purpose, sunflower production periods in Turkey were divided into two parts; unsupported (1988 to 1998) and premium supported (1999 to 2008). The dummy variable is taken to be "0" in 1988 to 1998 period and taken to be "1" in 1999 to 2008 period.

In the determination of effects of supports on producer welfare applied for sunflower production; Trakya Birlik's purchase price is taken as product price and the support unit price applied in 2009 by the Ministry of Agriculture and Rural Affairs is taken as the support unit prices. However, in the study, besides the purchase price of sunflower and premium support price, also the unit price variations of diesel fuel, fertilizer (20-20-0 compound fertilizer), seed and agricultural pesticides were researched between the years 1999 to 2010.

RESULTS AND DISCUSSION

Relationship between premium support, cultivated area and production amount of sunflower

In this study, in the period from 1988 to 2010, it is investigated whether there is a relationship between non-supported period of oil sunflower (1988 to 1998) and supported period (1999 to 2010) on sunflower cultivation area and the amount of production. For this purpose, the dummy variable for the unsupported period (1988 to 1998) is taken to be "0" and for the supported period (1999 to 2010) is taken to be "1".

The correlation between dummy variable related to support and sunflower production areas is found to be significant, and the correlation between the dummy variable related to support and the amount of sunflower production is found to be statistically insignificant in this study (Table 3). This situation can be explained with the

Table 3. The relationship between premium support periods and nonsupported periods with cultivation areas and production amount of sunflower in Turkey.

Variable	Criteria	Dummy variable
Sunflower production area (ha)	Correlation	-0.535(*)
	Significance level	0.008(*)
	Observation numbers	23
Sunflower production quantity (ton)	Correlation	0.092
	Significance level	0.677(**)
	Observation numbers	23

*P < 0.01; **Non-significant.

Table 4. Price variation occured in the factors that determine product cost, in premium support, and price of sunflower.

Years	Fuel (US$/lt)	Fertilizer (US$/kg)	Pesticide (US$/lt)	Seed (US$/kg)	Sunflower (US$/kg)	Premium support (US$/kg)
1999	0.369	0.170	2.948	5.528	0.268	0.029
2000	0.622	0.333	4.000	6.370	0.293	0.048
2001	0.415	0.311	3.111	3.457	0.278	0.022
2002	0.664	0.155	4.019	5.480	0.326	0.046
2003	0.998	0.203	4.706	10.338	0.426	0.080
2004	1.149	0.265	5.191	11.864	0.478	0.096
2005	1.445	0.277	4.817	12.597	0.504	0.130
2006	1.563	0.281	3.874	13.029	0.511	0.141
2007	1.948	0.421	5.127	18.799	0.876	0.171
2008	1.863	0.727	4.624	16.515	0.673	0.125
2009	1.639	0.348	4.684	18.066	0.642	0.141
2010	1.922	0.399	5.178	19.417	0.777	0.149

Resource: MARA, Costs of agricultural products in Trakya Region 1999-2010. Kirklareli Ataturk Soil and Water Resources Research Institute. Kirklareli.

decline at the ratio of 10.88% on average of sunflower cultivation areas to 559.456 ha before supported period (1988 to 1998) in comparison with the supported period (1999 to 2010). In other words, the premium support applied for oil sunflower since 1999 is inadequate in preventing the reduction of the cultivation areas which began before the support period. This circumstance also shows that the amount of support given to agricultural production is not in a level sufficient to promote the production. Therefore, the enhancement of sunflower plantations can be possible by providing continuous support and satisfactory reorganization to current conditions. As an example, in early 2000s the support policy applied to corn has successfully resulted in making Turkey self-sufficient over a period of 3 to 5 years whereas before 2000 it was foreign-dependent for corn production.

As a part of the research, attitude and the level of expectations of manufacturers to “Premium Support” are also analysed. 97.25% of manufacturers in the survey (536 Businesses) expressed their opinion that Premium Support applied by Agricultural and Rural Affairs should be continued. The expectation of 69.52% of companies (390 enterprises) for Premium Support unit prices of oil sunflower as a product of the year 2010 was 20.17 cent/kg. However, the expounded Premium Support unit price was 30% lower than the expected value of producers, that is to say 14.12 cent/kg. This shows that the value of stated unit price is far from the manufacturer expectations. In a study carried out on the subject, excepting premium support, diesel fuel and fertilizer support paid in Turkey within the scope of agricultural support is concluded to be insufficient (Ozcelik and Ozer, 2007).

The variation occured between 1999 to 2010 on the prices of sunflower production input and the premium support which has been granted since 1999 in Turkey is given in Table 4. When the data is analysed, it is conspicuous that especially the price of diesel fuel input has increased 5.2 times in the period 1999 to 2010, while

Table 5. Effects of sunflower support payments to producer welfare on enterprices in field of survey (2009).

Sunflower's income (excluding subsidizing payments)	
Number of enterprises (survey)	571
Sunflower production area (ha)	5.349.75
Production (ton)	9.487.08
Yield (ton/ha)	1.773
Price (US$/ton)	504.27
Gross production value (US$/ha)	894.24
Variable expenses (US$/ha)	575.54
Gross profit (US$/ha)	318.70
Sunflower's income (including subsidizing payments)	
Fertilizer support (US$/ha)	36.98
Fuel support (US$/ha)	36.98
Premium support(US$/ton)	140.00
Price (sales price+Premium support price) (US$/ton)	644.27
Gross production value (US$/ha)	1.142.29
Total supports (US$/ha)	322.18
Variable expenses(US$/ha)	575.54
Gross profit (US$/ha)	640.71
Share of subsidies in Gross Profit (%)	50.28

the price increase for the same period was recognized to be 3.5 times for seed, 2.3 times for fertilizer and 1.8 times for pesticides. Despite the 5.1 times increase on unit price of premium support in 2010 in comparison with 1999, the increase in purchasing price of the sunflower seed remained at the level of 2.9 times. Because of the diesel fuel and the product prices lower than the seeds which take important part of the costs, fluctuations are observed on sunflower cultivations areas. In the other studies carried in the region, in accordance with the evidences identified in the study, it is concluded that the purchase price of sunflower has significantly decreased in real terms in response to real increase especially in the price of diesel fuel (Aksoy and Gaytancioglu, 1996; Semerci and Kaya, 2010).

Effects of applied agricultural supports to producer welfare in sunflower production

Sunflower in Thrace is ranked as the second in the production pattern of agricultural enterprises with a ratio of 42.74% after the wheat. The contribution of sunflower to enterprises total agricultural income is calculated to be approximately 30%. These data clearly reveal the importance of sunflower for the producers in the field of research (Sahin et al., 2010). In this context, as well as the contribution of premium support and other support applications to welfare of sunflower producers, product cost reduction effect is also important.

The variations generated by support applications on Gross Production Values and Gross Profit of enterprises are examined by calculating the sunflower income both in per supported and unsupported unit area (ha) with the help of data acquired from 571 agricultural enterprises in scope of the research. In the study, the average sunflower yield of enterprises was calculated as 1.773 tons/ha. In case that the support is disregarded, the average GPV of sunflower is calculated to be 894.24 US$/ha, variable costs calculated to be 575.54 US$/ha and Gross Production Value is calculated to be 318.70 US$/ha (Table 5).

Conclusion

Among the farms of the survey, Gross Production Value of sunflower reaches on average the level of 1,142.29 US$ per hectare with an increase of 27.74% on the unsupported level in consideration of the average yield (1.773 ton/ha) and of the payments of oil seed supports. In conjunction with the support payments, sunflower Gross Profit reaches up to the level of 640.71 US$/ha with a growth of 98.87%. The share of subsidies in "Gross Production Value of New Sunflower" of which was calculated in consideration of supports is 50.28%. This rate clearly reveals the contribution and importance of support payments on Gross Profit and even in Gross Production Value of sunflower. Other studies in this issue reveal that although support payments (premium supports and field-based supports) are directly not efficient on increasing sunflower production, they play an important role on the formation of healthier sunflower purchase prices and on farmer's income. It also reveals that abolition of support payments will result in an absolute decrease on sunflower production (Erdal and Erdal, 2008; Koc, 2005).

Today, agriculture is a sector that should be supported because of its nature especially in developing countries. Turkey is one of the countries that have special emphasis on support of agricultural production. Agricultural support systems which are carried out in Turkey are in accordance with international rules. However, over time, due to the insufficiency of agricultural support policies, innovations of systems and fundamental changes have become inevitable. In the light of developments in recent years, a new agricultural production support system "Turkey Agricultural Basin Production and Support Model" was adopted since 2011 in Turkey. The basic expectations from the new system implemented are to provide an increase in producer income and availability to meet shortage of products supply (oil seeds etc.) with domestic production by supporting them in higher levels as well as an increase in agricultural production and productivity.

Agricultural subsidies for producers in Turkey comprise approximately 6% of total agricultural production value and 0.82% of Gross Domestic Product. In this case, as of now, the usage of support tools and field based supports and also the compliance with the rule of "de minimis" constitute a hindrance for the country. However, both the study findings and the evidences of previous researches demonstrate that premium supports and supports unit prices of diesel fuel and fertilizer are insufficient to increase sunflower production and they thereby reveal that unit support prices should definitely be increased.

This study carried out proves that the premium support applied in Turkey has no effect on increasing the cultivation areas of sunflower. It only shows that the supports of diesel fuel and fertilizers have significant increasing effects on producer income and decreasing costs for sunflower with premium supports. In consequence of the study, dissemination of production in watery conditions, encouragement usage of varieties with high oil content, usage of appropriate fertilizer according to results of soil tests, and the seed use should be proposed to be incorporated to agricultural subsidy system for sunflower production like in other products.

REFERENCES

Acar M, Bulut E (2009). Recent developments in agricultural support policies in Turkey and the world. S.U. IIBF. J. Soccal Econ. Res. 9 (17)1-19

Aksoy S, Gaytancioglu O (1996). Research on the effects of increases in agricultural input prices to product prices. I. Agricultural Economics Congress of Turkey, İzmir. p. 2.

Anderson K, Martin W, Valenzuela E (2006). The relative importance of global agricultural subsidies and market access. World Trade Rev. 5(3):357-376.

Anonymous (2009). Agriculture in Turkey with ecomonomical indicators. MARA. Agricultural Economics Research Institute. Project Ankara. 176:41-102.

Anonymous (2011a). (http:///www.fao.org/FAOSTAT-Agriculture /production/crops)

Anonymous (2011b). Available at: http:///www.tuik.gov.tr/agriculture/database

Erdal G, Erdal H (2008). The Effects of premium payment in Turkey agricultural supports system. Gaziosmanpasa Univ. J. Agric. Fac. 25(1):41-51.

Erkan O, Cicek A (1996). Research and sampling methods in agricultural economics. GOP University. Agricultural Faculty Publishing. 12(6). Tokat. Turkey. p.72.

Gaytancioglu O (1999). Economic anaysis of the agricultural politics applied on sunflower and the yield loss owing to warehousing. Trakya University. Tekirdag Agriclyural. Faculty, Tekirdag.

Green SB, Salkind NJ, Akey TM (2000). Using SPSS for windows, analyzing and understanding data. Second Edition. Prentice-Hall Inc. Upper Saddle River. New Jersey.USA. pp. 243-252.

Gunaydin G (2006). Agriculture sector in Turkey. J. Agric. Eng. 76-77:12-27

Ikdogan U (2008). Development of oilseed trade in the world and EU an evaluation in terms of Turkey. EU Expertise Thesis. Ministry of Agriculture and Rural Affairs of Turkey. Foreign Affairs and EU Coordination Department.Ankara. p. 128

Hennessy AD (1998).The Production effects of agricultural income support policies under uncertainity. Am.J.Agric. Econ.80:46-57.

Henningsen A, Kumbhakar S, Lien G (2009). Econometric analysis of the effects of subsidies on farm production in case of endogenous input quantities. AAEA & ACCI Joint Annual Meeting. Milwaukee. Wisconsin. USA.

Keeny R (2009). Transfer efficiency and distributional impacts of U.S. farm support: evidence from a macro–micro simulation. Am. J. Agric. Econ. 91(5):1289–1295.

Koc A (2005). Oilseed sector (seed, pulp and oil) status: Abatement of customs tariff rates and effects of premium supports. Vegetable Oils and Fats Industrialists Association, Hazar Advertising Printing, Publication Ankara 7:108.

MARA (2011a). 2010 Current data of Ministry of Agriculture and Rural Affairs of Turkey. General Directorate for Agricultural Production an Development. Ankara. p.29.

MARA (2011b). 2010 Annual Report of Tarsim Agricultural Insurance Pool. Ministry of Agriculture and Rural Affairs and Prime Ministry Undersecretariat of Treasury and Foreign Trade. Ankara. p. 7.

Mayrand K, Dionne S, Paquin M, Le Bell IP (2003). The Economic and environmental impacts of agricultural subsidies: An assessment of the 2002 US Farm Bill & Doha Round. Unisféra International Centre. Canada.

Ozcelik A, Fidan H (2003). Importance of sunflower regarding to economy of Turkey. Turkey First Oil Seeds, Vegetable Oils and Technologies Symposium. pp.92-102.

Ozcelik AO, Ozer O (2007). The Evaluation of supporting payments to farmers for fuel and chemical fertilizer in Turkey. J. Agric. Sci. 13 (1)1-8

Sahin I, Semerci A, Kaya Y, Citak N (2010). Determining efficiency of subsidizing policies and productivity on sunflower production. MARA. Tagem (08-R&D-6). p. 354.

Sahinoz A, Cagatay S, Teoman O (2007). Discussion of the Applicability of Premium Support System as Agricultural Support in Turkey and Economic Analysis of the System. MARA. Agricultural Economics Research Institute. Publ. Ankara. 155:75.

Semerci A, Kaya Y (2010). The Component of production cost in sunflower and its relationships with input prices. Int. Rev. Appl. Econ. 5(1-2):139-146.

Střeleček F, Zdeněk R, Lososová J (2009). Comparison of agricultural subsidies in the Czech Republic and in the selected states of the European Union. Agric. Econ–Czech 55(11):519-533

TURKSTAT (2010). Statistical indicators 1923-2009. Turk. Stat Inst. Publ. Ankara 3493:170.

Yamane T (1967). Elementary sampling theory. Prentice-Hall Inc.

Englowood Cliffs. N. Jersey.USA. (Please provide page number).
Zemheri O. (2010). Recent Developments in Agricultural Support Policies. The J. Minis. Agric. Rural Affairs. (Turk Tarım) 2010:195. Ankara. p. 18.

The potential and limitations of farmer participatory research in organic agriculture: A review

Carlo Ponzio, Ramesh Gangatharan and Davide Neri

Dipartimento di Scienze Agrarie, Alimentari ed Ambientali. Università Politecnica delle Marche, Via Brecce Bianche, 60131 Ancona, Italy.

Farmer participatory research (FPR) emerged in response to limitations of top-down Research and Development (R&D) approaches. The main purpose of this paper is to inspire and guide aspiring and new practitioners of participatory research and development (PR&D) to learn, reflect and constantly refine the way they work. The primary target users are field-based researchers involved in activities dealing with the interrelated issues of natural resource management, agriculture and rural livelihoods. They may have technical or social science backgrounds but share a common interest in drawing on the PR&D knowledge base. The paper is intended to enhance access to systematized information on field-tested PR&D concepts and practices among field practitioners and their organizations. This paper offers a conceptual model for participatory research projects that aim to improve the sustainability of organic agriculture.

Key words: Farmer participatory research, appropriate technology, agricultural research, on farm research, sustainability.

INTRODUCTION

Origin and emergence of farmer participatory research

Farmer participatory research emerged as a response to the generation of inappropriate technologies by scientists at research stations whose work was based on the transfer-of-technology model. Those working in this field began to develop a series of new research approaches that would result in technologies that would be beneficial to, and therefore adopted by, small farmers. The transfer-of-technology model was predominant in the 1950s and 1960s. The fact that small farmers did not adopt the technology packages developed at research stations led researchers to conclude that farmers were backward or ignorant, and that the key to success lay in creating a better extension service. Thus, the training and visit system (T&V) of Agricultural Extension was widely implemented. In the 1970s and early 1980s, non-adoption, still a problem, was attributed to constraints occurring at the farm level.

Farming systems research arose as a response, emphasizing research at the farm level to diminish constraints to the adoption of new technologies. It has received increased attention and recognition since the "Farmer First" (Chambers et al., 1989). Participatory technology development (Jiggins and De Zeeuw, 1992) concepts were introduced in the late 1980s. Finally, in the 1990s, some researchers came to believe that the problem was not the farmers, but the inappropriate technologies they were being encouraged to adopt. This marked the emergence and gradual evolution of farmer participatory research, an approach aimed at creating appropriate technology for small farmers (Chambers et al., 1989; Cramb, 2000) and for organic and biodynamic agriculture (Ponzio et al., 2013).

The concepts of "farmer participatory research" started to spread by the acknowledgement that the active participation of the final users of the research products to the research activity itself would have enhanced the efficiency of the identified technological solutions and facilitate their adoption, also enabling the farmers to become active proposers of key research objectives from their side, in line with their actual emerging needs (Lockeretz, 1987; Bachinger et al., 2000).

One key element of participation is an emphasis on developing the capacity of local people as an end in itself, as opposed to the purely mechanistic emphasis of participation as a means within the technology development flow that has often characterized research and extension programs. During the late 1980s and early 1990s, increasingly more field-based experiences emerged creating more space for methodological and institutional innovations for agricultural research and extension. Within these participatory approaches - as they became commonly known - a special emphasis was placed upon participation of local people and their communities, especially working with and through groups; and building upon the traditional or indigenous knowledge that they held (Chambers et al., 1989; Waters-Bayer, 1989; Haverkort et al., 1991).

Later on, with the onset of the innovative approaches of agro-ecology and rural sociology, a new "client-driven" concept was developed that requires decentralized technology development and devolves to farmers a high responsibility for adaptive testing (Ashby and Sperling, 1995). Nevertheless, consumers as well play a role in indirectly orientating agricultural research, as stated by (Guttman, 1978), being the primary beneficiaries of the research's products. Consumers demand more and healthier food, and they become aware of the negative externalities due to intensive agricultural and agro-processing practices. Evidence of this high attention by consumers to the way the food is actually produced is given by the fact that representatives of consumers associations actively participate to the works of the certification committees of the organic agriculture certification bodies, in order to closely monitor the conformity to the standards "from the field to the fork".

The challenge for development workers, researchers, and farmers is to design and use research methodologies that ensure the development and adoption of improved agricultural technologies to create sustainable agricultural production that will benefit the resource-poor farmers. Understanding the sustainability is the challenging question, with a continuous research for sustainable development which focuses on preserving or increasing the capacity of the systems being managed to produce desired benefits also in the future. Actor-oriented, integrative and participatory approaches (Cramb, 2000) are increasingly seen as a way to address the multiple and often conflicting social, environmental, and economic sustainability goals of different interest groups.

Education, research and extension in agriculture remain the vehicles to achieve sustainability in any modern food systems. The preparedness of future agriculturists requires a more holistic approach and this can be achieved through education reform in the college of agriculture through a promotion of innovative thinking and leadership (Borsari, 2012).

Education also serves to accentuate the need for policy change in the current management of our extremely centralized food systems and to foster transparency concerning information and regulation for anything concerned with food production and its distribution to markets and consumers (Borsari, 2011). Preservation of agrobiodiversity remains fundamental to any form of agriculture, while the mandate of lessening the dependence on non-renewable fossil-fuels to maintain production needs to shift toward more renewable energy sources (Borsari, 2012).

PARTICIPATORY APPROACHES TO AGRICULTURAL RESEARCH

Today, research in agriculture typically takes place under three modalities that in turn affect the choice of the physical site hosting the investigation activity (Figure 1): Basic research, applied research and on-farm research, this being principally executed within real operational farms (commercial farms). In the latter case, the researcher establishes a direct relationship with the farmer, which may occur at various levels of intensity: from a mere formal collaboration (the farmer limits himself to implement the experimental protocol) to a shared knowledge process to be developed through a full partnership: in this way, the farmer becomes totally involved in the phases of experimental design, assessment and discussion of the results.

Sutherland (1999) described four modes of participation which link farmers with researchers: (a) contractual, where the researcher is all-powerful; (b) consultative, where most of the key decisions are with the researcher but a certain emphasis is put on farmer consultation in problem identification and priority setting; (c) collaborative, when farmer and researcher exchange knowledge and share decision-making; (d) collegiate, where the farmer has greatest power, with the researcher responding to farmer specific requests of investigation.

About the quality of the researcher/farmer relationships, Okali et al. (1994) argued that "farmer participatory research, in principle, aims to operate at the interface between knowledge systems: it can be described as a people-centered process of purposeful and creative interplay between local individuals and communities on the one hand, with formal agricultural and research knowledge on the other - the collegiate interface".

According to the authors, the knowledge systems at

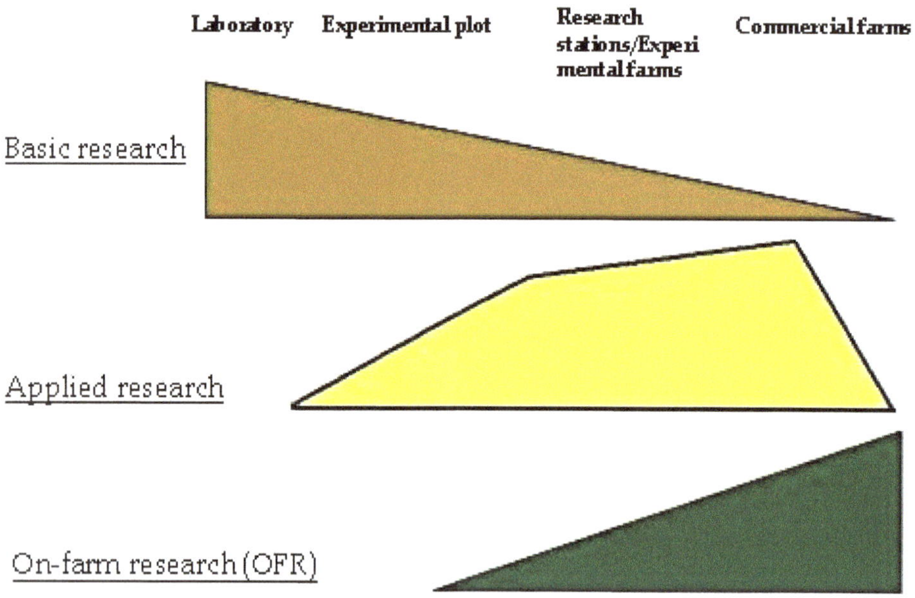

Figure 1. Agricultural research by site of operations.

stake are basically represented by the farmers' (local) knowledge on one hand and formal research (generic) knowledge on the other. Such knowledge systems may be therefore interfaced through 'creative interplay', so involving dialogue between different groups (local farmers and researchers): this dialogue will be respectful and serves to draw the two parties together in partnership ('collegiate interface'). While most of these assumptions may hold in optimal situations, however projects often face difficulties in linking ideas and actors in order to exemplify good practice (Sutherland, 1999).

Eksvard (2009) described the functions of a Participatory Learning and Action Research (PLAR) group in Sweden, composed by horticultural farmers, researchers and extensionists which was set up to test on-farm and evaluate the output of conventional research trials targeted at studying the best organic manuring options in organic horticulture. After pointing out the difficulties encountered inside the PLAR in harmonizing opinions and action plans among the farmers as well as between the groups of researchers and farmers, the author concludes that moving from conventional research approaches to trans-disciplinary approaches is not easy and strongly demands a common effort to relate the contextual knowledge of farmers to the abstract knowledge of scientists.

Sukkel et al. (2006) indicated in the Dutch Organic Farmers Network for Research, Development and Innovation (BIOM) a valid initiative for the improvement of the environmental and economic performance of the participating farmers. A wide range of practical experiments, resulting from specific bottlenecks highlighted by the network farmers, were carried out in 40 farms in cooperation with the agricultural scientists: the trials outputs mostly turned to increase yields, reduce labor input for hand-weeding and decrease nutrient surpluses. The authors strongly recommend the participatory approach to research through farmer's networks; however they state these tools also demand specific skills and attitudes from researchers, advisors and farmers.

Jones et al. (2006) reported about an on-farm research carried out to evaluate the performance of two soft wheat varieties in UK, by involving 14 organic farmers who grew the crops in their farms according to their standard cropping methodology. Measurements and laboratories analyses were carried out by the researchers but several field assessments were requested to the farmers (e.g. early and late crop groundcover, number and size of ears, straw length, etc.) who, however, showed a certain reticence in doing it. Farmers actually put forward the need for greater researcher-led assistance, which raised the issue whether the farmers well understood the concept of participatory approach and/or felt a poor ownership of the research project itself.

Yet the authors stated that new and valuable information was produced and both researchers and farmers considered useful the information on winter wheat variety performance under a range of organic systems; furthermore, farmers recognized the difficulty to reconcile the appearance of varieties in the field with their actual performance. However, authors recommended: (i) to spend more time in introducing the project and its objectives to the farmers; (ii) to discuss and develop the trial design in much closer link with them; (iii) to ascertain the full willingness and motivation of all the participants to

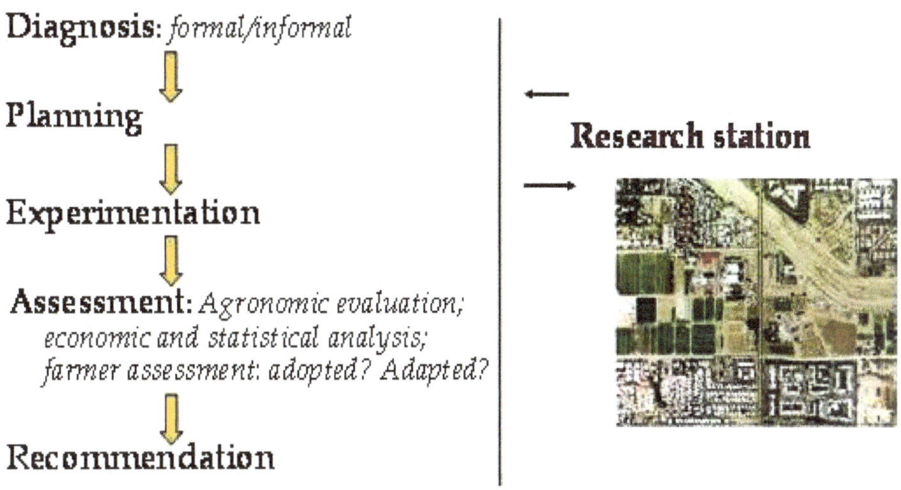

Figure 2. Typical stages for the execution of on-farm research: interactions with basic and applied research conducted inside laboratories and experimental stations are likely to occur (Tripp, 1991; Bachinger, 2000).

cooperate over the entire project (farmers and researchers).

The latter aspect was studied in depth by (Barreteau et al., 2010) who devised a conceptual analytic procedural framework to make participants' roles explicit in the implementation of different participatory research processes, thus preventing possible disappointment, reticence and project abandonment. The framework embraces three aspects: (i) the flow of information's among participants and the control over these flows for each step in the process; (ii) the timing and involvement of participants in the different steps of the research process; (iii) the modalities of communication among participants for each information flow (that is, bilaterally or as a group, mediated or face to face). The authors elaborated the framework from various experiences with participatory research; the framework is meant to be used from the very beginning of a participatory research process as a conceptual guide for researchers.

Basic, applied and on-farm research are interdependent and mixed approaches are very likely to occur; e.g. results from randomized block designs in other regions can be tested under local farming conditions (Figure 2) (Tripp, 1991; Bachinger et al., 2000).

According to the IAASTD report (McIntyre et al., 2009), participatory, collaborative methods and approaches have added value to the encounter between traditional/local knowledge actors and formal agricultural knowledge, science and technology (AKST) actors. Farmer-researcher groups in the Andes for instance brought together members of CIP (an international research institute) for the development and testing of measures and varieties to control late blight in potatoes, not only increasing productivity but also addressing issues for instance of inter-generational equity and the sustainability of soil management. Collaboration among knowledge actors in the commercialization and domestication of tree [and other] wild and semi wild species in participatory plant breeding (PPB) and in value-added processing are creating new value chains selling into both niche and mass markets. Other examples include efforts made in a number of countries to invite traditional/local knowledge actors into rural schools (e.g., Thailand) and universities (e.g., Peru, Costa Rica) as teachers and field trainers; to incorporate local AKST in the curricula and experiments run by village-based adult education and vocational training centers (e.g., India); and to expand opportunities for experiment-based, farmer-centered learning. Modern improved access to information and communication technologies (ICTs) shows large potential for extending and augmenting these developments.

Drinkwater (2002) put the emphasis on the relevance of studying "intact systems", in order to understand how a complex agroecosystem works as a whole in opposition to the typical factorial experiment approach, which aims at breaking down a complex system in order to isolate and study specific components and identify cause-effect relationships. The holistic participatory approach of on-farm research is fundamental because biodiversity performs key ecological services and if correctly assembled in time and space can lead to agroecosystems capable of sponsoring their own soil fertility, crop protection and productivity. There is consensus that at least some minimum number of species is essential for ecosystem functioning under constant conditions and that a larger number of species is probably essential for maintaining the stability of ecosystem processes in changing environments.

Research stages	*Stakeholders involved*
1. Livelihoods analysis and environmental assessment	Local farmers and soil scientists
2. Participatory nutrient budget, Semi-structured interviews and field visits	Representative farmers and key informants
3. Soil fertility feedback, discussions and Farm drives	Representative farmers, key informants and agricultural extension staff
4. Community meetings and Policy-Makers workshops	Community members, agricultural extension staff and policy makers
5. Participatory monitoring and evaluation of different soil nutrient management strategies	Agricultural extension staff with local communities

Figure 3. Stages of the methodological research framework followed by Dougill et al. (2002) to conduct interdisciplinary participatory research on land degradation in South Africa (redrawn from Dougill et al. (2002).

THE ROLE OF FARMERS AND RESEARCHERS IN ON-FARM RESEARCH AND PLANT BREEDING

Two different experimental approaches are discussed, both yielding meaningful results: i) field station trials, where simulated cropping systems are run in replicated plots, and ii) studies on whole agroecosystems in commercial farms. According to the author, an integrated research approach combining systems experiments with appropriately designed factorial experiments is highly recommended for a deep understanding of ecological processes in agricultural systems (Figure 2).

When approached from within, the investigated agroecosystem results more realistic in terms of scale of observation interconnected farming practices and management constraints to which the farmer is subjected: on-farm research therefore is likely to offer the opportunity to study it in a more integrated manner (Drinkwater, 2002). Through on-farm experiments farmers are given new skills, and confidence in problem-solving is enhanced (Bachinger et al., 2000).

Researches carried out by Dougill et al. (2002) in South-African small-scale farms allowed to study in depth the nutrients flow through the local agro-ecosystem and analyze in an interdisciplinary way the environmental, economic, political and social factors are influencing nutrient management, which is often the main cause behind the severe soil degradation occurring in the region.

The authors followed an original research methodological pathway (depicted in Figure 3) that started from holistic discussions on rural livelihoods, then turning to an in-depth participatory assessment of the key constraints likely to affect the natural resource management. It is stressed the importance to give to the farmers involved in the research process a good feedback on the research findings, that have to be discussed widely within local communities together with extension workers and, possibly, policy-makers.

Hard red spring wheat varieties were compared at six locations in organic farms in Minnesota and Nord Dakota, USA, over a three-year period. A basic scoring system was developed by researchers and farmers together. The farmers eventually indicated certain traits of the varieties under evaluation (e.g. grain yield, protein content, diseases resistance) as much more valuable than others (e.g. straw and stubble production, impact on succeeding crop), thus playing an active role in defining the variety prototype to be selected.

Ceccarelli and Grando (2007), researchers of the International Center for Agricultural Research in the Dry Areas (ICARDA), described a successful participatory plant breeding system (PPB), currently utilized in several Asian and North African countries. According to PPB, genetic variability is generated by the breeders in the experimental stations whereas the selection process of the most suitable varieties is carried out by the farmers and field extensionists in the farms (Figure 4).

In this way, the newly identified varieties are released faster with respect to the conventional breeding system, and most importantly the results are better adapted to the farmer's needs and environment.

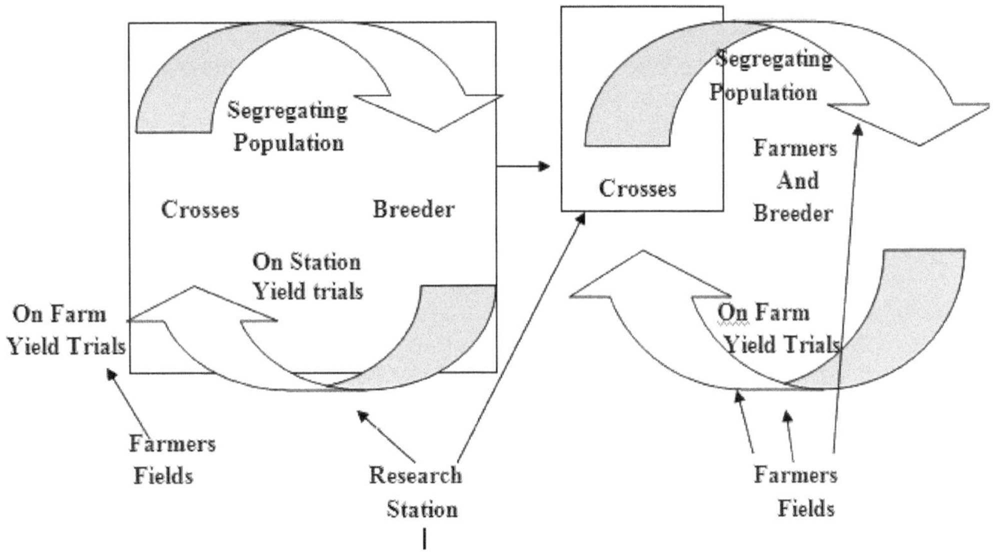

Figure 4. Conventional plant breeding is a cyclic process that takes place largely within one or more research stations (left) with the breeder making all decisions; decentralized participatory plant breeding is the same process, but takes place mostly in farmers' fields (right) and the decisions are taken jointly by farmers and breeders (redrawn from Ceccarelli and Grando, 2007).

The PPB advantages are particularly relevant in developing countries where multinational seed companies do not usually invest in big research programs because of low profitability and where many imported "improved" varieties are not in fact suitable for the local marginal cropping conditions. The authors conclude that on-farm breeding research, besides offering economic benefits, is characterized by psychological, moral and ethical added value stemming from the progressive empowerment of the agricultural communities, partners in the research activities.

A similar decentralized approach to assess new crop varieties was described by (Dorward et al., 2007) through the method of participatory varietal selection (PVS), conducted throughout Ghana in two agro-ecological zones, the Savannah and the Forest, by involving more than 2,000 small-scale farmers who evaluated, in several steps, in their own fields the performance of around 100 upland rice varieties, identified by the rice breeders. Once the farmers identified the most suitable varieties according to their own selection goals, the seeds of them were distributed to a small number of farmers, and the authors found that after a couple of years about 850 farmers in communities had already obtained the seeds from other farmers via informal mechanisms (gift, exchange or purchase), which gives the evidence of the good acceptance of the new seeds.

A semi-decentralized participatory approach was used by (Baidu-Forson, 1997) for the identification of the best farmers appreciated varieties of pearl millet (*Pennisetum glaucum L. Br.*). Fourteen varieties were comparatively tested in a research station and thirty farmers, from six villages across a north-south transect of western Niger, were selected to evaluate the varieties over two phases, by the support of structured questionnaires: the first, when the plants were at the reproductive stage (farmers visited the station and checked for the specific traits of millet plants and grain that were deemed by them more significant); the second, when post-harvest processing and food-quality traits of the grain were assessed at home, by the female sample farmers.

Unexpectedly for the author, the majority of farmers did not go for the highest productive varieties, rather they preferred the one characterized by early crop cycle; higher tillering capacity; large grain size and plant height > 2.5 m, all characteristics that offer higher probability of yield stability in the harsh environment of Sahel, thus indicating that farmer-led research objectives may somewhat differ from those of crop improvement programs devised by the scientists.

The participatory farmer approach to research is likely to present however negative facets, that have to be carefully assessed and addressed prior to begin the experiment. One aspect is about the difficulties to implement complex experimental designs and treatments due to the limited availability of time of the farmer, his/her lack of specific technical preparation and the structural inadequacy of the farm land: as a consequence, the potential of the analysis is drastically reduced (Selener, 2005).

The flexibility and the simplicity which are important traits of successful participatory research often lead to poor scientific validity of research results (Poudel et al., 2000; Wivstad and Natterlund, 2008).

Riley and Alexander (1997) reviewed the statistical

methods utilized in sixty participatory on-farm research papers and emphasized the complexity of analyzing quantitative and qualitative sets of data often coming from heterogeneous disciplines, which demands a very sophisticated statistical approach. Nevertheless, from the review it emerges that statistical methodology was often poorly defined and inadequately used.

Typically, the more farmer participation that was involved, the more complex the underlying design structure however followed by poor statistical analysis. Confounding of effects and inadequate sampling were encountered frequently due to lack of clear design structure. The authors conclude that the kind of statistical methodology suitable for use in participatory on-farm trials is in fact available, and is capable to add high value to the quality of modern, unstructured multidisciplinary design and to the summary of collected data, whether they are quantitative or qualitative. However, such powerful methodology is not documented in a form easily used by non-statisticians, nor it is easily accessible.

The participatory process is indeed very time-consuming; it demands a lot of commitment and hard work both from the researchers and farmer cooperators. In addition, participatory on-farm research is cost sharing, which means that farmer cooperators are usually expected to do their research at their own expense, and this results hard for farmers above all in the context of developing countries, especially when they are subjected to the risk of negative financial return from their farming enterprise (Poudel et al., 2000).

Another hindrance is represented by a possible conflict of interests which can arise, if not properly prevented, between the scientist - more oriented to identify technical and innovative solution of general value, suitable for more farming environments and "communicable" to the international scientific community - and the farmer, much more interested in specific, locally-adapted, solutions for her/his farm (Sutherland, 1999; Lockeretz and Stopes, 2000).

In a sociological study (Eshuis and Stuiver, 2005), analyzing the process of "learning in context" in a sustainable dairy-farming project that involved the participation of farmers and scientists, emphasized how difficult was the interaction between the two groups because of the differences between the heterogeneous forms of farmers' knowledge and scientific knowledge. However, such differences were progressively reduced during recurrent phases of alternating conflict and alignment over the validity of knowledge, highlighting the relevance of such phases for progress in learning and generation of innovation.

As underlined by (Sutherland, 1999) effective on-farm research - blending formal and farmer-led approaches - principally requires a cross section of expertise. The end result is likely to be a compromise of methods and approaches to fulfill the expectations of all stakeholders: Over time, there will be possibly iteration from formal to informal and back again. As the level of understanding improves, there may be scope for the further development of methods to improve research efficiency.

CONCLUSION

Experimentation by farmers cannot entirely replace conventional scientific research and conventional scientific research cannot replace farmers' on-farm research. There is a need for a new educational and experimental approach that favors a "symbiotic relationship" between the two through a decision-making process in research, defining innovative farmers' and scientists' roles, and the style of the research to be conducted in the farms. The result is the incorporation of the most important and valuable aspects of each into a new system which will both benefit the small resource-poor farmer and contribute to the scientific knowledge base. The final objective is to improve the quality of life of indigenous communities at the agriculture through participatory innovation based on local resources in sustainable and diversified agroecosystems.

ACKNOWLEDGEMENTS

The Italian Ministry of Agriculture (MIPAF) is acknowledged for sustaining the project "RADICI" in organic agriculture.

Abbreviations: FPR, Farmer participatory research; **R&D**, research and development; **PR&D**, participatory research and development; T&V, training and visit system; **PLAR**, participatory learning and action research; **PVS**, participatory varietal selection; **IAASTD**, international assessment of agricultural knowledge, science and technology for development; **AKST**, agricultural knowledge, science and technology.

REFERENCES

Ashby JA, Sperling L (1995). Institutionalizing Participatory, Client-Driven Research and Technology Development in Agriculture. Dev. Change 26:753-770.

Bachinger J, Stein-Bachinger K, Vögel R, Werner A (2000). Guide for on-farm experiments by organic farmers. Research Methodologies in Organic farming: On-Farm Participatory Research, Bari, Italy, 5-7 September 1999. In: Krell, R. and Zanoli, R. (Eds.) Research Methodologies in Organic farming: On-Farm Participatory Research: Proceedings, Food and Agriculture Organization of the United Nations, pp. 47-48.

Baidu-Forson J (1997). On-station farmer participatory varietal evaluation: a strategy for client-oriented breeding. Exp. Agric. 33:43-50.

Barreteau O, Bots PWG, Daniell KA (2010). A framework for clarifying "participation" in participatory research to prevent its rejection for the wrong reasons. Ecol. Soc. 15(2):1.

Borsari B (2011). Agroecology as a needed paradigm shift to insure food security in a global market economy. Int. J. Agric. Res. Gov. Ecol. 9(1/2):1-14.

Borsari B (2012). Curriculum framework for sustainability education. Acad. Exch. Quart. 16(1):74-78.

Ceccarelli S, Grando S (2007). Decentralized-participatory plant breeding: an example of demand driven research. Euphytica 155(3):349-360.

Chambers R, Pacey A, Thrupp LA (1989). Farmer First: Farmer Innovation and Agricultural Research. London: Intermediate Technology Publications.

Cramb R (2000). Processes influencing the successful adoption of new technologies by smallholders. In W. W. Stür, P. M. Horne, J. B. Hacker and P. C. Kerridge (eds.), Working with Farmers: The Key to Adoption of Forage Technologies. ACIAR Proc. 95:11-22.

Dorward P, Craufurd P, Marfo K, Dogbe W, Bam R (2007). Improving participatory varietal selection processes: participatory varietal selection and the role of informal seed diffusion mechanisms for upland rice in Ghana. Euphytica. 155:315-327.

Dougill AJ, Twyman C, Thomas DSG, Sporton D (2002). Soil degradation assessment in mixed farming systems of southern Africa: Use of nutrient balance studies for participatory degradation monitoring. Geogr. J. 168:195-210.

Drinkwater LE (2002). Cropping systems research: Reconsidering agricultural experimental approaches. Hortechnology, July-September. 12(3).

Eksvard K (2009). Is conventional agricultural research fit for the purpose of supporting ecological agriculture? A case study of an attempted transition in Sweden. Renew. Agric. Food Syst. 25(1):55-68.

Eshuis J, Stuiver M (2005). Learning in context through conflict and alignment: Farmers and scientists in search of sustainable agriculture. Agric. Hum. Values 22:137-148.

Guttman JM (1978). Interest groups and the demand for agricultural research. J. Polit. Econ. 86(3):467-484.

Haverkort B, Kamp J. van der, Waters-Bayer A (1991). Joining Farmer's Experiments: Experiences in Participatory Technology Development. London, Intermediate Technology Publications. P. 269.

Jiggins J, De Zeeuw H (1992). "Participatory technology development in practice: Process and methods." In C. Reijntjes B. Haverkort and A. Waters-Bayer (eds.), Farming for the Future: An Introduction to Low-externalinput and Sustainable Agriculture London: MacMillan. pp. 1350-1362.

Jones H, Hinchsliffe K, Clarke SM, Pearce B, Gibbon D, Lyon F, Harris F, Thomas J, Wolfe M S (2006). A participatory methodology for large scale field trials in the UK. Joint Organic Congress, Odense, Denmark, May 30-31.

Lockeretz W (1987). Establishing the proper role for on-farm research. Am. J. Altern. Agric. 2:132-136.

Lockeretz W, Stopes C (2000). Issues in on-farm research. In Research Methodologies in Organic Farming: On-farm Participatory Research. R. Zanoli and R. Krell Eds., REU Technical Series 63. FAO, Rome.

McIntyre BD, Herren HR, Wakhungu J, Watson RT (2009). Agriculture at a crossroads. A Synthesis of the Global and Sub-Global International Assessment of Agricultural Knowledge, Science and Technology for Development (IAASTD) Reports. Island Press, Washington, DC. pp. 71-74.

Okali C, Sumberg J, Farrington J (1994). Farmer participatory research: Rhetoric and reality. London: IT Publications.

Ponzio C, Gangatharan R, Neri D (2013). Organic and Biodynamic Agriculture: A Review in Relation to Sustainability. Int. J. Plant Soil Sci. 2(1):95-110.

Poudel DD, Midmore DJ, West LT (2000). Farmer participatory research to minimize soil erosion on steepland vegetable systems in the Philippines. Agric. Ecosyst. Environ. 79:113-127.

Riley J, Alexander CJ (1997). Statistical literature for participatory on-farm research. Exp. Agric. 33:73-82.

Selener D (2005). Participatory Research and Development for Sustainable Agriculture and Natural Resource Management, A Sourcebook. Vol. 1: Understanding Participatory Research and Development. CIP-UPWARD and IDRC Publ.

Sukkel W, van Leeuwen Haagsma W, van der Wel C (2006). BIOM, Organic farmers Network for Research, Development and Innovation. Joint Organic Congress, Odense, Denmark, May 30-31, 2006.

Sutherland A (1999). Linkages between farmer-oriented and formal research and development approaches. Agricultural Research and Extension Network. Network Paper No. 92. January 1999. UK Dept. for International Development (DFID). P. 24.

Tripp R (1991). Planned Change in Farming Systems: Progress in on-farm research. John Wiley & Sons publications.

Waters-Bayer A (1989). Participatory Technology Development in Ecologically-Oriented Agriculture: Some Approaches and Tools. Agricultural Administration (Research and Extension) Network Paper No. 7:63.

Wivstad M, Nätterlund H (2008). Learning in context – improved nutrient management in arable cropping systems through participatory research. Paper presented to16th IFOAM Organic World Congress, Modena, Italy, June 16-20.

Marketing performance and efficiency of evaporative-preservation cooling system for fresh tomato marketing in Ondo State, Nigeria

Taye S. Mogaji[1]*, Awolala D. Olufemi[2] and Olorunisola P. Fapetu[1]

[1]Department of Mechanical Engineering, Federal University of Technology Akure, Ondo State, Nigeria.
[2]Department of Agricultural and Resource Economics, Federal University of Technology Akure, Ondo State, Nigeria.

This paper presents the marketing performance and efficiency of evaporative-preservation cooling system for fresh tomato as an update of the performance evaluation of the evaporative cooling system developed at the Federal University of Technology Akure, Nigeria. A total of one hundred and five (105) fresh tomato marketers were sampled. Marketing margin analysis was performed to determine the performance of the preserved tomato vegetable during dry season based on their mode of preservation. Shepherd's efficiency formula result analysis shows that fresh tomato market was efficient with average marketing efficiency of 17.65%. The two-independent samples t-test results (t-value = 74.014) further reveal that profit margin for users of evaporative cooling system is significantly different from users of traditional mode of preservation, at 5% level of significance. However, disaggregated wholesaler's profit margin of ₦6,261.6 per month for users of traditional preservation method implies that they are not efficient compared with users of evaporative cooling system averaging ₦12,381.4 per month. Given the wide tropical variability in temperatures and relative humidity, evaporative cooling system is of economic importance towards commodity marketing development in Ondo State, Nigeria.

Key words: Evaporative cooling system, tomato, marketing performance, marketing efficiency, marketing margin, wholesale market, Ondo State, Nigeria.

INTRODUCTION

Evaporative cooling system has prospective use for the short term preservation of vegetables soon after harvest with higher market value given a long term market supply. Most of the post-harvest losses incurred on fruits and vegetables in developing countries are due to lack of adequate storage facilities and perishable nature (Mitra and Baldwin, 1997). Hence, quality depends on post-harvest handling, transportation and storage (Haidar and Demisse, 1999). Kader (1992) estimated the extent of post-harvest losses in fresh fruit and vegetables at 5 to 25% in developed countries and 20 to 50% in the developing countries. Low temperature handling and storage have been described as the most important physical method for post-harvest loss control which can be achieved using less expensive method of evaporative cooling (Seyoum and Woldetsdik, 2004). This cooling system has been reported for achieving a favourable environment in the storage structure for fruit and vegetables (Helsen and Willmot, 1991; Umbarker et al., 1991). One of the cooling methods employed for extending shelf life and minimizing post-harvest losses of perishable produce is mechanical refrigeration, controlled atmospheres, and adiabatic cooling which is considered less expensive for small scale peasant farmers, retailers and wholesalers (Mogaji and Fapetu, 2011). The study was carried out in Ondo State, Nigeria with an

*Corresponding author. E-mail: mogaji_taye@yahoo.com.

approximated land area of 14,788.72 km^2 and living about 3.44 million citizens. It has a temperature range of 21 to 29°C and relatively high humidity throughout the year (Ondo State Government, 2011). The State is largely agrarian with existence of many rural and urban market centers designated for food commodity market, namely roots and tubers, fruits, and vegetables in commercial quantities. Ondo State is administratively sub-divided into Northern, Central and Southern Senatorial districts. Each Senatorial district has few major specialized wholesale produce markets available to it. Wholesale markets are usually located at the districts main urban centre, and are the main collection points for surplus produce from the area. Wholesale markets have better storage, transportation, communication for buyers and sellers.

Market efficiency and performance

In reality, given that market will not self-regulated in terms of efficient use of resources and that no excess profits will get to any operator under perfect competitive condition, economists face difficulty in defining the criteria for evaluating economic efficiency of markets. Agricultural economists have applied two approaches for analyzing market performance. The first approach analyzes the productive efficiency of well-defined marketing sub-systems while the second approach focused on analyzing the market structure and the resulting performance. The methods and criteria used by both approaches are instrumental in identifying actual and potential inefficiencies in the marketing system and characteristically, lead to the development of marketing policy recommendations to government (Kilmer and Armbruster, 1999).

Marketing performance

Performance is a measure of pricing and operational efficiency. Individual producers as well as the public have a stake because the degree of efficiency attained affects producer's prices and profit, costs to the consumer and their real income, and the general resource utilization. Improved marketing efficiency is a common goal of farmers, marketing organizations, consumers and society (Kohls and Uhl, 1985). Higher efficiency means better performance, while lower efficiency denotes poor performance. Markets are efficient when the ratio of the value of output to the value of input throughout the marketing system is maximized (Abbot and Makeham, 1981). Performance of the market is reflection of the impact of structure and conduct on product price, costs and the volume and quality of output (Cramers and Jensen, 1982). If market structure resembles monopoly rather than pure competition, one can expect poor market performance.

Marketing margin

Marketing margin is the most commonly used measure of market performance. It describes price differences between other points in the marketing chain, for example between producer and wholesale, wholesale and retail, prices. It measures the share of the final selling price that is captured by a particular agent in the marketing chain (Mendoza, 1995). Marketing margins consist of marketing functions such as grading, packing, loading, unloading, transportation, storage, and levies. They are the same as returns to all factors of production, that is, land, labour, capital and entrepreneurship involved in marketing (Adeyokumnu, 1973). The data on marketing costs are needed to disaggregate the gross marketing margin of an enterprise at different marketing stages. This provides information on the costs of particular marketing functions, which can be compared with costs incurred by other enterprises to assess the operational efficiency (Scarborough and Kydd, 1992).

Large gross margins may not necessarily express high profit, but increased qualities and quantities of service, low labour, capital and management productivity, leaving producers and consumers better off while small gross margins may co-exist with inefficient resource use, poor coordination and consumer satisfaction, and disproportionate profit elements due to low productivity (Trotter, 1992). This study therefore focused on the marketing performance and efficiency of evaporative-preservation cooling system for fresh tomato relative to the traditional mode of preservation.

METHODOLOGY

This study analyzed the marketing margin and performance of tomato wholesalers for users of evaporative cooling system and the traditional method of preserving fresh tomato vegetables. Figure 1a, b and c show the developed evaporative cooling system and the corresponding control samples. The work of Mogaji and Fapetu (2011) is suggested for further details about the developed system and methods.

A cross-sectional field survey through focus group discussions (FGD) and personal interviews were used to obtain relevant information from fresh tomato marketers in Ogbese wholesale market (Northern district), Iloro wholesale market (Central district), and Owena wholesale market (Southern district) which were purposively selected for the study.

The bulk of sales in these markets dominate food commodity marketing activities taking place in Ondo State. Data were collected in the dry season of November, 2010 to March, 2011, an appropriate period to capture the effect of marketing margin efficiency when supply of fresh tomato vegetable are usually associated with high temperatures, perishability rate and cost of storage. In each wholesale market, 10 wholesalers and 15 sub-wholesalers were randomly sampled among those who used traditional preservation method, while 5 in each category of wholesalers and sub-wholesalers were purposively selected as users of the evaporative cooling. Seventy-five (75) respondents from users of traditional preservation method and thirty (30) marketers from users of evaporative cooling system were selected for the study. The population for the study therefore numbered 105 fresh tomato

Figure 1a. Evaporative cooling system with the fresh tomatoes fruit.

Figure 1b. Test products inside cabinet after 14 days.

Figure 1c. Test Products in Ambient Air after 14 days (Control Sample).

marketers. The wholesalers were interviewed at one weekly interval to allow for sub-wholesaling and retailing activities during the week.

The marketing performance of tomato marketers was determined to measure the economic benefit of using evaporative cooling system for fresh tomato preservation over the traditional method in the study. Marketing margin analysis was performed to determine the efficiency of the tomato vegetable marketing system during dry season. Given that marketing margin equals the difference between what the consumer pays and the farm gate price per unit of commodity item (Kohls and Uhl, 1985), and the assumption that wholesalers buy directly from the farmers, marketing margin are estimated as follows:

$$Marketing\ margin = (Selling\ Price - Buying\ cost)\ per\ unit \quad (1)$$

$$Net\ marketing\ margin = (marketing\ margin - marketing\ cost\ )\ per\ unit \quad (2)$$

Table 1. Socioeconomic characteristics of tomato vegetable marketers in the senatorial districts.

Characteristics	Dominant indicators at aggregate level					
	Northern Sen. district	**Mean value**	**Central Sen. district**	**Mean value**	**Southern Sen. district**	**Mean value**
Age of marketers	68.6% between 41 and 50 years	43 years	55.2% between 31 and 40 yrs	38 years	47.7% between 41 and 50 years	46 years
Level of education	42.8% completed basic education	6 years	78.1% completed basic education	6 years	50.0% completed basic education	6 years
Marketing experience	40.4% between 21-30 years	22 years	38.6% between 11-20 years	13 years	52.2% between 21-30 years	23 years
Start-up capital	52.0% less than ₦50,000	₦46,500	36.8% between ₦40,000 and ₦80,000	₦65,000	45.5% between ₦40,000 and ₦80,000	₦68,000

Source: Field data, 2010.

Hence,

$$Wholesaler\ net\ marketing\ margin = (Wholesaler\ margin - marketing\ cost) \quad (3)$$

It should be noted that marketing costs were mainly from transportation, handling/labour, rent, security fees, association membership and local government levy as reported in the study area.

Marketing efficiency is given by:

$$Marketing\ efficiency = \left(\frac{value\ added\ by\ marketing\ activities}{marketing\ cost}\right) \times 100 \quad (4)$$

Sherpherd formula technique was used to estimate wholesaler marketing efficiency (Sherpherd, 1993):

$$Marketing\ efficiency = \left(\frac{Consumer\ price}{Total\ marketing\ cost}\right) - 1 \quad (5)$$

$$Profit\ margin, \pi = (pQ) - \alpha + \beta X \quad (6)$$

where, Total revenue (pQ) equals price by quantity; Total cost ($\alpha + \beta X$) equals total fixed and variable costs

Prices and margins were valued at the average prices in naira per basket (40 to 45 kg) of fresh tomato vegetable during the 2010/11 dry season.

The two-independent samples t-test was performed to analyze if the profit margin of tomato marketers using evaporative cooling system is not significantly different from marketers who use traditional method for fresh tomato preservation. The hypothesis is stated as:

$$H_0 = \overline{\mu_1} = \overline{\mu_2} = 0 \quad (7)$$

The t- test relationship was specified as:

$$t_c = \frac{\overline{\mu_1} - \overline{\mu_2}}{\sqrt{\frac{s_1^2}{n_1}} - \sqrt{\frac{s_2^2}{n_2}}} \quad (8)$$

μ_1, means of profit margin of marketers who used evaporative cooling system; μ_2, means of profit margin of marketers who used traditional preservation; n_1, number of sampled tomato marketers who used evaporative cooling system; n_2, number of sampled tomato marketers who used traditional preservation; s_1, sampled variance for marketers who used evaporative cooling system; s_2, sampled variance for marketers who used traditional preservation.

The H_0 is accepted when the decision criterion holds that $t_c < t_{(\alpha,\ (n_1+n_2-2))}$ at α =5% level of significance, but rejected if otherwise.

RESULTS AND DISCUSSION

Socioeconomic characteristics of the respondents

Table 1 reveals that an average marketer of fresh tomato vegetable is middle aged. In the Northern and Southern districts, 41 and 50 years was the model age bracket, while it was 31 and 40 age bracket in the Central district. Most of the respondents are basically literate, having completed a 6-year basic education. Average marketing experience was between 21 and 30 years per respondent in both Northern and Southern districts (≥22 years), while 13 years was reported in the Central. The implication is that tomato vegetable trade requires adequate understanding of the market before entry. It was further revealed in Table 1 that start-up capital averaged over ₦65,000 in the Central and Southern districts while

Table 2. Net marketing margin and marketing efficiency at wholesale level in Ondo State, 2010/2011 dry season.

Variable	Estimates at senatorial district level			
	Ogbese market (North)	Iloro market (Central)	Owena market (South)	Average
Farm gate	4110.0	4570.0	4230.0	4303.3
Wholesale price	5040.5	5610.0	5100.0	5250.2
Consumer retail price	5720.8	6550.4	5840.2	6037.1
Marketing Cost	329.4	341.6	335.0	335.3
Marketing margin	930.5	1040.0	870.0	946.8
Net marketing margin	601.1	698.4	535.0	611.5

Costs and returns are valued in ₦ per basket (40 to 45 kg); Source: Field data, 2010.

Table 3. Sherpherd formula marketing efficiency of wholesale fresh tomato market.

Variable	Estimates at senatorial district level			
	Ogbese market (North)	Iloro market (Central)	Owena market (South)	Average
Consumer retail price	5720.8	6550.4	5840.2	6037.1
Selling price	5040.5	5610	5100	5250.2
Total marketing cost	329.4	341.6	335.0	335.3
Marketing efficiency (%)	16.37	19.16	17.43	17.65

Estimates are valued in ₦ per basket. Source: Field data, 2010.

<₦50,000 is reported in the Northern district. It can be concluded that start-up capital for fresh tomato vegetable market is considerably high, and volume of sales required to breakeven in the wholesale market.

Marketing performance analysis

Net marketing margin

As presented in Table 2, on aggregate, ₦698.4 wholesaler net marketing margin per basket (highest) was reported in Iloro market (Central) and ₦601.1 in Ogbese market (North) while the lowest was found in Owena market (South) districts. The average marketing cost at ₦335.3 per fresh tomato basket, resulted in average net market margin of ₦611.5 for wholesale market. However, a very high average farm gate price of ₦4303.3 per basket was observed due to the seasonal scarcity usually associated with tomato vegetable marketing especially during the dry copping season. The net marketing margins of Iloro and Ogbese wholesalers markets might be due to their lower purchasing prices due to their benefit of economies of scale. Major supplies usually come from the Northern part of Nigeria as local productions are not enough to meet consumer demands. A small difference between the highest and lowest net margins in the wholesale market was likely from similar economic characteristics of key players in the market, especially those that concerned distribution. The lowest net marketing margin of Owena market wholesalers reflects higher transportation and handling costs. This result agreed with previous study that inadequate marketing services such as transport, packing and handling represent major obstacles that face marketing activities (Altoum, 2008).

Marketing efficiency

Market efficiency of 19.16% was estimated for Iloro market (highest), Owena market recorded 17.43% efficiency and Ogbese market recorded 16.37% (lowest) as indicated on Table 3. Average marketing was 17.65% efficiency. A market that is efficient does not only bring sellers and buyers together, it enables entrepreneurs to take advantage of opportunities, to innovate and improve in response to demand and price changes (Fakayode et al., 2010). The high market efficiency implied that wholesalers in Iloro market got higher coefficients of marketing efficiency because they run marketing activities more efficiently than others, perhaps due to their closeness to the urban retailers and efficient price control. This result could therefore be seen as highest pricing efficiency in fresh tomato vegetable marketing in the study area.

Disaggregated wholesaler's profit margin by preservation techniques of fresh tomatoes

The profit margin of tomato wholesalers is presented in

Table 4. Averagely monthly cost, revenue and gross margin from sales of fresh tomatoes by preservation techniques.

Cost variable (₦)	Traditional system (₦)	Evaporative cooling (₦)
	Mean value of users (N=75)	Mean value of users (N=30)
Transportation costs	1020.00	680.00
Labour cost	294.00	195.20
Rent	400.00	400.00
Security levy	200.00	200.00
Association membership fee	200.00	200.00
Local government levy	80.00	80.00
Total marketing cost	2194.00	1755.20
Total buying cost	34426.40	60246.20
Total cost	36620.40	62001.40
Average revenue	7575.20	13256.60
Profit margins	6261.60	12381.40

Source: Field data, 2010.

Table 5. t-Test result of profit margin between users of evaporative cooling method and traditional system.

Null hypothesis	$t_{(0.05, 103)}$	t_{cal}	Decision
$H_{0:}\ \overline{\beta}_1 = \overline{\beta}_2 = 0$	1.980	74.014	Reject null

Source: Computed from Field Data, 2010.

Table 4. The average profit margin for users of primitive traditional system was ₦6,261.6 per month while that of users of evaporative cooling monthly reported average of ₦12,381.4. The higher average profit margin obtained by users of evaporative cooling was attributed to the scale of business transactions, in terms of number of tomato baskets that marketers were able to buy (at farm gate) during the dry season, preserved to retain its market value (Figure 1b), and sold, thereby resulted into higher profit over the traditional preservation system which allows for quick perishability. Users of traditional mode of preservation were unable to buy in large quantities for fear of loss in value (Figure 1c), over time. The positive influence of scale of market operations thus signify the higher returns to scale over traditional preservation among users of evaporative cooling system, which tend to further increase the margin.

Hypothesis testing

The summary result of the two-independent samples t-test is presented as shown in Table 5. The result presented in Table 5 reveals that profit margin of marketers who used evaporative cooling system is significantly different from those who use traditional method. The hypothesis tested shows that the computed t-value of 74.014 is greater than t-tabulated of 1.98 at 5% level of significance. The set hypothesis is rejected that profit margin of wholesalers using evaporative cooling system is not significantly different from marketers who use traditional method. This can be attributed to the higher quantities of tomato baskets that users of evaporative cooling were able to purchase at farm gate and store for longer days before being sold. This testifies to the improved marketing performance of wholesalers who store with evaporative cooling system over traditional preservation.

Conclusions

The study showed that fresh tomato wholesale market is efficient in Ondo State, Nigeria. The evaporative preservation cooling system developed is more efficient compared with the traditional preservation method as a cost effective means of preserving fresh vegetables through reduction of post harvest losses and increasing of income generated from agricultural produce. Given the same operational conditions, the average monthly profit margin for users of evaporative cooling method is twice that obtained by users of the traditional method. Marketing efficiency between users of traditional presser-vation and evaporative cooling system revealed that scale of market transactions among users of evaporative cooling system were higher, thereby resulted into higher profit margin over the traditional preservation system. Consequentially, evaporative cooling system is found suitable for application in the farms at both retail and wholesale level along the market flow channel until the product reaches the consumer. Marketers are thus encouraged to take advantage of opportunities, to innovate and improve in response to demand and price changes.

The present analysis showed that the evaporative preservation cooling system is more efficient as compared to the traditional preservation method from both marketing efficiency and performance points of view. Subjecting to the same operational conditions of averagely monthly cost, revenue and gross margin from sales of fresh tomatoes, the average profit margin per month for the wholesalers who were users of evaporative cooling method is two times that obtained by the users of the primitive traditional system. However, average marketing efficiency of 17.65% was obtained in fresh tomato wholesale market by the users of the evaporative cooling system means. It can thus be concluded that the marketing performance and efficiency of the evaporative-preservation cooling system elucidated a cost effective means of preserving fresh vegetables through reduction of post harvest losses and increasing of income generated from agricultural produce.

REFERENCES

Abbot JC, Makeham JP (1981). Agricultural Economics and Marketing in the Tropics. Wing Tai Cheung Printing Co. Ltd, Rome. p. 58.

Adeyokumnu TO (1973). The Marketing of Rice in Egba Division, Western Nigeria. Bull. Rural Econ. Sociol. 8:243-253.

Altoum YA (2008). Evaluation of the Factors Affecting the Production and Marketing of Tomato Crop in Khartoum State. PhD. Thesis. Faculty of Agriculture University of Khartoum, Sudan.

Cramer GL, Jensen W (1982). Agricultura Economics and Agribusiness, 2nd Edition. McGraw Hill Book Company, USA. p. 222.

Fakayode SB, Omotesho OA, Babatunde RO, Momoh AA (2010). The Sweet Orange Market in Nigeria - How Viable? Res. J. Agric. Biol. Sci. 6:395–400.

http://www.ondostate.gov.ng/ondo_state.html. Land Area and Population of Ondo State, Nigeria, 2011

Haidar J, Demisse T (1999). Malnutrition and Xerophthagma in rural communities of Ethiopia. East Afr. Med. J. 10:590-593.

Helsen A, Willmot JJ,(1991). Wet air cooling of fruits, vegetables and flowers. Current practice in Europe. Technical innovation in freezing and refrigeration of fruits and vegetables, International Institute of Refrigeration, Paris, France. pp. 169-77.

Kader AA (1992). Postharvest Technology of Horticultural Crops. (2nd ed), Publication No. 3311, University of California, Berkeley, CA., U.S.A. pp. 228-345.

Kilmer LR, Armbruster WJ (1999). Economic Efficiency in Agricultural and Food Marketing. Iowa State University, Press. http://agris.fao.org/agris-search/search/display.do?f=1989/US/US89063.xml;US8908012

Kohl's RL, Uhl JN (1985). Marketing of Agricultural Products. Fifth .edition., Coiler Macmillan Publishing Company, New York, USA. p. 83.

Mendoza G (1995). A Primer on Marketing Channels and Margins. In G.J.Scott(eds.). Prices, Products, and People: Analyzing Agricultural Markets in Developing Countries. Lynne Reinner Publishers, Boulder, London. pp. 257-275.

Mitra SK, Baldwin EZ (1997). Post harvest physiology and storage of tropical and subtropical fruits. CAB International. West Bengal. India, pp. 85-122.

Mogaji TS, Fapetu OP (2011). Development of an evaporative cooling system for the preservation of fresh vegetables, Afr. J. Food Sci. 5(4): 255-266.

Scarborough V, Kydd J (1992). Economic Analysis of Agricultural Marketing. A Manual Marketing Series 5, Chartham United Kingdom: Natural Resources Institute, UK.

Seyoum TW, Woldetsadik K (2004). Forced ventilation evaporative cooling of fruits: A case study on Banana, Papaya, and Orange. Lemon and Mandarin. Trop. Agric. J. 81(3):179-185.

Shepherd AW (1993). A guide to marketing costs and how to calculate them". Marketing and Rural Finance Service. Agricultural Service Division, Food and Agriculture Organisation (FAO), Rome.

Trotter BW (1992). Applying Price Analysis to Marketing Systems: Methods and Examples from the Indonesian Rice Market. Marketing Series 3. Natural Resource Institute. Chatham. UK.

Umbarker SP, Bonde RS, Kalase MN (1991). Evaporatively cooled storage structure for oranges (citrus). Indian J. Agric. Eng. 1(1):26-32.

Knowledge and perception of extension agents on information and communication technologies (ICTs) use in extension service delivery in Ondo State, Nigeria

Ajayi, A. O., Alabi, O. S. and Akinsola, T. O.

Department of Agricultural Extension and Rural Development, Obafemi Awolowo University, Ile-Ife, Osun State, Nigeria.

This study determined the knowledge and perception of Extension Agents (EAs) on information and communication technologies (ICTs) use in extension service delivery in the two Agricultural zones of Ondo State. Fifty EAs were randomly selected from each zone. Validated questionnaire was used for data collection and appropriate statistical tools were used to analyze the data collected. The results showed that half (50.0%) of the EAs were highly aware of the various ICT while 50.0% had high knowledge on ICT use. Furthermore, majority (72.0%) of the EAs had indifferent perception towards ICT use. The results also showed that there were significant association between EAs awareness on ICT and EAs membership of professional association (X^2=53.341: df=1: p≤0.05) and highest educational level (X^2=99.96: df=4: p≤0.05). Furthermore, significant relationship existed between EAs knowledge on ICT use (r=0.656: p≤0.01), their perception towards ICT use (r=0.301: p≤0.01) and their awareness on ICT. The results of this study show that programmes designed to improve EAs use of relevant ICTs should address their education and professional association participation.

Keywords: Knowledge, perception, awareness, information communication technology.

INTRODUCTION

Global attention directed at agriculture due to emerging challenges of food security in recent years, resulting partly from age long negligence of dissemination of appropriate technology. Increasing production is a major challenge facing present agriculture. Smallholder farmers which dominate the landscape of developing world need to improve farming by acquiring adequate knowledge and information (UN, 2005). The role of public agricultural extension service has traditionally been to provide the important link between agricultural research and farming communities, especially for technology transfer in support of agricultural and rural development (FAO, 2004). Several authors (Anderson and Feder, 2004; Anandajayasekeram et al., 2008; Aker, 2010) posited that public agricultural extension service over the years has been working via different approaches, methodologies and programmes to ensure farmers adoption of improved technologies with little success. Agricultural Development Project is the agency charged with the responsibility of Agricultural Extension in Nigeria and it is no longer financed by World Bank (Omotayo, 2005).

Given the urgent need for current Agricultural Knowledge and Information System (AKIS) by farmers, the use of conventional communication channels such as farm/home visit, personal letters, and use of contact farmers, for disseminating agricultural information as entrenched in the T and V extension approach is becoming less effective in the face of changing government policies and dwindling financial resources as a result of the withdrawal of the World Bank funding. One way to address this is through encouraging the adoption of Information and Communication Technologies (ICTs) by both researchers and extension workers to transmit relevant information to farmers in a most efficient way (Salau and Saingbe, 2006). ICT has many potential applications in agricultural extension most especially in accessing required information and knowledge (Richardson, 1997; Chapman et al., 2004; Mcnamara, 2009). ICT can bring new information services to rural areas where farmers, as users, will have much greater control. It follows from the foregoing that ICT are information transmission technology built on the potential of electronic communication devices such as computers and telecommunication equipment, for connecting and accessing various ends in the information pathway (Aboh, 2008).

Recent happenings across the globe are testimonies to the tremendous influence of technology in various sectors such as health, communication, entertainment and so on. Global landscape has changed tremendously as a result of these technological innovations. The influence of technology in agriculture has been established over the years with the introduction of chemicals, fertilizers, labour saving equipment, improved seeds and better farming methods and systems. The introduction of various relevant ICTs in agricultural information dissemination could help farmers' access market information, land resources and services, management of pests and diseases, rural development programmes (Meera et al., 2004) and help in broadening the orientation of farmers in production activities thereby causing a major turnaround in the agricultural sector as it is doing in many other sectors. If extension agents are crucial to agricultural and rural development and use of various ICTs promises a boost for agricultural and rural development, then a study on the knowledge and perception of extension agents on ICTs use in extension service delivery is important.

Objectives of the study

The specific objectives of the study are to:

1. Describe the socio-economic characteristics of extension agents in Ondo State;
2. Determine the level of knowledge of extension agents on ICT use in extension service delivery in Ondo State;
3. Determine the perception of extension agents on ICT use in extension service delivery in Ondo State; and
4. Determine the extension agents' awareness on ICTs in Ondo State.

Hypotheses of the study

Three hypotheses were set in the null form to provide direction for the study. These are:

1. There is no significant relationship between ICT awareness and socio-economic characteristics of EAs.
2. There is no significant relationship between ICT awareness and the knowledge of extension Agent on ICT use.
3. There is no significant relationship between ICT awareness and the perception of Extension Agent on ICT use.

METHODOLOGY

The study was conducted in Ondo State which is in the southwest geopolitical region of Nigeria. The state population according to 2006 census was 3,441,024 people, with an estimated land area of 15,500 km^2. Ondo state Agricultural Development Project (ODSADEP) divides the state into two zones namely, Owo and Ondo. The 60 Extension Agents in Owo zone and 63 Extension Agents in Ondo zone formed the sample population for the study with 50 extension agents being randomly selected from each zone.

Data collection instrument

The instrument used for the study was questionnaire with five sections: Socio economic characteristics, awareness of various ICTs, knowledge on ICT use, perception towards ICT use, and the constraints on ICT use.

Measurement of variables

Independent variable such as age, number of years of formal education, length of service, number of farm families/farmer they are responsible to, membership of professional association; monthly income and cosmopoliteness – the degree of external orientation were measured by direct questions. EAs knowledge on ICT use in extension service delivery as well as their perception on ICT use in extension service delivery were measured on a Likert scale and were categorized by using the mean plus or minus one standard deviation. The dependent variable was awareness on various ICTs. This was measured as the total awareness score on ICT use using the following divisions; have you heard about it; have you seen it; do you have skills to use it; have you used it; and do you have it. This was coded as 1 for yes and 0 for no in all the five categories. Data collected were summarized using frequency, percentage, mean and standard deviation while Pearson product moment correlation, chi square and regression analyses were used to test the hypotheses stated.

RESULTS AND DISCUSSION

Socio-economic characteristics of extension agents

Result in Table 1 shows that majority (71.0%) of Extension Agents were male. The average age of the Extension Agents was 41.31 years and the standard deviation was 6.67. The analysis shows that more than half (51.0%) of the respondents were between the active ages of 41 and 50 years. It further revealed that more

Table 1. Distribution showing the socio-economic characteristics of Extension Agents n=100.

Socio-economic characteristics		Frequency	Percentage	Central tendency
Sex	Male	71	71.0	
	Female	29	29.0	
Age(years)	≤30	8	8.0	
	31-40	37	37.0	X= 41.31
	41-50	51	51.0	SD= 6.69
	≥51	4	4.0	
Years of formal education	11-15	25	25.0	X= 16.89
	16-20	71	71.0	SD= 1.93
	≥21	4	4.0	
Length of service(years)	≤25	97	97.0	X=12.74
	26-30	2	2.0	SD= 6.43
	≥31	1	1.0	
Farmers/farm family covered	≤300	63	63.0	
	301-500	16	16.0	X= 296
	501-700	5	5.0	SD= 283.98
	≥701	16	16.0	
Membership of professional association	Executive	4	4.0	
	Committee	6	6.0	
	Ordinary	24	24.0	
	No	66	66.0	
Professional cadre	VEA	58	58.0	
	BEO	27	27.0	
	ZEO	6	6.0	
	SMS	7	7.0	
	WIA	2	2.0	
Monthly income(Naira)	≤100,000.00	74	74.0	
	100,000.01-120,000	8	8.0	X= N81,865
	120,000.01-140,000	6	6.0	SD= N33,741
	≥140,000.01	12	12.0	
Cosmopoliteness	Biannually	5	5.0	
	Annually	6	6.0	
	Monthly	14	14.0	
	Weekly	64	64.0	
	Daily	11	11.0	
Reason for cosmopoliteness	For other business	39	39.0	
	To visit friends	17	17.0	
	For conferences	17	17.0	
	To acquire more knowledge on ICT	27	27.0	

Source: Field survey (2012).

than half of the agents are adults with requisite experiences which could be useful in dealing with fellow adults (farmers) in the course of their work. Only 4.0% of the respondents were 51 years and above. The average year of formal education was 16.9 years with a standard deviation of 1.9. The analysis shows that majority

Table 2. Distribution of extension agent on their awareness of various ICTs.

List of ICTs	Heard		Seen		Used		Have		Skills	
	Frequency	Percentage	Frequency	Percentage	Frequency	Percentage	Frequency	Percentage	Frequency	Percentage
Television	100	100.0	100	100.0	100	100.0	81	81.0	77	77.0
Radio	100	100.0	100	100.0	100	100.0	93	93.0	94	94.0
Video recorder	77	77.0	78	78.0	29	29.0	21	21.0	32	32.0
Digital camera	81	81.0	75	75.0	53	53.0	27	27.0	47	47.0
Print media	80	80.0	64	64.0	35	35.0	25	25.0	21	21.0
Video player	84	84.0	75	75.0	65	65.0	61	61.0	78	78.0
Telegram	58	58.0	35	35.0	2	2.0	1	1.0	11	11.0
Projector	70	70.0	55	55.0	25	25.0	9	9.0	34	34.0
Web camera	64	64.0	45	45.0	34	34.0	20	20.0	38	38.0
Mobile phone	100	100.0	100	100.0	100	100.0	89	89.0	83	83.0
Global positioning system	63	63.0	44	44.0	27	27.0	14	14.0	35	35.0
Fax	61	61.0	36	36.0	20	20.0	3	3.0	9	9.0
Computer	100	100.0	100	100.0	72	72.0	55	55.0	62	62.0
Internet	83	83.0	59	59.0	56	56.0	49	49.0	56	56.0

(71.0%) of the respondents had between 16 and 20 years of formal education while only 4.0% of the respondents had formal education greater than or equal to 21 years.

Also 40.0% of the respondents had Agricultural Extension as their major area of study while 4.0 percent had soil science but all areas of agriculture was ably represented. The average length of service was 12.7 years with a standard deviation of 6.4. The analysis shows that majority (97.0%) of the respondents had been in service for less than or equal to 25 years. Most (63.0%) of the respondents were responsible to less than 300 farmers/farm family and the average farmer/farm family responsible to was 296. Only 34.0% of the respondents belonged to professional associations with 24.0% as ordinary members, 6.0% as committee members and only 4.0% as executive members.

Many (58.0%) of the respondents were Village Extension Agents while 2.0% were in charge of Women in Agriculture. Most (74.0%) of the Extension Agent earned less than or equal to ₦100,000 monthly with their average monthly income put as ₦81, 865 with a standard deviation of ₦33, 741. Many (64.0%) of the Extension Agents travel out of their community weekly and 39.0% of the Extension Agents left their community for fortnightly training meetings.

Awareness of various ICTs

Result in Table 2 shows that all of the Extension Agents had heard of television while all of them had seen television. Also, majority (81%) had television set while all had used a television set. Furthermore, 77.0% of the Extension Agent had skills on how to operate television set. However, all of the Extension Agents had heard of Radio while all had seen Radio. Furthermore, all had used Radio while 93.0% had Radio and 94.0% of the Extension Agents had skill on how to operate Radio. Also, 81.0% of Extension Agents had heard about Digital camera, 75.0% had seen digital camera while 53.0% had used digital camera, 27.0% had digital camera and 47.0% had skills on how to operate digital camera.

Result in Table 2 also revealed that majority (64.0%) of the Extension Agents had heard about web camera while 45.0% had seen web camera and 34.0% had used web camera. Furthermore, 20.0% had web camera while 38.0% had skills on how to operate web camera. All of the Extension Agents had heard about mobile phone, had seen mobile phone and had used mobile phone respectively. Also, 89.0% had mobile phone while 83.0% had skills on how to operate mobile phone. Furthermore, all of the Extension Agents had heard about and seen computer respectively while

Table 3. Mean score of extension agents on knowledge of ICT use.

Statements	Mean	Standard deviation	Ranking
GSM for making calls or text messages to colleagues or farmers	4.45	1.10	1
GSM use as electronic diary and reminders for meetings and events	4.30	1.16	2
Receiving and disseminating information through email	3.62	1.52	3
Computer for word processing	3.32	1.39	4
Membership of social networks on internet	3.25	1.83	5
Assessing information for research on the internet	3.19	1.52	6
Computer for data processing	3.17	1.44	7
Power point preparation and presentation	2.97	1.46	8
Use of projectors in lecture and seminar	2.92	1.39	9
Organizing and setting up various ICT equipments	2.45	1.34	10
Presenting television programme	2.35	1.15	11
Presenting Radio Programme	2.20	1.11	12
Scripting of radio programme	2.07	1.27	13
Scripting of television programme	2.04	1.13	14
Documenting events through digital camera	1.96	1.21	15
Documenting events through video recording	1.83	1.12	16

72.0% had used computer. Also, 55.0% had computer while 62.0% had skills on how to operate computer. The awareness of Extension agents on other ICTs were also presented in Table 2.

Extension Agents' awareness score was further analyzed and the summary presented in Table 5. Results in Table 5 reveals that half (50.0%) of the Extension Agents had high awareness on various ICTs with a score range of ≥54.96 while 12.0% had moderate awareness with a score range of 54.96 to 21.82 and 38.0% had low awareness with a score range of ≤21.82.

Knowledge on ICT use

Result in Table 3 shows that the knowledge of Extension Agents was highest in using Global System of Mobile communication (mobile phones) for making calls or text messages to colleagues or farmers with a mean score of 4.45. This was followed by GSM use as electronic diary and reminders for meetings and events with a mean score of 4.30. Also the Extension Agent had moderate knowledge in using computer for word processing with a mean score of 3.32 and accessing information for research on the internet with a mean score of 3.19. However, the result in Table 3 also shows that the Knowledge of the Extension Agents was lowest in documenting events through video recording with a mean score of 1.83. This showed that Extension Agent had more knowledge on the use of mobile phone, e-mail and computer which is in line with Adesope et al. (2005) whose finding showed that mobile phone, computer for word processing and electronic mail were the commonest information technology used by extension agents.

Extension Agents' knowledge score was further Analyzed and the summary presented in Table 5. Result in Table 5 reveals that half of the Extension Agents had high knowledge on ICT use with a score range of ≥60.76 while 14.0% of the Extension Agent had moderate knowledge on ICT use with a score range of 60.76 to 31.42 and 36.0% of the Extension Agents had low knowledge on ICT use with a score range of less than or equal to 31.42.

Perception towards ICT use

Result in Table 4 shows that the mean values of the various perceptional statements used to measure the perception of Extension Agents about ICT use. The statement "knowledge of ICT use has a great significance in Agricultural development" ranked highest with mean score of 4.47. This was followed by the statement "it's a waste of time bringing ICT into agriculture" with a mean score of 4.39. However, the statement "infrastructural facilities to support ICT use are not available" ranked lowest with a mean score of 2.60. This signifies that Extension Agent perception towards the use of ICT in Agriculture was mostly influenced by statements "knowledge of ICT use has a great significance in agricultural development, it's a waste of time bringing ICT into agriculture, ICT use influences rate of adoption and extension work can be greatly enhanced by ICT".

Extension Agents' perception score was further analyzed and the summary presented in Table 5. Result in Table 5 shows that majority (72.0%) of the Extension Agent had indifferent perception towards ICT use with score range of 45.59 to 35.97 while 15.0% of the Extension Agent had unfavourable perception towards ICT use with a score ≤ 35.97. However, 13.0% of the

Table 4. Mean score of extension agents on perception towards ICT use.

Perceptional statements	Mean	Standard deviation	Ranking
Knowledge of ICT use has a great significance in Agricultural development	4.47	0.66	1
It's a waste of time bringing ICT into agriculture	4.39	1.06	2
ICT use influences rate of adoption	4.38	1.09	3
Extension work can be greatly enhanced by ICT	4.21	0.98	4
The benefits of ICT use outweigh the financial burden involved	3.85	0.90	5
ICT removes a lot of cost, barriers and saves time	3.80	0.93	6
ICT is a cost effective and practical tool for facilitating extension service delivery	3.61	0.93	7
ICT use is good but not necessary in extension service delivery	3.55	1.33	8
The financial burden of using ICT is unbearable	3.01	1.04	9
ICTs are too complicated and hard to use	2.91	1.16	10
Infrastructural facilities to support ICT use is not available	2.60	1.16	11

Table 5. Distribution showing the awareness level, knowledge level and perception level towards the use of ICT (n=100).

Levels on ICT use	Frequency	Percentage
Awareness		
High(≥54.96)	50	50.0
Moderate (54.96-21.82)	12	12.0
Low(≤21.82)	38	38.0
Knowledge		
High (≥60.76)	50	50.0
Moderate (60.76-31.42)	14	14.0
Low (≤31.42)	36	36.0
Perception		
Favourable (≥45.59)	13	13.0
Indifferent (45.59-35.97)	72	72.0
Unfavourable (≤35.97)	15	15.0

Source: Field survey 2012.

Extension Agents in the study area has a positive perception towards ICT use with a score of greater than 45.59.

Correlation analysis

The variables studied were subjected to correlation analysis using Pearson product moment correlation to find out the relationship between the dependent variable (awareness on various ICTs) and independent variables studied. Result in Table 6 reveals that there was no significant relationship between the age of Extension Agents, length of service, and number of years of formal Education, number of farmer/farm family covered, monthly income and Awareness on various ICTs but there was positive and significant correlation between Knowledge of various ICT ($r=0.656$, $p\leq 0.01$) and awareness level of various ICT as well as Perception towards ICT use ($r=0.301$, $p\leq 0.01$) and awareness on various ICT. This implies that the higher the Knowledge of extension agents on ICT use, the higher their awareness on it and also, the better their perception towards ICT use, the higher their awareness on various ICTs.

Chi square analysis

The Chi square analysis carried out presented the following results. There is significant association between membership of professional association ($X^2=53.341$, df=1, $p\leq 0.05$), highest educational level ($X^2=99.96$, df=4, $p\leq 0.05$), frequency of leaving the community ($X^2=196$, df=4, $p\leq 0.01$), reason for leaving the community ($X^2=149.1$, df=3, $p\leq 0.01$) and awareness level of various

Table 6. Correlation analysis showing the relationship between some independent variable and awareness level score.

Independent variable	Correlation coefficient (r)	Coefficient of determination (r^2)
Age	0.051	0.002601
Length of service	-0.52	0.2704
Number of years of formal education	-0.063	0.003969
Number of farmer/farm family covered	0.017	0.000289
Monthly income	0.089	0.007921
Knowledge on ICT use	0.656**	0.430336
Perception towards ICT use	0.301**	0.090601

* = 0.05; **= 0.01; Source: Field survey 2012.

Table 7. Chi square analysis showing association between some socio-economic characteristics and awareness level of various ICTs.

Characteristics	Chi square (X^2)	Df	Level of significance
Major Area of study	205.7	6	0.206
Membership of professional Association	53.341	1	0.050*
Professional cadre	157.9	3	0.354
Gender	41.661	1	0.314
Highest educational level	99.960	4	0.034*
Frequency of leaving the community	196.0	4	0.009**
Reason for leaving the community	149.1	3	0.015**

*= 0.05; ** = 0.01; Source: Field survey, 2012.

Table 8. Regression analysis of some independent variables and awareness level of various ICT.

Parameter	Unstandardized coefficients		Standardized coefficients	t	Sig.
	B	Std. error	(Beta)		
(Constant)	-2.639	16.891		-0.156	0.876
Age in years	-0.181	0.293	-0.073	-0.617	0.539
Number of years of formal education	0.029	0.762	0.003	0.038	0.970
Length of service	-0.034	0.321	-0.013	-0.105	0.916
Number of farmer/ farm family	-0.006	0.005	-0.097	-1.214	0.228
Monthly Income	0.000	0.000	0.205	2.066	0.042
Knowledge total	0.752	0.104	0.666	7.265	0.000
Perceptional total score	0.177	0.304	0.051	0.581	0.563

R = 0.682; R^2= 0.466; Adjusted R^2 = 0.425; Standard error of estimate = 12.56; F value =11.455; Sig. = 0.000. Source: Field survey, 2012.

ICT (Table 7). This implies that being a member of a professional association, being literate, frequency of leaving the community and the major reason for leaving the community contributes to extension agents' awareness of various ICT.

Regression analysis

Result in Table 8 shows the result of the multiple regression analysis carried out. The Adjusted R-square was 0.425, which implies that the significant independent variables accounted for about 42.5% of the variation in the extension agents' awareness of various ICTs. Hence, the result shows a positive and significant relationship between awareness of various ICT and Knowledge on ICT use (b = 0.666, p≤0.01) and monthly income (b= 0.205, p≤0.05). This implies that any programme designed to improve the awareness of Extension Agents on ICT use should be planned with emphasis on the two variables. This is because, the more extension agents become knowledgeable on various ICTs, the better their interest and desire that will eventually improve their awareness on these ICTs. Furthermore, the more income available to the extension agents, the easier it becomes to acquire these ICT and use them. In addition, the analysis of variance for the regression analysis yields an F – value of 11.455, which is significant at $p < 0.01$ level of significance.

CONCLUSIONS AND RECOMMENDATIONS

Although the extension agency in Nigeria cannot be said to be gender biased, yet majority of the agricultural extension agents were males. This might be due to the nature of the job – the stress and rigours involved. The average age of the extension agent is 41.31 years. This established that extension agents are still in their active years and can still cope with the nature of the work. Agricultural extension was the major area of study of the Extension Agents in the study area. This is expected since most of the knowledge and skill required for success in extension delivery are taught as part of the curriculum for a degree in agricultural extension. The agents studied have spent an average of 12.74 years in service and are responsible to an average of 296 farmer/farm family. Majority of the Extension Agent does not belong to professional association despite the many benefits they stand to gain and they earn an average of ₦81, 865 monthly. There was high level of awareness of ICT among the agricultural extension agents, especially in computer, mobile phone, radio, television and video player. These ICTs are relatively the cheapest on the list of the ICTs listed and some of these are household items. Mobile phones are everywhere in the nation and personal laptop is gaining prominence among civil servants because of its various uses. The Extension Agent perception of ICT use in extension service delivery was mostly indifferent. This might be due to the cost implication of ICT use, the complicated nature of the ICTs and the needed infrastructural facilities not available.

The study recommends that adequate information on various relevant ICT suitable for extension service delivery should be given to extension agents so that they can improve their knowledge on it and also help them to develop more positive perception that will enhance the future use of these ICT in extension service delivery. Also, the study recommends that anytime programme that will enhance the awareness about utilization of various ICTs by extension agents is being planned, variables such as knowledge on ICT, monthly income and perception towards ICT use that showed significant relationship with extension agents' awareness in the study should be given preference. The policy implication of this study therefore is that government needs to reorient its policies in order to harness ICT potential for contribution to agricultural development in the country.

REFERENCES

Aboh LC (2008). Assessment of the frequency of ICT tools usage by agricultural extension agents in Imo State, Nigeria. J. Agric. Social Res. (JASR) 8:2.

Aker JC (2010). Dial 'A' for Agriculuture: Using Information and Communication Technologies for Agricultural Extension in Developing Countries. Tuft University, Economics Department and Fletcher School, Medford MA02155.

Anderson JR, Feder G (2004). Agricultural extension: Good intentions and hard realities. The World Bank Research Observer 19(1):41-60.

Anandajayasekeram P, Puskur R, Workneh S, Hoekstra D (2008). Concepts and practices in Agricultural extension in Developing countries: A source book. IFPRI (International Food Policy Research Institute), Washington, DC, USA, and ILRI (International Livestock Research Institute), Nairobi, Kenya. p. 275.

Chapman R, Slaymaker T, Young J (2004). Livelihoods Approaches to Information and Communication in Support of Rural Poverty Elimination and Food Security. Overseas Development Institute, UK Department of International Development and Food and Agricultural Organisation.

Food and Agriculture Organization of the United Nations (2004). *Institute building to strengthen agricultural extension*. 27th FAO regional conference for Asia and the pacific Beijing, china, May, 17th – 21st.

McNamara K (2009). Improving Agricultural Productivity and Markets: The Role of Information and Communication Technologies. Agriculture and Rural Development Notes, Issue 47, April, The World Bank, Washington DC.

Meera SN, Jhamtani A, Rao DUM (2004). Information and Communication Technologies in Agricultural Development: A Comparative Analysis of three Projects from India. Network, p. 135. Agricultural Research and Extension Network.

Omotayo AM (2005). Information and Communication Technology (ICT) and Agricultural Extension: Emergency issues in transferring Agricultural Technology in Developing countries. In: S.F. Adedoyin (Eds.): Agricultural Extension in Nigeria. Ilorin: Agricultural Extension Society of Nigeria.

Richardson D (1997). The Internet and Rural and Agricultural Development: An Integrated Approach. Rome: FAO.

Salau ES, Saingbe ND (2006). Access and utilization of ICTs among Agricultural Researchers and Extension workers in selected institutions in Nassarawa State of Nigeria. http://patnsukjournal.net/vol4No2/pl.pdf, pp. 1-6.

United Nations (2005). Global E-government Readiness Report: From E-Government to E-Inclusion. UNPAN/2005/14, United Nations, New York.

Sorghum production for food security: A socio-economic analysis of sorghum production in Nakuru County, Kenya

Robert M. Ogeto[1], Erick Cheruiyot[2], Patience Mshenga[3] and Charles N. Onyari[4]

[1]Kenya School of agriculture, P. O. Box 1909-10100, Nyeri, Kenya.
[2]Department of Crops, Horticulture and Soil Sciences, Egerton University, P. O. Box 536, Njoro, Kenya.
[3]Department of Agricultural Economics and Agribusiness Management, Egerton University, P. O. Box 536, Njoro, Kenya.
[4]Embu University College, P. O. Box 6-60100, Embu, Kenya.

Kenya's Vision 2030 identifies agriculture as one of the key sectors to deliver sustainable economic growth and improved livelihoods for the poor in the rural areas. However, the sector continues to face several endemic and emerging constraints that require special attention. During the first two decades after independence, Kenya's economy grew at an average rate of 6 percent per year and this was substantially driven by a robust agriculture sector. However, until about five years ago, the overall economy barely grew, partly as a result of a decline in agricultural activities. The agriculture sector continues to face challenges in production due to frequent and prolonged drought hence the need for drought resistant crops like sorghum. The utility of sorghum is in its climatic adaptability and household as well as industrial use. However, there is a remarkably low production as well as consumption among Kenyan communities against the food security challenges. This paper provides insights into the socio-economic characteristics of farmers and the factors that influence their participation in its production. Simple random and snowball sampling methods were applied in collecting data from 207 farmers using a questionnaire. Data collected was analyzed using the double hurdle model. Of the factors under study, only land tenure, gender, farm size and educational level were significantly influencing sorghum production in the study area. It was recommended that more women should be encouraged to participate in sorghum production as it was found that they participated in sorghum production more than men.

Key words: Sorghum production, development, food security, Nakuru County.

INTRODUCTION

Kenya's Vision 2030 has identified agriculture as one of the key sectors to deliver sustainable economic growth and improved livelihoods for the poor in the rural areas. However, the sector continues to face several endemic and emerging constraints at the global, regional and national level that require special attention. During the first two decades after independence, Kenya's economy grew at an average rate of 6% per year and this was substantially driven by a robust agriculture sector (MOA, 2009). However, until about five years ago, the overall economy barely grew, partly as a result of a decline in agricultural activities. Despite experiencing mixed results over the years, agriculture still remains a major contributor towards the Kenyan economy though its share of the Gross Domestic Product (GDP) declined from 23 percent in 2007 to 22 per cent of real GDP in 2010 and 24% in 2011 (KNBS, 2010). Besides, the sector is the growth engine for the non-agricultural sector with a multiplier effect of about 1.64. Furthermore, agriculture provides 75 percent of the country's industrial raw materials and employs over 80 percent of Kenya's workforce (MOA, 2009).

Food security and development in Kenya

Food security has moved to the forefront of the development debate, and it is likely to remain a chief development concern in Sub-Saharan Africa in the foreseeable future. The compounding effects of sharp increases in food prices in 2007 up to mid-2008 and the global economic downturn of 2009 are estimated to have reversed the steady decline experienced from the late 1960s to 2004-2006 in the proportion of undernourished population in developing countries (FAO, 2009). Estimates suggest that the share of the population in developing countries suffering from hunger increased in both 2008 and 2009, reaching close to 20 percent (FAO, 2009). In Sub-Saharan Africa, estimates indicate an increase in the proportion of undernourished from 28 percent in 2004-2006 to 29 percent in 2008 (UN, 2009). This implies a reversal in progress towards achievement of Millennium Development Goal 1, to halve the population living in hunger decreased by 2015 from 1990 levels, both globally and in Africa.

Climate change will compound these challenges. Africa is the region most at risk of hunger linked to climate change. Globally, the number of people at risk of hunger is projected to increase 10-20 percent by 2050 as a consequence of climate change (IFPRI, 2009). This is due to weak coping capacities and the acute vulnerability of agricultural production to climate shocks and stresses.

In Kenya, declining performance of the sector in terms of its growth has been one of the major concerns facing policy makers and those having interests in the sector. A decline in agriculture has thus, far reaching implications in terms of employment and income inequality as well as food security for the country (UNDP, 2002). Agricultural productivity is needed for growth since the actual yields of major crops in Kenya are far below the potential yield (Karugia, 2003). There is need to expand areas under crop production and this could mean venturing into dry areas hence the need for drought tolerant crops like sorghum to bridge this gap. Sorghum is a very important source of food and farm income for smallholder farmers, which can be enhanced especially if linked to new markets (Hamukwala, 2010). Sorghum is important to a substantial number of farm households in the East African region, especially in the semiarid areas. Most cultivars of these drought resistant grains represent food for the household, and cash income if there is a market for the sale of surplus production. Some cultivars are also grown for brewing, which offers another source of cash (Schmidt, 1988).

The utility of sorghum as a cereal crop

Despite its major contribution, the agriculture sector continues to face challenges in production due to frequent and prolonged drought both locally and internationally. Drought is perhaps the most prevalent abiotic stress affecting plant growth, survival and productivity in the world (FAO and ICRISAT, 1996). It is reported that the effect of drought is more pronounced in the Semi-Arid Tropics (SAT), where rainfall is generally low, erratic and poorly distributed. The drastic effect of drought can be reduced by growing drought tolerant crops such as sorghum. Sorghum is the world's fifth most important cereal, in terms of both production and areas planted. It is primarily grown in areas experiencing low rainfall. Most of these areas are unsuitable for the production of other grains unless irrigation is available (FAO and ICRISAT, 1996). Sorghum is unique in its ability to produce under a wide array of harsh environmental conditions where other crops grow or yield poorly. It is grown with limited water resources and usually with low fertilizer supply or other inputs by a multitude of smallholder farmers in many countries (FAO, 1995).

Total consumption of sorghum closely follows the global pattern of output, since most of it is consumed in the countries where it is grown. It is used for human food and as animal feed (FAO, 1995). This report further indicates that, although in the early 1990s a very large part of the sorghum output was used directly as human food, its share has continuously declined in quantity since then. On the other hand sorghum consumption as animal feed has more than doubled from 30 to 60% since the early 1990s. While total food consumption of all cereals has risen considerably during the past 35 years, world food consumption of sorghum has remained stagnant, mainly because it is regarded in many countries as an inferior grain, although the crop compares nutritionally well with other grains.

As human food, sorghum has many uses with some of its products being: dehulled boiled sorghum, sorghum stew, sorghum *pilau*, sorghum green grams *pilau*, sorghum *ugali*, sorghum ginger biscuits, sorghum bread, sorghum queen cakes, sorghum cake, sorghum *chapatti*, sorghum porridge and sorghum beverage (MOA, 2007). Other uses of sorghum include use of stalks as dry season fodder and as fencing materials in addition to industrial uses as animal feed, making of industrial starch and fuel for alcohol production.

There are good prospects for the expansion of the industrial market for sorghum if its yields can rise fast enough to catch up with yields of competing cereals (FAO and ICRISAT, 1996). In spite of these prospects, sorghum tends to be grown in traditional farming systems. As a result, the yields generally average less than 1 ton/ha and vary considerably from year to year with majority of sorghum growers not producing enough to meet family requirements in most years.

In Kenya, sorghum is grown principally in the often drought-prone marginal agricultural areas of Eastern, Nyanza and Coast Provinces. Farm production went up from 118,227 t in 2002 to 126,433 t in 2003 (EPZA, 2005). Production of sorghum increased by a dramatic 75 percent from 602,910 bags in 2008 to 1,055,051 bags in 2009 with some slight improvement on the yield per ha to 6.09 bags; much lower compared with the 14.0 bags/ha recorded in 2005 and 10.5 bags/ha recorded in 2007. The area under the crop also registered an increased acreage to 173,172 ha in 2009 from 104,041 ha in 2008 a 67% increase. The acreage further increased to 254, 125 ha in 2011 from 225,762 ha in 2010 a 13% increase. This increase is attributed to increased area dedicated to the crop on account of being drought resistant and hence a primary poverty eradication vehicle especially in marginal areas (MOA, 2012).

Research data on the influence of socio-economic factors on participation and extent of participation in sorghum production in Kenya is not available. The objective of this research was to study the influence of socio-economic factors on participation and extent of participation in sorghum production in Njoro and Rongai of Nakuru County.

Sorghum production in Nakuru County

Over the years, Kenya has experienced diminishing rainfall amounts probably due to the effects of the global climate change. This has led to decreased food production making her a food deficit country with most areas receiving relief food supplies in many of the years. This has necessitated the government to come up with programmes to encourage the growing of drought tolerant crops like sorghum as a means of ensuring food security and improved incomes. The demand for sorghum is gradually increasing owing to industrial needs such as brewing by East African Breweries Limited. However, sorghum production remains low despite the added agribusiness benefit.

Though Nakuru County is among the high potential areas for sorghum, indications are that only a few farmers are engaged in commercial production and that the farmers are not able to produce sufficient quantities for consumption and get surplus for sale. The current status of sorghum cultivation in the County is not clear and underlying factors to lack of response to emerging trends in sorghum demand needs to be determined if the crop's promotion is to be effectively effected. However, there has been paucity of research with regard to the socio-economic and institutional characteristics of sorghum producers and the acreage under sorghum production. There is therefore need to analyze the socio-economic factors influencing participation and extent of sorghum production as an initial step to finding the solution to increasing sorghum cultivation among smallholder farmers in Nakuru County.

METHODOLOGY

Study area

The study was conducted in Njoro and Rongai Districts in Nakuru County. Nakuru County lies within the Great Rift Valley and borders eight other districts namely, Kericho, and Bomet to the West, Koibatek and Laikipia to the North, Nyandarua to the East, Narok to the south West and Kajiado and Kiambu to the south. The County covers an area of 7,235.3 Sq Km and is located between longitudes 35" 28" and 35" 36" and latitudes 0" 12" and 1" 10" South (NCAPD, 2005). It lies about 2100 m above sea level.

The area surrounding Nakuru town is known for its agricultural potential with numerous small farms and also vast agricultural enterprises. The main crops grown and marketed in Nakuru County include coffee, wheat, barley, maize, and beans.

The crops provide the primary raw material for the manufacturing industries found in Nakuru. These industries include flour milling and grain ginneries. However some areas of Nakuru County are marginal, hence conducive for production of sorghum in such areas.

Rongai district comprises four divisions, eighteen locations and thirty seven sub-locations. It falls within an altitude range of between 1650 and 1850 m above sea level (a.s.l) with a temperature range of between 17 and 29°C. There are three agro-ecological zones namely Lower Highland 3 (LH_3), Upper Midland 2 (UM_2) and Upper Midland 3 (UM_3). The annual rainfall is between 600-1000 mm. As per the 2009 census, the district population is 142,127 with 33,868 farm families which form 24% of the population (Jaetzold and Schmidt, 2010). The major crops in the district include maize, wheat, sorghum, finger millet, cow peas, beans, Irish potatoes, sweet potatoes, fruits, vegetables and cassava whereas the major livestock include dairy cows, dairy goats and poultry.

Njoro district comprises four divisions, twenty one locations and forty four sub-locations. There is a wide variation in altitude with Mauche lying within 2100- 2500 m a.s.l, Mau Narok 1700-2850 m a.s.l and Lare 1650-2200 m a.s.l. with a temperature range of between 11 and 24.5°C. The divisions lie within varied agro ecological zones with Mauche lying within UH1, UH2, and LH, Mau-Narok lies within UH1, UH2,LH, Njoro lies within UH2, UH3 and LH3 while Lare lies within LH2, LH4 and UM3. The annual rainfall is between 600-1800 mm as per the 2009 census, the district population is 287,648 with 35,012 farm families which form 12.17% of the population (Jaetzold and Schmidt, 2010). The major crops in the district include Maize, Beans, Wheat, Irish potatoes, Cabbages, Kales, Carrots, Pyrethrum and Sorghum, whereas the major livestock include dairy cows, dairy goats, sheep and poultry. The study area particulars are shown in Figure 1.

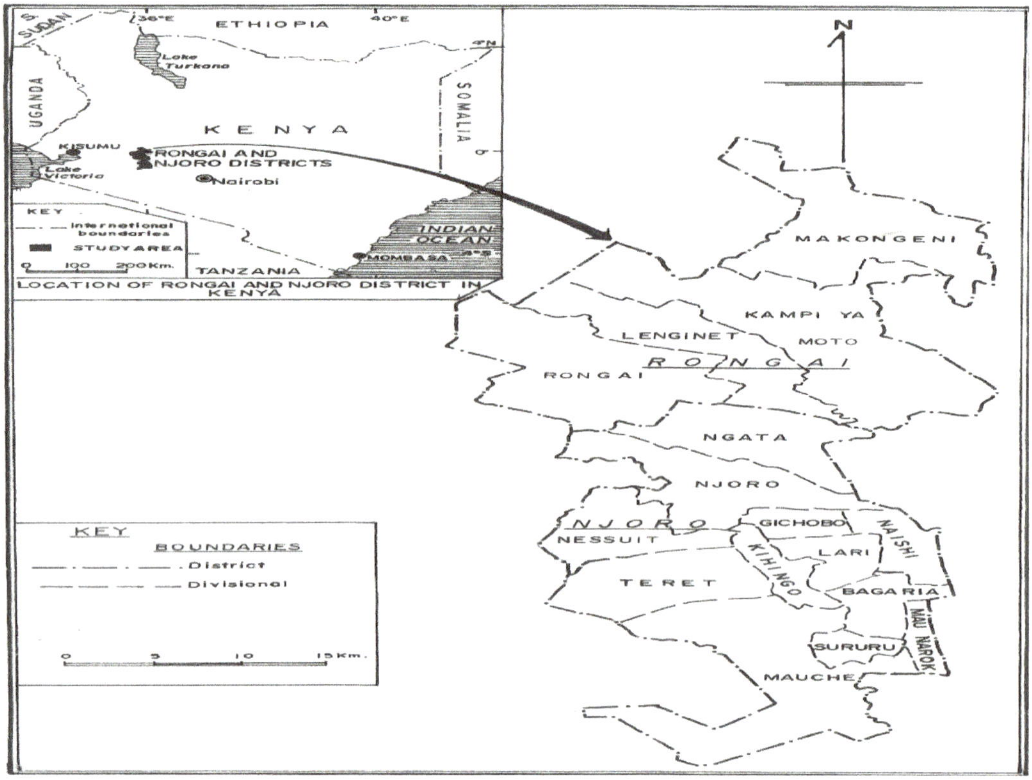

Figure 1. The study area particulars.

Data collection

In Kampi Ya Moto Division, simple random sampling was applied in order to choose a sample of 105 farmers from a sample frame that was provided by the District Agricultural office. This method of sampling involved giving a number to every member of the population in the sampling frame, placing the numbers in a container and then picking any number at random. The subjects corresponding to the numbers picked were included in the sample.
In Njoro Division, both purposive and snowball sampling were applied. In the division, the number of sorghum farmers was so small as compared to non sorghum farmers and not well known, hence simple random sampling alone could not be used. Snowball sampling is useful when the population that possesses the characteristics under study is not well known and there is need to find subjects. Purposive sampling was first used to identify sorghum farmers in the division who will form part of the sample. Snowball sampling was then applied using the purposively identified cases to identify more sorghum farmers to form part of the sample. In snowball sampling, initial subjects with the desired characteristics (sorghum farmers) were identified using purposive sampling technique. The few identified subjects were then asked to name others that they know to be producing sorghum until the list was exhaustive. Then simple random sampling was applied to select non sorghum farmers from the sampling frame until the desired sample size of 102 farmers was attained.

Analytical technique

The double hurdle regression model was used. It models non-participation and potential participation apart from the observed participation. It is a two tier decision or selection equation and the outcome depends on the selection to participate. The empirical model is as shown as follows:

Double hurdle empirical model:

$y_{i1}^{*} = Q_i^{\prime}\alpha + v_i$	Participation decision
$y_{i2}^{*} = X_i^{\prime}\beta + u_i$	Extent of participation
$y_i = X_i^{\prime}\beta + u_i$	If $y_{i1}^{*} > 0$ and $y_{i2}^{*} > 0$
$y_i = 0$	Otherwise

Where: y_{i1}^{*}: Latent variable describing the household's decision to participate in sorghum production; y_{i2}^{*}: Latent variable describing households' extent of sorghum production; y_i : Observed dependent variable; Q_i : Vector of variables explaining the participation decision; X_i : Vector of variables explaining the extent of sorghum production; v_i, u_i: respective error terms assumed to be independent and distributed as $v_i \sim N[0, \alpha^2]$ and $u_i \sim N[0, \alpha^2]$.

The first equation defines the participation decision and non-participation decision model where y_{i1}^{*} takes the value of 1, if a household made a decision to produce sorghum and 0 if no production. This equation was used for analysis for objective one. The second equation defines the intensity of participation where y_{i2}^{*} is the acreage under sorghum production. Q_i and X_i define socio-economic factors that affect the discrete probability of participation or non-participation and intensity of participation respectively. The socio-economic factors under consideration were: Age of household head, gender of household head, household size,

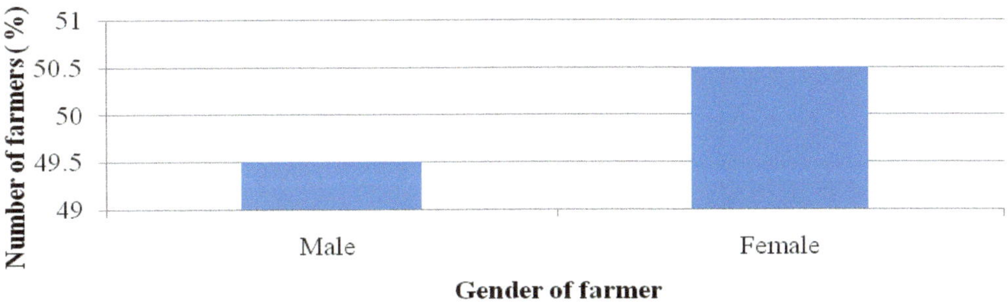

Figure 2. Participation in sorghum production in Nakuru County by gender.

storage facilities, household income and educational level attained by the farmer. v_i and u_i are the error terms of estimation in the participation and intensity of participation functions respectively.
According to Gujarati (2004) and Pallant (2007), the coefficients measure the expected change in the model for a unit change in each independent variable, all other independent variables being equal. The sign of the coefficients shows the direction of influence of the socio-economic variable and it therefore follows that in this study, a positive value in tier 1 indicates an increase in the likelihood that a household will participate in sorghum production and a positive value in tier 2 indicates an increase in the likelihood that the household will put more land under sorghum production. On the other hand, a negative value in tier 1 shows that the household is less likely to participate in sorghum production and in tier 2 that the household is less likely to increase the acreage under sorghum.

The significance values (P-values) show whether a change in the independent variable significantly influences the dependent variable at a given level. In this study, the variables were tested at the 5% and 10% significance levels. Thus, if the significance value is greater than 0.1, then it shows that there is insufficient evidence to support that the independent variable influence the dependent variable. If the significance value is equal to or less than 0.1, then there is enough evidence to support a claim presented by the coefficient value. If the significance value is less than or equal to 0.05, then the variable is significant at 5% significance level. If the significance value is greater than 0.05 but less than or equal to 0.1, the variable is significant at 10% significance level. The standard error measures the standard deviation of the error in the value of a given variable (Hill et al., 2001; Gujarati, 2004).

RESULTS AND DISCUSSION

Socio-economic characteristics of sorghum farmers

Gender of farmers

The results on demographic characteristics showed that there were a larger proportion of females (54.6%) as opposed to males (45.4%). This was as expected given that farming in the region is dominated by women who are involved in such enterprises as sorghum farming, which is grown in the two regions for subsistence purposes. This is in agreement with findings of Doss (1999) that women are more involved in subsistence crop production unlike men. The phrase 'the African farmer and her husband' also expresses the importance of women farmers in Africa, where some 80% of all those engaged in food production are women. On average, women work more hours than men in producing food as they produce the bulk of the food for local or family consumption (Commonwealth Secretariat, 2001). The findings of the study further revealed that a larger proportion of women participate in sorghum production (50.5%) as opposed to males (49.5%) of the total number of farmers who participated in sorghum production as shown in the Figure 2. This implies that women are more involved in sorghum production than men. This agrees with Doss (1999) who concludes that a frequently made distinction is that cash crops and export crops are "male crops," while subsistence crops are "female crops". The standard explanation for the division of crops by gender is that women are responsible for feeding the family and thus prefer to grow small scale subsistence crops for household consumption at the expense of commercialization, whereas men are responsible for providing cash income.

Age distribution of farmers

Age of the household head is an important aspect in agriculture because it determines experience one has in a certain type of farming. Household head's experience further influences household members' farming activities since they usually provide guidance (Ngqangweni and Delgado, 2003). The results of the study revealed that majority of the farmers were of young ages with 36.7 and 26.6% of the farmers falling in the 30-39 and 40-49 age ranges respectively which is the most active age in life in terms of enterprise development and management. Majority of the famers were in the range of between 30-59 years with a few farmers over 60 years. The average age of the farmers was 43 years which implies that majority of the farmers in the study area are of medium age groups. The study further revealed that 52.9% of the farmers who were below 30 years were participating in sorghum production. On the other hand, 57.9% of the

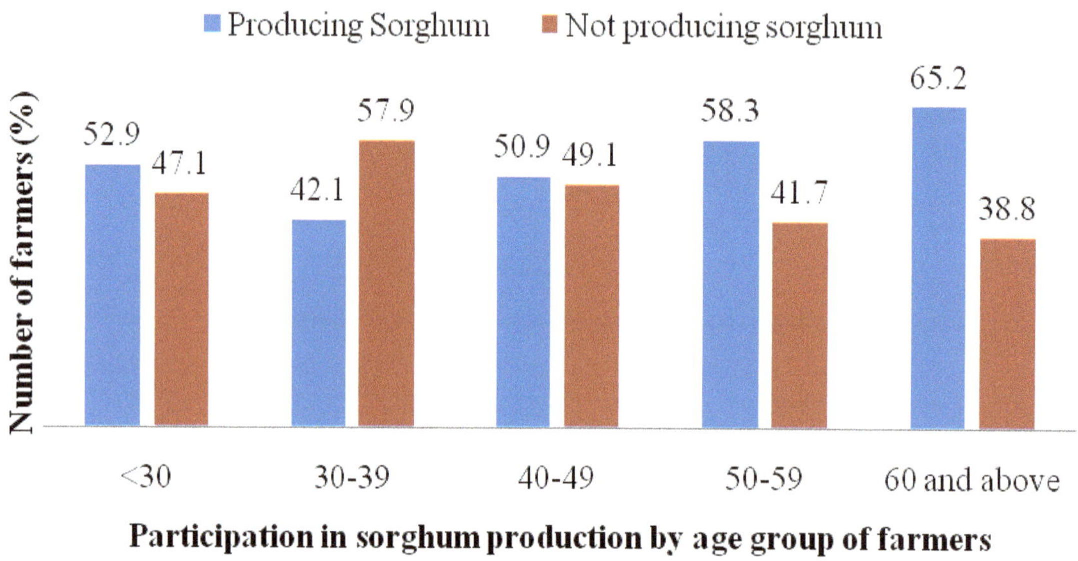

Figure 3. Participation in sorghum production in Nakuru County as per age group of farmers.

farmers who were between 30 and 39 years did not participate in sorghum production with only 42.1% participating as shown in the Figure 3. In the age group of 40 to 49 years, 50.9% were participating in sorghum production. In the age group of 50 to 59 years, 58.3% of the farmers were participating in sorghum production whereas in the age group of farmers who are 60 years and above, 65.2% participated in sorghum production. The results implied that the probability of participation in sorghum production increased as the age of the farmers increased. This means that older farmers in the study area were participating in sorghum production more than the younger farmers. This could be because sorghum becomes a good substitute for grains like maize in the diet of most elderly people. It is often recommended for feeding by elderly people especially those with diabetic complications.

Education level of farmers

Education is an important aspect in enterprise development and farm management. The results showed that 41.1, 42.5 and 10.1% of the respondents had primary, secondary and tertiary level of education, respectively. Only 6.3% of the respondents reported to have no formal education. Majority of the farmers in the study area had primary and secondary education. This generally means that most of the respondents are literate. These results were in agreement with the findings of UNESCO that showed that about 38% of the adult population in sub-Saharan Africa, or 153 million adults cannot read or write and that over 60% of these are women (UNESCO, 2010). The findings further showed that 69.2% of the farmers without education were participating in sorghum production as shown in the Figure 4. Out of the 85 farmers with primary education, 54.1% were producing sorghum whereas out of the 88 farmers with secondary education 47.7% of the farmers were producing sorghum. Only 38.1% of the farmers with tertiary education were producing sorghum. The results generally indicate a declining number of farmers who produce sorghum as the level of education increases. This implies that famers with low level of education (no education and primary level) in the study area were participating more in sorghum cultivation than those with higher education (secondary and tertiary levels). This gave an implication that the probability of participation in sorghum production in the study area decreases as the level of education of the farmer increases. As the level of education of the farmers increase, their probability of getting off-farm engagements increases (Weir and Knight, 1999). They will also in most cases have a tendency of producing high value crops with big margins as compared to sorghum. These could be the reason for the declining production trends in sorghum production as education levels increased.

Land tenure system

Land tenure provides the legal and normative framework within which all agricultural as well as other economic activities are conducted. When tenure rights are certain, they provide incentives to use land in a sustainable manner or invest in resource conservation whether for

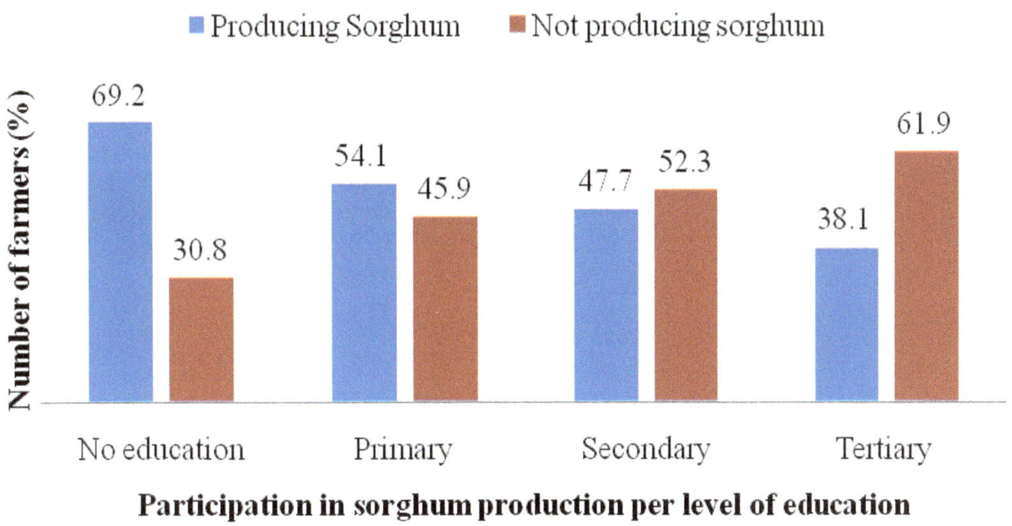

Figure 4. Participation in sorghum production in Nakuru County as per level of education of farmer.

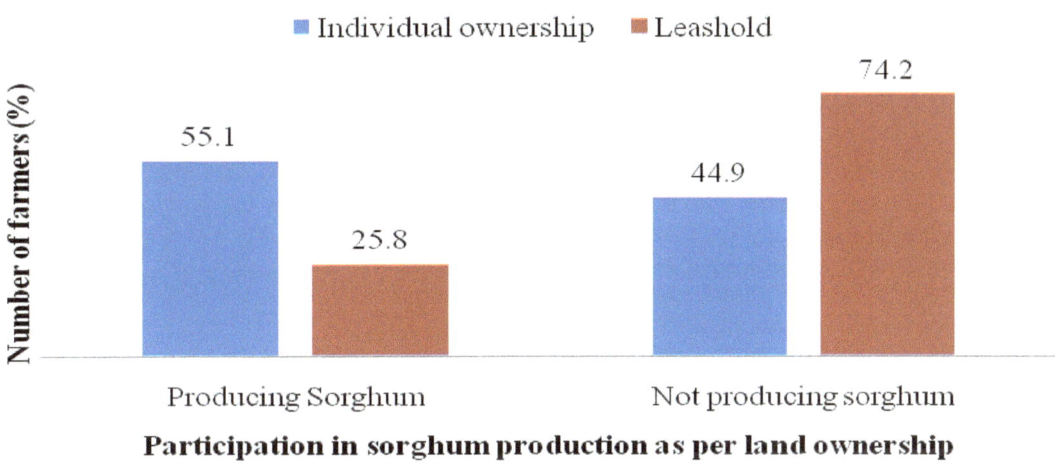

Figure 5. Participation in sorghum production in Nakuru County as per land tenure systems.

the individual or group of individuals (Ogolla and Mugabe, 1996). Land tenure types and policies also tend to determine the nature of agriculture and influence other land use practices (Waiganjo and Ngugi, 2001). It was noted that 84% of the farmers in the study area had individual land ownership while 16% owned land on leasehold. This implied that a majority of the farmers in the study area owned land on individual basis as is often the case in Kenya (Kameri-Mbote, 2005). This is important because it gives farmers an opportunity to invest in the land for productive purposes. The results further indicated that 55.1% of the farmers with individual ownership of land were cultivating sorghum as shown in the Figure 5. On the other hand, only 25.8% of the farmers with leased land were cultivating sorghum. The results show that farmers in the study area on individual ownership of land participated more in sorghum production as compared to those with leased land. This gave an indication that majority of the farmers with leased land were giving priority to other high value crops for higher returns as opposed to sorghum which was produced majorly for subsistence by majority of the farmers even with those on individual land ownership.

Household size

The household is the major source of farm labour in small-scale agriculture, although most sorghum farming operations in the study area are gender specific. Women

Table 1. Sorghum production in Nakuru County as influenced by household sizes of farmers.

Household size distribution			Participation		Total
			Yes	No	
	<4	Count	17	16	33
		Household size distribution (%)	51.5	48.5	100.0
	4-6	Count	57	66	123
		Household size distribution (%)	46.3	53.7	100.0
	7-9	Count	22	15	37
		Household size distribution (%)	59.5	40.5	100.0
	>9	Count	9	5	14
		Household size distribution (%)	64.3	35.7	100.0
Total		Count	105	102	207
		Household size distribution (%)	50.7	49.3	100.0

Source: Sorghum survey, Njoro and Kampi Ya Moto divisions, December 2011 – June 2012.

play a major role especially during harvesting, threshing, winnowing and bagging of sorghum. Nganga (2009) attributes large household sizes to the desire to have enough family labour and hence farmers keen on using family labour instead of hired labour will in most cases have more children. In his study, the use of free child labour was also positively associated with family size implying that families that rely on child labour tend to have more children. This is especially true for women who use older children as baby-sitters as they perform other chores both inside and outside the household. The findings of this study showed that about 59.4% of the farmers had between 4 and 6 members in their household, while about 17.9% of the farmers had between 7 to 9 persons in their households. Only 6.8% had over 9 persons in their household. The average household size was 6 persons and this was almost in tandem with Kenya's mean household size of 5.1 persons (GOK, 1994). Only about 24.7% had more than 6 family members of the household which are fairly large sizes. This means that majority of the farmers in the study area generally had large household sizes. Generally, as family sizes increase, household demands increase which constrains family income unless there is a steady source of income a reason for a positive relationship between household size and poverty levels in most developing countries. It was further noted that 51.5% of the farm households with less than 4 family members were not producing sorghum whereas 46.3% of the farmers with household sizes of between 4 and 6 family members were producing sorghum as shown in the Table 1. However, the percentage of those producing sorghum increased as the household sizes increased, as about 59.3% of the famers who had between 7 and 9 family members were producing sorghum. The percentage of those producing increased further to 64.3% for farmers with over 9 family members. These results gave an implication that the likelihood of farmers in the study area of participating in sorghum production increased as the household sizes increased. As household size increases, the demand for various household items increases which in turn reduces the disposable income for production. This means little income is available for purchase of farm inputs and this could lead to production of sorghum by majority of the big households since sorghum requires very little input investment.

Farm size

Farm size plays an important role in deciding what enterprises farmers undertake and the land size to be put under each of those enterprises. The study results showed that majority of the farmers in the study area, that is, 72% of 207 farmers, had less than 4 acres. Overall, 19.3% of the farmers had between 4 and 8 acres and 3.9% having between 9 and 13 acres, with only 4.8 per cent having more than 13 acres of land for farm operations as shown in the Table 2. The results showed that majority of the famers had limited sizes of land for farm production.

The findings also indicated that more farmers with large farm sizes participate in sorghum production than those with small farm sizes. Out of the 149 farmers with less than 4 acres of land, 48.5% participate in sorghum production and 50.9% of 40 farmers with 4-8 acres produce sorghum as shown in Figure 6. On the other hand, 50% of the farmers with between 9 and 13 aces participate in sorghum with an equal number not participating. About 80% of the farmers with over 13 acres produce sorghum. Farmers with larger farm sizes

Table 2. Ranges of farm sizes in Nakuru County.

Farm size in acres	Number of farmers	Percentage of farmers
<4	149	72.0
4-8	40	19.3
9-13	8	3.9
>13	10	4.8

Source: Sorghum survey, Njoro and Kampi Ya Moto divisions, December 2011 – June 2012.

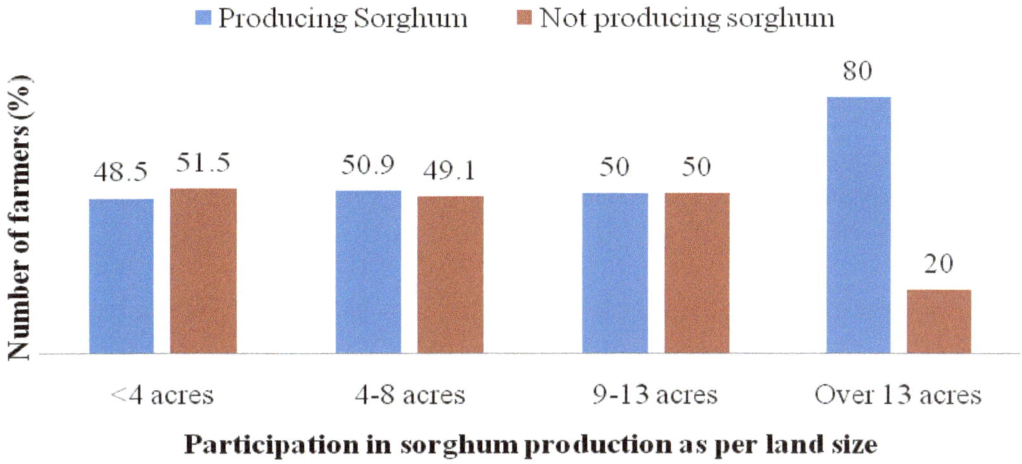

Figure 6. Participation in sorghum production in Nakuru County as per farm size.

are participating more in sorghum production as compared with those with small farm sizes. This means that the probability of participation in sorghum production in the study area increases fairly as the farm sizes increase which gives a positive relationship between farm size and participation in sorghum production. This could be occasioned by the fact that as land size increases, it gives room for more alternative enterprises.

Household income

Household income is important in determining the types and scale of enterprises to be undertaken by farmers. From the findings, there were two main sources of income namely farm income and off-farm income. Sources of farm income mainly include revenues from farm crop and livestock produce and revenues earned from farm assets. Sources of off-farm income included mainly off-farm employment. Out of the 207 farmers, 56.7% had no off-farm income and wholly relied on farm income while 43.3% of the farmers had access to both farm and off-farm sources of income. It was also noted that out of the 207 farmers, 64.3% had incomes ranging from Kshs 800 to Kshs 10, 000 per month which indicate that income could be a limiting factor of production among farmers in the study area as only 12.6% had incomes of over Kshs 20,000 per month. This gave an implication that majority of the farmers in the study area may not be having access to adequate disposable income for farm investment hence limiting and prioritizing the types of enterprises undertaken at farm level. About 51.6% of the farmers with income of less than Kshs 10,000 did not participate in sorghum production whereas in the income category of between Kshs 10,000 and Kshs 20,000, 56.4% of the farmers participated in sorghum production as shown in Figure 7.

On the other hand, 50% of farmers with income of over Kshs 20,000 participated in sorghum production. This showed that farmers with higher incomes participate more in sorghum production as compared to those with low incomes. This also gives an implication that as farm incomes increase in the study area, the likelihood of participation in sorghum production also increases hence a positive relationship between household income and participation in sorghum production in the study area. This could be due to the fact that farmers with more income have enough resources for alternative enterprises. They could also be investing in sorghum production to make use of the available demand by EABL

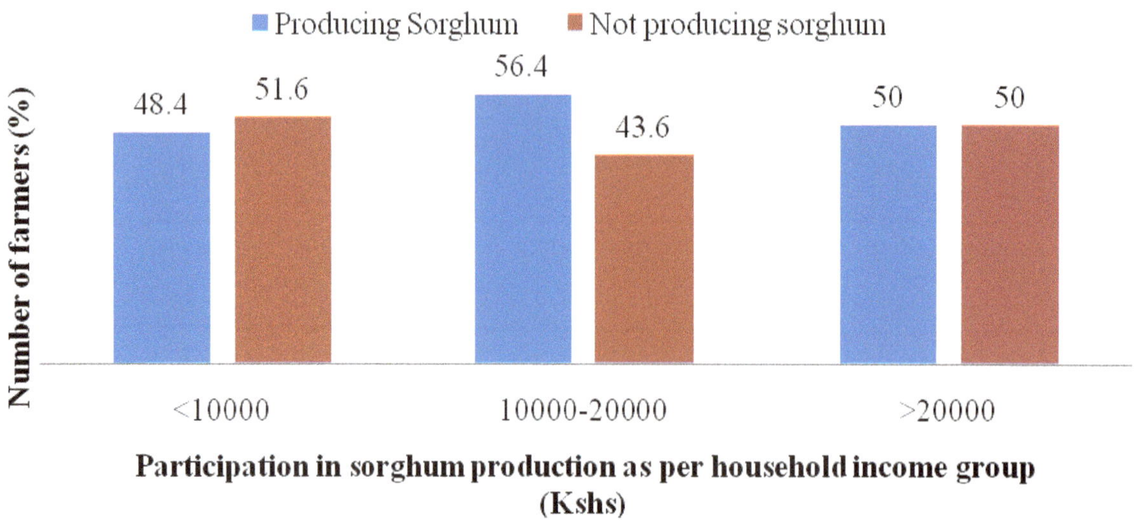

Figure 7. Participation in sorghum production in Nakuru County as per category of household income.

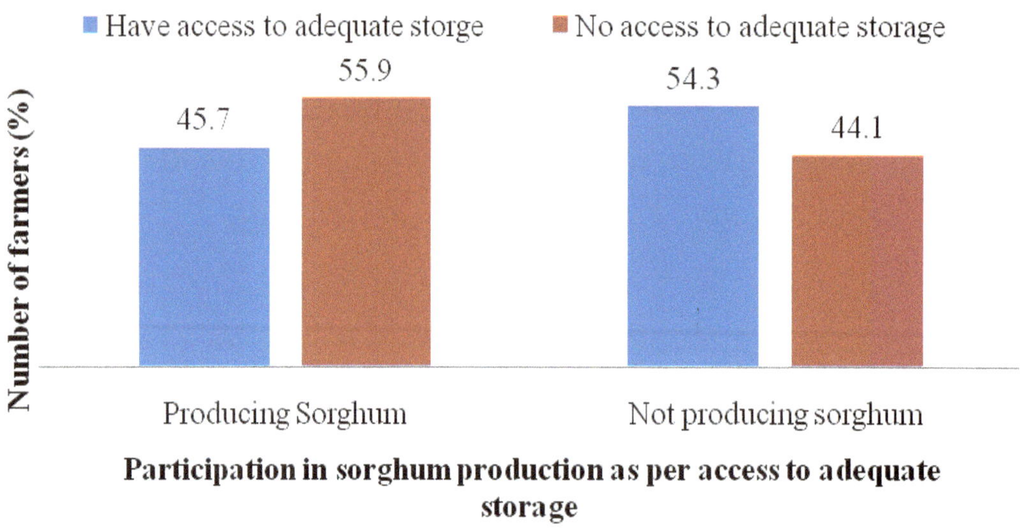

Figure 8. Farmer participation in sorghum production on the basis of access to storage.

and other markets.

Access to adequate storage facilities

Storage facilities are important in decision making on cereal enterprise selection depending on varying storage requirements. About 50% of the farmers in the study area have access to adequate storage facilities. This means that half of the farmers in the study area have got access to adequate storage facilities for their produce. About 55.9% of the farmers without adequate storage were participating in sorghum production whereas only 45.7% of farmers with adequate access to storage were participating in sorghum production as shown in Figure 8.

The results indicate that more farmers without access to adequate storage were producing sorghum than those with access to adequate storage. This showed that storage might not be a major consideration in sorghum production in the study area. Both those with or without access to adequate storage have almost equal chances of participating or not participating. Low sorghum production needs little storage space and thus storage facility is not a key factor influencing participation in sorghum cultivation.

Model empirical results

The main objective was to identify the socioeconomic

Table 3. Double hurdle results for participation in sorghum production

Variable	Tier 1 (Participation in sorghum production)					
	Coef.	Std. Err	z	Significance (p>|z|)	[95% Conf. Interval]	
AGE	-0.0024	0.0107	-0.23	0.822	-0.0235	0.0186
HHSIZE	-0.0064	0.0572	-0.11	0.910	-0.1187	0.1058
GENDER	-0.3601	0.1922	-1.87	0.061**	-0.7367	0.0165
EDLEVEL	-0.0418	0.0300	-1.40	0.163	-0.1006	0.0169
LANDTENURE	-0.5726	0.2567	-2.23	0.026*	-1.0757	-0.0695
FARMSIZE	0.0473	0.0252	1.87	0.061**	-0.0021	0.0968
HHINCOME	-0.00001	0.00001	-0.93	0.353	-0.00003	0.00001
ADQTSTORE	0.1574	0.1874	0.84	0.401	-0.2099	0.5247
_cons	2.6955	1.4653	1.84	0.066**	-0.1764	5.5674

N = 207; * and * * statistically significant at 5% and 10% significance level respectively. Land tenure is significant at 5%. Gender and farm size are significant at 10%.

factors that influence participation and extent of participation in sorghum production among the smallholder farmers in the study area. The data collected was analyzed using the double hurdle model. The double hurdle regression results of participation and the extent of participation in sorghum production are presented in Tables 1 and 2. The tables show the estimated coefficients (α value), standard error and significance values of independent variables in the model. As indicated in Tables 1 and 2, some predictor variables influence decisions in participation in sorghum production significantly in tier 1 and some variables also influence extent of sorghum production significantly as shown in tier 2. Out of the 8 independent variables used in the model in tier 1, only land tenure was statistically significant at the 5% significance level whereas gender of the household head and farm size were statistically significant at the 10% significance level as shown in Table 3.

Gender of the household head had a negative sign. It had been expected that the variable could have a positive or negative sign. The significance values of 0.061 showed that there was enough evidence to support that if the number of households headed by males increases then the likelihood of participation in sorghum production diminishes and vice versa. From the results on gender, it had been shown that there were slightly more women producers of sorghum than men. This agrees with Doss (1999) and Curran and Cook (2009) who indicated that women are more likely to participate in production of crops which are more of a subsistence nature than men. They argue that women are usually responsible for cultivating food for home consumption on household plots whereas men usually grow the main cash crop on personal plots. In many areas, women are also allowed to cultivate and gain income on a personal plot but because sorghum is not often used as a cash crop, women are the primary cultivators.

Land tenure had a negative sign. It had been expected that the variable could have a positive or negative sign. The significance value of 0.026 indicates that there was enough evidence to support that land tenure affects participation in sorghum production. A close examination showed that majority of the farmers not participating in sorghum production were those on leasehold. This is what could be the cause of the negative value of the correlation coefficient in statistical terms. Farmers who are on individual ownership are expected to participate more than those on leasehold but only 45% of them were not participating and this could also be contributing to this scenario. Waiganjo and Ngugi (2001) noted that tenure insecurity whether customary or statutory tenure regimes undermine the effectiveness of farming activities. They conclude that land tenure types and policies tend to determine the nature of agriculture and influence other land use practices. Farm size had a positive sign and a significance value of 0.061. This means that there was enough evidence to support that as farm size increases; the likelihood of participation in sorghum production also increases. The positive sign indicated a positive relationship between farm size and participation in sorghum production in the study area. This indicates that as land size increases the likelihood of participation in sorghum production increases. This could be probably because, as farm size increases, more enterprises can be undertaken simultaneously. Farmers with larger farm sizes are also more likely to adopt modern varieties of sorghum for higher returns which is in agreement with Nambuya et al. (2005) who found that adoption of improved varieties is positively correlated to farm sizes, and that most farms with small land holdings were not growing improved varieties. This means that most small scale farmers grew indigenous varieties and had a lower likelihood of participating in sorghum production as compared to large scale farmers. Out of the 8 independent variables used in the model in tier 2, only educational level and land tenure were statistically significant at the 5% significance levels as shown in

Table 4. Double hurdle results for extent of participation in sorghum production

Variable	Tier 2 (Extent of Participation in sorghum production)					
	Coef.	Std. Err	z	Significance (p>|z|)	[95% Conf. Interval]	
AGE	-0.0056	0.0074	-0.75	0.455	-0.0202	0.0090
HHSIZE	0.0602	0.0419	1.44	0.151	-0.0220	0.1424
GENDER	0.0846	0.1289	0.66	0.512	-0.1682	0.3373
EDLEVEL	0.0446	0.0219	2.04	0.041*	0.0017	0.0875
LANDTENURE	0.3648	0.1800	2.03	0.043*	0.0120	0.7176
FARMSIZE	0.0091	0.0093	0.97	0.332	-0.0092	0.0274
HHINCOME	5.79e-06	7.24e-06	0.80	0.424	-8.40e-06	0.00002
ADQTSTORE	-0.0539	0.1300	-0.41	0.679	-0.3087	0.2010
_cons	-0.8409	0.9659	-0.87	0.384	-2.741	1.0522
sigma						
_cons	0.4346	0.0631	6.89	0.000*	0.3111	1.5582

N = 207; * statistically significant at 5% significance level. Educational level and land tenure are significant at 5%.

Table 4.

Educational level of the household head has a positive sign and a significance value of 0.041. This means that there is enough evidence to support that as the educational level of the household head increases the likelihood of more acreage under sorghum cultivation increases. Therefore as the farmers' education level increased there were greater chances of increasing the acreage under sorghum in the study area as it had been hypothesized. This agrees with the findings of Wakili (2012). He asserted that more years of formal education is imperative for better understanding and adoption of new technologies which subsequently make it possible to move closer to the frontier. Furthermore, educated farmers are expected to be more receptive to improved farming techniques and therefore should have higher level of technical competencies than farmers with less education. Farmers with low level of education would be less receptive to improved farming techniques as they have low adaptive capabilities.

Land tenure has a positive sign and a significance value of 0.043. There is enough evidence therefore to support that as the number of households with individual land tenure increases, the likelihood of more acreage under sorghum production increases. From the earlier results on land tenure, majority of the households with individual land ownership were producing sorghum. The tenure system is important in decision making in terms of choice of enterprises. It is much more convenient to make production decisions when land is individually owned than when leased and this could be the reason for the positive influence of land tenure on extent of sorghum production.

CONCLUSION AND RECOMMENDATIONS

It was concluded that land tenure, gender of household head, farm size and educational levels of household head were significantly influencing sorghum production in the study area. It was also concluded that if more women were encouraged to participate in sorghum production, then this could lead to tremendous increase in sorghum production in the study area.

To fully tap the potential of increased participation and acreage under sorghum production in the study area, more women should be encouraged to participate in sorghum production as it was found that they participated in sorghum production more than men. Emphasis should also be given to the significant variables by maximizing on their influence especially those that influence sorghum production positively. This study only focused on socioeconomic factors of sorghum production. There is need for further research on the influence of other factors, such as economic and agronomic factors, and their influence on sorghum production.

REFERENCES

Commonwealth Secretariat (2001). Gender Mainstreaming in Agriculture and Rural Development. United Kingdom: Commonwealth Secretariat.

Curran S, Cook J (2009). Gender and Cropping: Sorghum in Sub-Saharan Africa. Washington: University of Washington.

Doss CR (1999). Twenty-Five Years of Research on Women Farmers in Africa: *Lessons and Implications for Agricultural Research Institutions; with an Annotated Bibliography.* CIMMYT Economics Program Paper No. 99-02. Mecico: CIMMYT.

EPZA (2005). Grain Production in Kenya. Export Processing Zones Authority.

FAO (1995). Sorghum and Millets in human nutrition. Rome: FAO.

FAO (2009). The State of Food Insecurity in the World: Economic crises – impacts and lessons learned. FAO.

FAO and ICRISAT (1996). The world sorghum and millet economies Facts, trends and outlook. Italy, India: FAO and ICRISAT.

GoK (1994). *Sessional paper No. 2 of 1994 on National Food Policy.* Nairobi: Government of Kenya.

Gujarati D (2004). Basic Econometrics, Fourth Edition. The McGraw–Hill Companies, New York.

Hamukwala PT (2010). Sorghum and pearl millet improved seed value chains in Zambia: Challenges and opportunities for smallholder farmers. The University of Zambia, Lusaka, Zambia.

Hill RC, Griffiths WE, Judge GG (2001). Econometrics, Second Edition. John Wiley & Sons, New York.

IFPRI (2009). Climate Change – Impact on Agriculture and Costs of Adaptation.

Jaetzold R, Schmidt H (2010). Farm Management Handbook of Kenya, Volume II, Part B: Natural Conditions and Farm Management Information, Central Kenya (Rift Valley and Central Provinces) 121-151:379-404.

Karugia JT (2003). A micro level analysis of agricultural intensification in Kenya: The case of food staples. Department of Agricultural Economics - University of Nairobi.

KNBS (2010). Gross Domestic Product; Third Quarter 2010. Kenya National Bureau of Statistics, Nairobi.

MOA (2007). Recipe Manual for Traditional and Underutilized Foods. Government of Kenya.

MOA (2009). Strategic Plan 2008-2012. Nairobi: Government of Kenya.

MOA (2012). Economic Review of Agriculture. Nairobi: Ministry of Agriculture.

Nambuya IS, Bashasha B, Adipala E (2005). Factors affecting adoption of improved sorghum varieties in eastern Uganda and their implication for policy intervention. *African Crop Science Conference Proceedings*, Kampala, pp. 7: 697-703.

NCAPD (2005). District Strategic Plan 2005 - 2010. National Coordinating Agency Population Development.

Nganga TKW (2009). Family Size, Economics and Child Gender Preference in the Nyeri District of Kenya. 18th Annual IAFFE Conference at Simmons College, Boston, Massachusetts, USA 26th – 28th June 2009 University of Nairobi. pp. 1-40.

Ngqangweni S, Delgado C (2003). Decisions on livestock keeping in the semi-arid areas of Limpopo Province. Working Paper, Department of Agricultural Economics, Extension and Rural Development, University of Pretoria, Pretoria.

Ogolla BD, Mugabe J (1996). Land Tenure Systems, In Land We Trust; Initiative Publishers, Nairobi Kenya.

Pallant J (2007). SPSS: A Step by Step Guide to Data Analysis using SPSS for Windows, Third Edition. New York: Open University Press.

Schmidt OG (1988). Priorities in post harvest research support for sorghum and millet in Eastern Africa. International Development Research Centre (IDRC).

UN (2009). The Millennium Development Goals Report.

UNDP (2002). Kenya Human Development Report, UNDP, Nairobi.

UNESCO (2010). Education for All Global Monitoring Report. Retrieved February 1, 2012, from UNESCO: www.efareport.unesco.org.

Waiganjo C, Ngugi EN (2001). The effects of existing land tenure systems on land Use in Kenya today. *International Conference on Spatial Information for Sustainable Development, 2–5 October 2001*, (pp. 1-10). Nairobi, Kenya.

Wakili AM (2012). Technical efficiency of sorghum production in hong local government area of adamawa state, Nigeria. Russian J. Agric. Socio-Econ. Sci. 6:1-6.

Weir S, Knight J (1999). The Effects of Education on Farmer Productivity in Rural Ethiopia. UK: Centre for the Study of African Economies, Department of Economics, University of Oxford.

The sustainability of cotton production using genetically modified seeds by emerging smallholder farmers in South Africa: a global and African pesrpective

Magicson Mthembu[1], Cliff S. Dlamini[2] and Banele Nkambule[3]

[1]Faculty of Natural and Agricultural Sciences (Centre for Sustainable Agriculture), University of the Free State, South Africa.
[2]WWF South Africa, C/O Table Mountain Fund, Centre for Biodiversity Conservation, Kirstenbosch National Botanical Gardens, P. O. Box 23273, Claremont, 7735, South Africa.
[3]World Vision Swaziland, P. O. Box 2870 Mbabane Swaziland.

This review focuses on the sustainability of adoption of genetically modified seeds by developed countries like United States, Australia and developing countries such as China and India. It then zooms into the African perspective in relation to countries that adopted Bt cotton. The review indicates that the Bt technology has played a critical role improving cotton yields and reduce the cost of regular spraying for bollworm insects. South Africa is one of the few developing countries, and the only one in Africa that has adopted genetically modified crops for commercial production.

Key words: Sustainability, Bt cotton, *Bacillus thuringiensis*, smallholder farmers, genetically modified.

INTRODUCTION

International overview

Cotton production in the world

Cotton is grown mainly for the fiber although a small quantity of the seed is used as a source of food, feed and oil for both human and animal consumption. Cotton seed is refined before it is utilized for human consumption to remove gossypol which is toxic to humans and monogastric animals (Jia, 1998). Table 1 shows a list of top ten countries ranked as the top producers of cotton lint.

Cotton engineered to produce insecticidal proteins of *Bacillus thuringiensis* referred to as Bt cotton has been one of the widely adopted and effective transgenic seed grown worldwide. In 2001 roughly 4 million hectares of land was reported to be cultivated with Bt cotton globally (Brookes and Barfoot, 2006). According to Williams (2001), the United States of America accounted for nearly 2.4 million hectares (60%) of Bt cotton planted that year. China recorded plantings of nearly 1.5 million hectares and Australia roughly 0.1 million hectares (James, 2001). Other countries that were the first adopters of Bt cotton includes Argentina, Indonesia and Mexico (Bennet et al., 2006).

ADOPTION OF GENETICALLY MODIFIED SEEDS IN THE WORLD

Bt cotton was officially approved for sale in India in 2002 (Qaim, 2002). Prior to this time, India had been battling to control bollworm infestation that halved the yields. Within a decade after the approval, Bt cotton was adopted by nearly 7 million farmers and today they cover 97% of the area planted with the cotton in India (Qaim, 2002). The high adoption rates were mainly due to simplification of regulatory procedures and well established bio-safety records by the Ministry of Environment and Department of Biotechnology hence leading to expeditious commercial release (Karihaloo and Kumar, 2009).

As a result, cotton production in India has risen with the use of the hybrid Bt cotton seeds, benefitting farmers and helping the country to become net exporter of the cash crop. A study jointly undertaken by the Council for Social Development (CSD) and Bharat Krishak Samaj (BKS), showed that overall production of cotton has grown by 9.25% since introduction of Bt cotton in 2002/2003 and farmers income jumped up to nearly 375% (Karihaloo and Kumar, 2009). Of all the crops and countries engaged in GMO (genetically modified organisms controversies today, only a few are as important as Bt cotton in India. In India, the cultivation of hybrid Bt cotton seeds which began in 2002/2003, has seen the average income of farmers increasing by almost 575% with average net returns at Rs 64, 113, 96 per hectare. The average nationwide cotton yields went from 302 kg/ha in the 2003/2004 season to a projected 481 kg/ha in 2011/2012 by 59% overall (Abhiyan and Abhiyan, 2012).

From China's perspective, current estimates suggest that as much as 95% of all cotton grown in that country is of the genetically modified variety. Vast planting of transgenic crops producing insecticidal proteins from the bacterium have helped to control several major pest and reduce the need for insecticide sprays (Abhiyan and Abhiyan, 2012)

In 1991, the biotechnology research centre of the China Academy of Agriculture Sciences (CAAS) initiated a major research program to develop cotton varieties that would contain a gene that would produce a Bt toxin which would control cotton bollworm (Alston et al., 1995). The first successful genetically engineered cotton plant was produced in 1993; by 1999, twenty new cotton varieties containing the Bt gene had been produced. In 1995, CAAS started testing these varieties in experimental fields regulated by the Ministry of Agriculture. In 1997, the Chinese bio-safety committee approved four CAAS varieties for commercial use in nine provinces, in 1998 farmers planted 10, 000 ha of CAAS Bt cotton (Deng, 1999). CAAS had difficulty selling more of it in 1998 because the government seed companies, which have regional monopolies on cotton seed sales, were not interested in distributing it. As a result CAAS formed a joint venture to commercialize Bt cotton called Bio-century Transgenic Corporation Ltd. This partnership greatly increased Bt seed cotton production in the country (Jia, 1998).

In the United States, the gene technology for managing insect pests was approved for commercialization by the US Environmental Protection Agency (EPA) in October 1995 and is now available from several seed companies in the country, as well as from many other cotton growing countries around the world (Rabobank, 1996). Preservation of Bt technology is critical for cotton producers across the US cotton Belt because of increasing insecticide resistance and production cost (Abhiyan and Abhiyan, 2012). Frequent introduction of new transgenic cotton varieties creates a need to continuously evaluate their cost effectiveness and to develop efficient plans for their deployment.

All Bt cotton plants contain one or more foreign genes derived from the soil-dwelling bacterium, *Bacillus thuringiensis*, thus, they are transgenic plants (Hardee et al., 2000). The insertion of genes from Bt causes cotton plant cells to produce insecticide proteins, often referred to as cry proteins. These insecticide proteins are effective in killing some of the most injurious caterpillar pest of cotton, such as the larvae of tobacco budworms and bollworms.

OVERVIEW OF AFRICAN CONTEXT

Cotton growing is an important livelihood strategy for large numbers of rural household in Africa. The crop is grown mainly on small family farms with a scale less than 4 ha (Gouse et al., 2002). Burkina Faso is by far the biggest cotton producer in Africa, where about three million people are involved in growing the crop (Hillocks, 2009). Cotton farming provides an income to small holders living in drier areas, where there are no other alternative crops that can be produced. Cotton is also a demanding crop in terms of crop husbandry and pest management. Hillocks (2009) indicated that the average yields in most Sub-Saharan Africa countries are well below the yield potential of current varieties under rainfed conditions. In Eastern and Southern Africa, yields of 500 to 700 kg/ha of seed cotton are typical for varieties with a yield potential closer 3,000 kg/ha (Abhiyan and Abhiyan, 2012).

According to Abate et al. (2000), cotton is one of the most important cash crops in West Africa and is a vital catalyst to economic development in that region. Cotton is the most important agricultural export and constitutes a major share of export earnings in Africa. Cotton has been produced in West Africa since the colonial era, where it was concentrated in the semi-arid regions. The occurrence of diseases and insect pressure limited agricultural development in wetter, higher potential areas. The area under cotton production keeps expanding; therefore production principles need to be improved for crop protection (Abhiyan and Abhiyan, 2012).

Insects are a major threat to cotton, about 15% of the world cotton production is lost due to insect attack every year (Oerke, 2005). In West Africa, the numbers are higher, with about 23% of cotton production lost to insect

Table 1. Top 10 International cotton growing countries.

Rank	Country	Production (million bales)
1	China	33.0
2	India	27.0
3	United states	18.0
4	Pakistan	10.3
5	Brazil	9.3
6	Uzbekistan	4.3
7	Australia	4.2
8	Turkey	2.8
9	Turkmenistan	1.6
10	Greece	1.4

Statista (2013).

Table 2. Genetically modified crops approved for commercial release in South Africa.

Crop	Year first approved	Season first approved
Insect resistant cotton	1997	1997/1998
Insect resistant maize	1997	1998/1999 (yellow)
Herbicide tolerant cotton	2000	2001/2002
Herbicide tolerant soybeans	2001	2001/2002
Herbicide tolerant maize	2002	2003/2004
Stacked gene cotton	2005	2005/2006
Insect resistance		
Herbicide tolerance		

Source: Gouse et al. (2008).

(Vognan et al., 2002). Among insects, the cotton bollworm complex is the most damaging to cotton yields in West Africa (Vaissayre and Cauquil, 2000). In particular, the major bollworm pest throughout West Africa is American bollworm, which is found throughout the region. The other bollworm varies from country to country, and includes the pink bollworm, spotted bollworm, spiny bollworm and red bollworm. Damage to cotton plants is characterized by feeding activity on squares (flower buds), flowers, and cotton bolls. Flower and boll damage is the most severe as it results in the shedding of the plant's reproductive parts and drastically reduces potential yield (Abhiyan and Abhiyan, 2012).

Chemical insecticides are used extensively in cotton production to control insect pest, with the primary target being bollworms. The number of sprayer per crop season varies between locations and from one year to the next. Cotton producers spray about six times per year, although as many as ten sprayings can be required.

OVERVIEW IN THE SOUTH AFRICAN CONTEXT

South Africa is one of the few developing countries, and the only one in Africa that has adopted genetically modified crops for commercial production (Gouse et al., 2002). The first field trials of transgenic crops on the African continent were initiated in 1989 with an application to the South African Department of Agriculture from the US seed company Delta and Pine land, which used South Africa as an over wintering haven for field trials of cotton. This application was approved according to guidelines established by the South African Committee for Genetic Experimentation (SAGENE), a scientific advisory committee set up in 1979 to monitor and advice on the development of GMOs in the country.

Insect resistant cotton has been produced since the 1997/2002 period, herbicide tolerant cotton has been made available for commercial production and tolerant soybeans have been introduced on a small scale. Table 2 indicates genetically modified seeds approved to be produced in South Africa.

A survey conducted in 2006 estimates that the area planted under GM crops now stands at 609 000 hectares (Gouse et al, 2008). According to this survey, the GM share of soybean crop in South Africa stands at 59%, while GM maize holds a 29% market share of maize and GM cotton makes up about 90% of the area planted with the crop.

In general this modern technology is well suited for high technology farming, but now it has also made an appearance in the low input management agriculture system. South Africa is marked by a strong system of intellectual property rights (IPR) and a dualistic agricultural system with large and small holder farmers operating under similar but not identical market access conditions. Genetic modification technologies have been characterized by an impressive adoption rate in South Africa.

According to Gouse et al. (2002), large scale cotton farmers reported pesticide and application cost savings and peace of mind about bollworms as the major benefits of Bt cotton. Small holder farmers indicated financial savings on pesticides and yield increases as the major benefit and reason for adoption of Bt cotton.

ADOPTION BY SMALL HOLDER FARMERS

According to Gouse et al. (2002), the rapid adoption rate of Bt cotton earlier on by small holder farmers in the Makhathini flats of Kwazulu Natal has been the focus of considerable international attention since it is one of the earliest examples of widespread planting of GM crops by resource poor farmers. The experience with GM cotton brought benefits initially, and this story became frequently quoted as testimony that GM crops can indeed have a positive impact on small holder farmers in developing countries.

The project of introducing Bt cotton to small holder cotton farmers was further roll on to Mpumalanga in the Nkomazi area. The Nkomazi area has always been the biggest contributor of cotton production in the country. The Bt technology was received with great hope as it addresses the challenges of insects and weed control. According to DARDLA (2011) about 259 smallholder farmers planted Bt cotton in Nkomazi area over nine villages. A total of 600 hectares was reported planted in the 2011 season with the Bt seeds. The successful yields gains received by farmers in the 2011 season opened the negotiations for prospects of building a ginner in the future.

EMPLOYMENT IN THE COTTON INDUSTRY

According to Department of Agriculture, Forestry and Fisheries (DAFF) in 2003, it was estimated that the primary cotton industry employed on average an estimated 58, 950 seasonal and 6, 550 permanent workers. This numbers were expected to increase to 154, 215 seasonal and 17, 135 permanent workers should local cotton production increase to 74, 000 tons annually, or in a scenario where small holder farmers produce 27% of the crop at 74 000 tons of cotton lint, an increase of 112, 534 seasonal and 12, 504 permanent workers, as well as an additional 11, 533 small holder farmers was expected.

In the 2008/2009 production year, employment figures were estimated at around 7, 000 cotton farm workers and for the 2009/2010 production year, employment figures increased to 8, 000 cotton farm workers due to increase in the area planted. According to the strategic plan developed by cotton South Africa (2004), the cotton industry was planning on broadening participation to enable small holder farmers to contribute on average 25% of the national crop by 2007 and 35% by 2014, to raise productivity which will improve the social life of the rural poor who depend on cotton farming for survival.

COTTON MARKETING STRUCTURE

Local cotton and cotton lint are marketed in the following three ways. Firstly, the seed cotton is sold by the producer to a ginner who gins the cotton and in turn sells the cotton lint for his own account to spinners and the seed to processors, either directly or by making use of agents. The ginner determines the price the producer obtains for his seed cotton and the producer relinquishes ownership of the cotton. Secondly, the producer may decide not to sell his seed cotton to the ginner, but contracts the ginner to gin it on his behalf on payment of a ginning fee. In this case, the cotton lint and seed remains the property of the producer who then either markets these products himself or contracts the gin or someone else to market the cotton lint on his behalf. Lastly, the producer can also gin. He then either markets the cotton lint and seeds himself or contracts someone else to do it on his behalf.

In the ginning process seed cotton is separated from the fibre. The purchase of seed cotton takes place in terms of the grading standards applicable to handpicked and machine picked cotton and linked to the South African grading standards for lint. The latter serves as a norm for the sale of the cotton lint. In the event of any disputes arising during the delivery of the cotton that cannot be resolved to the satisfaction of both parties, the Quality Control Department of cotton, South Africa is accepted as an impartial body in the role of arbitrator.

During the same period of declining cotton production, the industry experienced improvement in cotton producer prices which began increasing in 2007/2008 and at the same time a peak was attained at approximately R4 700 per ton in response to improvement in international cotton prices. In 2008/09 market seasons, there was a slight decline in prices of seed cotton due to low production levels and low demand for South African seed cotton. In the 2009/2010 and 2010/2011 marketing season, average producer prices of seed cotton started to increase again. The change between 2009/2010 and 2010/2011 period represent 11% increase in the average

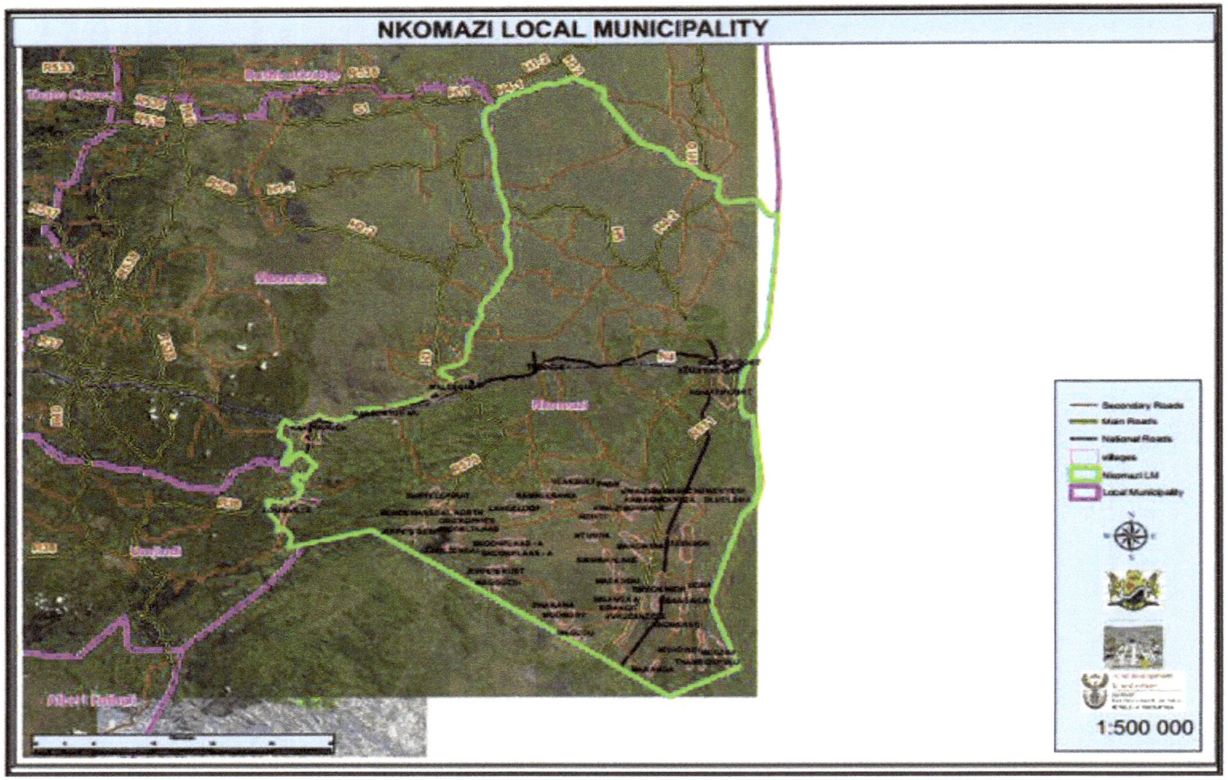

Figure 1. Map of Nkomazi municipality. Source: DARDLA (2013).

producer prices of seed cotton.

COTTON PRODUCTION IN THE NKOMAZI AREA OF SOUTH AFRICA

Nkomazi is located in the eastern part of the Lowveld area in the Mpumalanga province. It lies 350 km east of Gauteng bordering Mozambique and Swaziland. The Nkomazi area has a potential of about 3, 500 hectares that can be cultivated with cotton. The leading tribal village in cotton production includes Steenbok, Brink, Albersnek, Mangweni, Mbangwane and Tsambokhulu. Figure 1 shows a map of the Nkomazi area in the Mpumalanga province of South Africa.

PERSPECTIVE OF NKOMAZI AREA IN RELATION TO COTTON PRODUCTION

Cotton production has been a culture in this part of the province and it has been produced for many years. However, it must be mentioned that the scale of production has decreased tremendously of late as a result of factors such as, poor market prices, high production costs, and uncertainty of rain distribution due to climate change and development of the sugar industry which resulted in conversion of some potential cotton lands for sugarcane production.

It is for this reason the Mpumalanga Department of Agriculture, Rural Development and Land Administration (DARDLA) is implementing a strategy to revive cotton production in the area where farmers are supplied with production inputs such as seed and chemicals.

The aim of the Department is to assist farmers by providing them with a cotton ginner which will buy all cotton produced by farmers in the region. It also affirms that building a ginner will require volumes of cotton in order to run efficient (DARDLA, 2011).

Cotton planted in this part of the province is mainly on dry land and the farmers are planting Bt cotton sponsored by the Department with some other production inputs.

CONCLUSION

This review indicates that genetic modified cotton seeds brought hope in the South African cotton industry. The adoption of Bt cotton comes after many studies have demonstrated its benefits. These include significant yield gains as well as reduced insecticide and labour usage in different countries. Before the introduction of Bt technology, farmers relied on a full time spray program to control the problematic bollworm complex. The types of insecticides that farmers used on bollworms are among the most toxic, killing even beneficial insects and endangered the environment.

The information gathered from regional and international reports further reveals that the use of genetic modified seeds has benefited farmers by reducing sprayings for insect pest and easy control of weeds by spraying herbicides. However the global challenge remains the preservation of seed technology by following the correct and sustainable production principles by farmers.

ACKNOWLEDGEMENTS

The University of the Free State, WWF South Africa and World Vision Swaziland are highly appreciated for their invaluable support throughout the preparation of this review paper. Special thanks go to all authors cited in the text.

REFERENCES

Abate T, van Huis A, Ampofo J (2000). Pest management strategies in traditional agriculture: An African perspective. An. Rev. Entomol. 45:631-659.

Abhiyan HB, Abhiyan BS (2012). A decade of BT cotton in Madhya Pradesh: A report.

Alston JM, Norton GW, Pardey PG (1995). Science under scarcity principles and practices of agricultural research evaluation and priority setting. Ithaca: Cornell University Press.

Bennet R, Morse S, Ismael Y (2006). The economic impact of genetically modified cotton on South African smallholders: yield, profit and health effects. Agric. Econ. 28(1):59-71.

Brookes G, Barfoot P (2006). Global impact of biotech crops: socio-economic and environmental effects in the first ten years of commercial use. Ag. Bio. Forum. 9(3):139-151.

Deng W (1999). Monsanto, Beijing. Bollguard in herbicides. Beijing: Monsanto.

Department of Agriculture, Rural Development and Land Administration (DARDLA). (2011). Operational plan for 2011/12. Nelspruit: DARDLA.

DARDLA (2013). Nkomazi map. Department of Agriculture Rural Development and Land Administration, Mphumalanga, South Africa.

Gouse M, Kirsten JF, Van der Walt WJ (2008). Bt Cotton and Bt maize: an evaluation of direct and indirect impact on the cotton and maize farming sectors in South Africa. Pretoria: University of Pretoria.

Gouse M, Kirsten JF, Jenkins L (2002). Bt cotton in South Africa: Adoption and the impact on farm incomes amongst small-scale and large scale farmers. Pretoria: University of Pretoria.

Hardee DD, Van Duyn MB, Layton MB, Bagwell RD (2000). Management of the Tobacco budworm-bollworm complex. Retrieved on 3 August 2012, from http://www.ars.usda.gov/is/np/btcotton/btcotton.pdf.

Hillocks RJ (2009). GM cotton for Africa. Outlook Agric. *38*(4):311-316.

James C (2001). Preview: Global review of commercialized transgenic crops. Ithaca: The International Service for the Acquisition of Agric-biotech Applications.

Jia SR (1998). Development of resistance management strategies for commercial cultivation of Bt cotton in China. In: Proceedings of 5th International Symposium. The biosafety results of field tests of genetically modified plants and microorganisms, Braunschweig, Germany, 6-10 September 1998.

Karihaloo JL, Kumar PA (2009). Bt cotton in India: a status report. Second Edition.

Oerke EC (2005). Centenary review crop losses to pests. J. Agric. Sci. 144:31-43.

Qaim M (2002). Small farmers benefit from insect killing cotton. Delhi: Nature Publishing Group.

Rabobank (1996). The international cotton complex. Utrecht: Rabobank.

Statista (2013). The leading statistics portal.

Vaissayre M, Cauquil J (2000). Destructive insects and diseases of the cotton plant in Africa to the South of the Sahara. Montpellier: The French Ministry of Agriculture.

Vognan G, Ouédraogo M, Ouédraogo S (2002). Description of the cotton system in the Burkina Faso region. Intermediary report. Bobo Dialasso: Institut de l'Environnement et de Recherches Agricoles (INERA).

Williams MR (2001). Cotton insect database. Proceedings of the Beltwide cotton conference.

The impact of cassava processing on the livelihoods of women processors in Central Gonja district of the northern region of Ghana

Yidana J. A., Osei-Kwarteng M. and Amadu Y.

Department of Horticulture, Faculty of Agriculture, University for Development Studies (UDS), Tamale, Ghana.

The study was conducted in the Central Gonja District located at the southern end of the Northern region of Ghana which lies between latitude 9° 15′ and 9° 05′ N and longitude 0° 45′ and 1° 0′ W. The study focused on the contribution of cassava processing to the standard of living of women processors. Information on the level of income of processors, cost of processing, level of food security as well as challenges facing processors were obtained with structured and semi-structured questionnaires from 50 selected women processors. Data was pooled, collated and summarized into frequency distribution and percentages using Microsoft Excel. The average monthly revenue was 88.1, Ghana cedis and the average monthly total cost was 44 Ghana cedis implying a monthly net profit of 44.1 Ghana cedis for an average cassava processor. Challenges faced by the processors include: Inadequate funds (100%), inadequate storage facilities (98%), low price of processed products (90%), poor packaging materials (84%) and health hazards (64%). Cassava processing is profitable and contributes significantly to the standards of living of women, however, the challenges processors encounter could drastically reduce the level of processing and economic returns affecting the standards of living of the processors.

Key words: Ghana, cassava, processors.

INTRODUCTION

Cassava *(Manihot esculenta* Crantz) is a root and tuber crop that has been identified as important for food security, especially in Africa and serves as a staple food for over 200 million people in the African continent (CTA, 2005). In areas where cassava is a main staple, people have developed ways for its processing into storable products such as tapioca, starch, dough and gari. It plays a major role in efforts to alleviate the African food crisis because of its efficient production of food, year-round availability and tolerance to extreme stress conditions (Hahn and Keyser, 1985; Hahn, 1987). Women have played a significant role in food production and processing. For instance women provide 60 to 80% of the labour requirements (UN/ECA, 1974). A study conducted by IITA in 1985 showed that, women contribute about 82% of the labour requirements for cassava production and processing (Akorada, 1992).

In many rural areas in Ghana and several developing countries, women play a crucial role in providing and improving household food security (CTA, 2005). Women also provide most of the labour for harvesting and post-harvest activities (FAO, 1996). Cassava is important, not only as a food crop but even more so as a major source of income for rural households (Davies et al., 2008). As a cash crop, cassava generates cash income for the largest number of households in comparison with other staples

(Nweke, 1996). It is produced at relatively low cost and more frequently than other staples for sale (Davies, 1991). Cassava has some inherent characteristics which makes it attractive, especially to the smallholder farmers in Ghana (Bokanga, 1992). Some of these are: It is rich in carbohydrates, available all year round and resistant to drought. Furthermore cassava is more tolerant of low soil fertility and more resistant to pests and diseases than grains (Ekong, 2003). These attributes combined with other socioeconomic considerations are therefore what IFAD has recognized in the crop as lending itself to a commodity-based approach to poverty alleviation (Akorada, 1992).

Cassava is largely consumed in many processed forms ("Gari", Dried and Dough) in Ghana. Cassava production and processing have been one of the major occupations of women in Central Gonja District for several years. However, there has not been much improvement in the standard of living of these women and their families (Asinobi et al., 2009). Also the contribution of cassava processing to the living standards of these women has not been documented or is not known. This study is relevant in terms of the valuable information it has provided on the level of income and food security as well as other challenges mitigating against a better standard of living for these women processors. This study was carried out to find out the perceived contribution of cassava processing to the standard of living of women in Central Gonja District.

MATERIALS AND METHODS

The Central Gonja district is located at the southern end of the Northern Region of Ghana on latitude 9° 10' N and longitude 1° 10' W covering a total land area of 8353 km^2 representing 12% of the total landmass of the Northern Region (Google map.com).

Research instrument and data collection

A content validated interview schedule with components to address the objectives of the study was used to collect data from the sampled respondents. Both primary and secondary data were sourced. The primary data was taken from the sampled respondents while the secondary data was gathered from the library, internet, Ghana MoFA and other published literature. Personal visits were made to the area of study to carry out a face to face interview with the women processors.

The questions were designed to obtain information from respondent based on the following parameters:

(1) Demographic characteristics of processors;
(2) Level of income of cassava processors;
(3) Cost of cassava processing;
(4) Level of food security of cassava processors;
(5) Challenges facing women in cassava processing.

Target population

The study targeted women in the Central Gonja District who were engaged in cassava processing ("gari' and dough) on a small scale basis.

Sampling procedure and sample size

Purposive sampling was used to identify women involved in small- scale cassava processing in the area for interview. A total of Fifty questionnaires were administered.

Cost of production

Cost of production is the average cost of producing one unit of commodity (Adegeye and Dittoh, 1985). In this work, cost of production of processed cassava includes all costs involved in the production process. Therefore, total cost of processing cassava is the sum of total fixed cost and total variable cost.

$$TC = TFC + TVC$$

Returns to investment

$$TR = PQ + CB$$

Where TR = total monthly revenue; P = price per output; Q = quantity of output; PQ = revenue from sale of processed cassava, and CB = revenue from sale of by-product.

Profit

$$P(\Pi) = TR - TC$$

Where TR = total revenue; TC = total cost, and P (∏) = profit.

Analysis of data

Manual analysis was used in computing the profit realized from the processing of cassava which included the following; cost of production, returns to investment and profitability. Microsoft excel was also used for analysis and tables and bar charts were used for results presentation.

RESULTS

Demographic characteristics

The majority (44%) of the women was in the age group of 31 to 40 years and a majority (48) also had no formal education (Table 1). Women in the age group of 51 to 60 were not many in the cassava processing venture. Most of the women (28%) had 2 dependants and 10% of them had 5 dependants (Table 2).

Marital status

Married women dominated the processing venture. Women who were single, divorced and widowed were also involved in the processing representing 12, 8, and18%, respectively.

Table 1. Frequency of age of respondents.

Age (years)	Frequency	Percent	Cumulative (%)
21-30	5	10.0	10.0
31-40	22	44.0	38.0
41-50	14	28.0	88.0
51-60	9	18.0	100.0
Educational			
No formal	24	48.0	48.0
Primary	8	16.0	64.0
J.H.S	12	24.0	88.0
S.H.S	6	12.0	100.0

Source: Field survey, 2011.

Table 2. Frequency of number of dependants of processors.

Number of defendant	Frequency	Percent	Cumulative (%)
1	9	18.0	18.0
2	14	28.0	46.o
3	12	24.0	70.0
4	10	20.0	90.0
5	5	10.0	100.0

Source: Field survey, 2011.

Profitability analysis of cassava processing

All the women after processing sold in pans. The average weight of a pan of processed cassava was 14 kg however the prices quoted for a pan of processed cassava varied from seller to seller and ranged between 10 and GH¢14. From the study, processors did not necessarily incur the same expenditure on the same items they used in cassava processing. The industry did not require the use of complex capital items. The only items used were pans, baskets, knives, jute sacks among others and they were valued to get the total fixed cost.

From the study, cassava processors incurred an average monthly variable cost of over GH¢40 and spent over GH¢20 on raw materials. Also these women paid an average of GH¢11.6 for labour. They also calculated their total average monthly cost to be the sum of the total monthly fixed cost and total monthly variable cost and that was GH¢44.

Returns from cassava processing were realized through the sales of the finished cassava product and the by-product such as cassava peels. The by-product generated additional average revenue of GH¢3.3 per month. Due to differences in selling prices, different levels of revenue were realized. The study revealed that 15, 5, 48, 18, and 14% of the processors sold their processed cassava at GH¢14, GH¢13, GH¢12, GH¢11, and GH¢10 per pan (14 kg) respectively. 50% of the women processed two pans (28 kg) in a week and 8% of the women processed four pans (56 kg) per week. In a week, an average of two pans (28 kg) was processed. This implies that, the average revenue per month from the sale of processed cassava was GH¢84.8. Nonetheless the average selling price of a pan of processed cassava was GH¢12.

The average net profit per month was computed as GH¢44.1 (Table 3) which was about 50% of the total monthly revenue.

Challenges facing cassava processors

The result of challenges facing cassava processor is given in Table 4.

DISCUSSION

The cassava processing industry is dominated by women. It is seen as a preserve for women in Central Gonja District. Cassava processing is dominated by the most active labour group. This was the young population between 31 and 40 years old. The low population of cassava processors who were between the ages of 50 to 60 years could be attributed to the labour-intensive nature of cassava processing which does not favor old

Table 3. Profit analysis of cassava processing.

Item	Monthly average value(GH¢)
Revenue	-
By-product	3.3
Finished product	84.8
Total revenue (TR)	88.1
Cost VC FC	44
Total cost (TC)	44
Net profit	44.1

Source: Field Survey, 2011.

Table 4. Challenges facing cassava processors.

Challenges	Percent
Inadequate funds	100
Inadequate storage facilities	98
Low prices of processed products	90
Poor packaging materials	84
Health hazards	68

Source: Field survey, 2011.

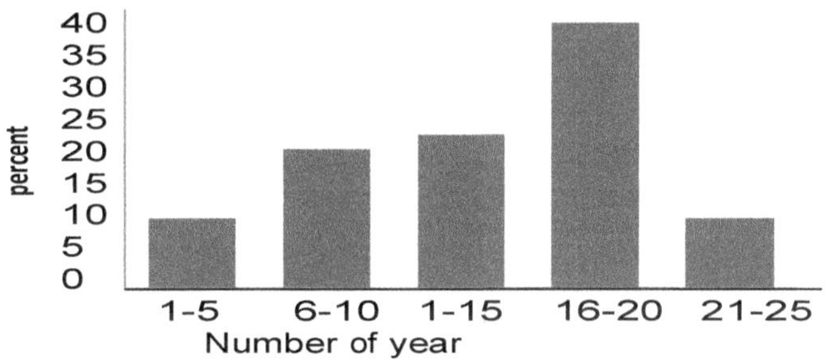

Figure 1. A bar chart illustrating the number of years in cassava processing. Source: Field survey, 2011.

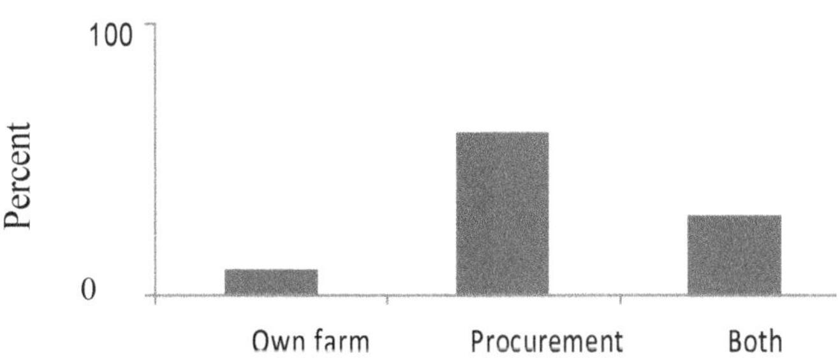

Figure 2. A bar chart illustrating the source of raw materials for cassava processing.Source: Field survey, 2011.

keeping. Since most of them found it difficult to state the age (Figure 1). The observed low level of formal education (48%) might have effect on their record-

unit of processed pan of cassava in kilograms. This confirms the findings of an earlier research that the food processing industry of Ghana just like any other African country is dominated by women who have little or no formal education (Asiedu, 1989). Married women (62%) dominated the processing venture; the reason is that, before cassava is processed, men are always needed to assist in uprooting and peeling of the cassava. Moreover, the complication involved in processing cassava cannot be done single-handed. Also the study observed that, married women sometimes gained the opportunity to use cassava from their husband's farm (Figure 2). Few women who were single, divorced and widowed as compared to married women were also involved in the processing. This could be attributed to the fact that, cassava processing is the main economic activity in the study area.

One would have expected that, the higher the number of dependants one had, the higher the probability of engaging in the processing. However, the relationship between the number of dependants and participation was inverse. Thus, as the number of participants increase the number of dependants decreases and vice versa. One possible explanation is that, for women who had more than five dependants, some of the dependants were matured and have possibly become financially independent depriving the women of their services in the processing. Women who had two dependants were more inclined to processing cassava since their dependants (probably of youthful) were a labor force for them.

The persistence of women in cassava processing could be attributed to the fact that, the women get profit (GH¢44.1 per month) from the business which is about 50% of their total monthly revenue. This supports the reason why the women have been in the business for so many years (Figure 2).

The industry did not use complex capital items. The only items used were pans, baskets, knives, jute sacks among others. The study found that cassava processors incurred an average monthly variable cost of GH¢44. On the average, cassava processors spent over GH¢20 on raw materials. It also came to light that the women pay on average GH¢11.6 for labour. The reason for high cost for labour was that family labour was costed during the study.

Inadequate funds were a major challenge for the cassava processors, so consequently they produced in small quantities. The processors also complained that, their processed products were low priced and this was attributed to inadequate storage facilities which made them to sell at any price to prevent spoilage. Packaging adds values to the products, prolong the shelf life and when not done properly, could lead to contamination and poor patronage of the products (Asinobi et al., 2009). Health and health related issues were not also left out. It was observed that, the women sometimes sustained injuries, cuts, shoulder and chest pains during the

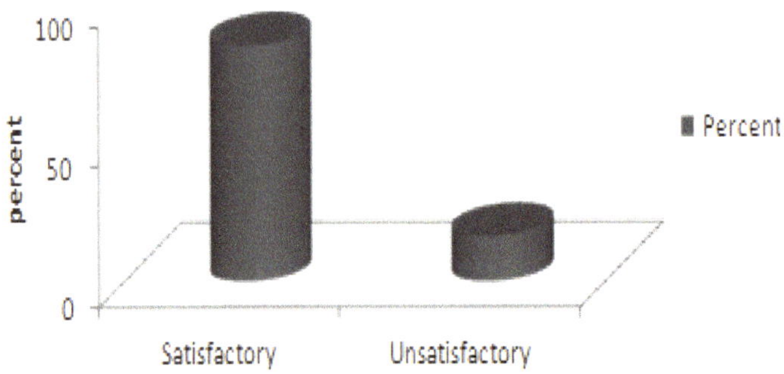

Figure 3. Level of family food security of processors. Source: Field survey, 2011.

processing which sometimes force them to use large portion of their income for medication. Due to this challenge, the processors are not able to process on a large scale and this confirms earlier findings by Asinobi et al. (2009) which reported that cassava processors in Ghana are faced with challenges like low prices, inadequate funds and poor packaging materials. Majority of the processors (84%) described their family food security to be satisfactory whiles the remaining 16% were of the view that their food security is unsatisfactory (Figure 3). These findings therefore agreed with Al-hassan and Irene (2005) that cassava is grown in Ghana mostly by smallholders as a food-security crop.

Conclusion

The survey showed that, cassava processing is profitable and contribute significantly to the standards of living of women cassava processors, in terms of income generation and family food security. However, cassava processors in the study area were constrained with inadequate storage facilities, poor packaging materials, health problems, low price of processed products and inadequate fund for the expansion of their scale of processing. The consequences of all these could drastically reduce the level of processing and economic returns affecting the standards of living of the processors.

REFERENCES

Adegeye AJ, Dittoh JS (1985). Essentials of Agricultural economics. Impact publishers Ibadan. Nigeria.

Akorada (1992). Roots Crops for Food Security in Africa, Collaborative study of cassava in Africa (COSCA). Work Intel printers limited Ibadan. P. 8.

Al-hassan R, Irene E (2005). The Potential For Farm/Non-Farm Linkages in The Cassava Subsector in Ghana. P 171. Asiedu JJ (1989). Processing Tropical Root Crops. A Technological Approach Macmillan Publishers Ltd.

Asinobi CO, Ndimantang B, Nwajiuba CU (2009). Cassava Production, Processing Trends and Constraints Imo State University, Owerri.

Bokanga M (1992). Cassava Fermentation and Industrialization of Cassava Food Production. Proceedings of the Fourth Symposium, International Symposium for Tropical Root Crops African Bulletin ISTR.C- AB: pp. 197- 201.

CTA Spore (2005). Information for Agricultural Development in ACP Countries. P. 120.

Davies RM (1991). A Survey of Cassava Project on Technology for W omen, Institute for processing machinery in Oyo State. B.Sc. Thesis. Agriculture Research. Ahmadu Bello University, Nigeria.

Davies RM, Olatunji MO, Burubai W (2008). A Survey of Cassava Processing Machinery in Oyo State © IDOSI Publications.

Ekong EE (2003). An Introduction to Rural Sociology. Dove Educational Publishers, Uyo, Nigeria.

FAO (1996). Rural women and food security; current status and perspectives, P. 4.

Hahn SK (1987). An Overview of African Traditional cassava Processing and Utilization, Outlook on Agriculture, pp. 110-118.

Hahn SK, Keyser J (1985). Cassava: A Basic Food of Africa. Outlook Agric. 4:95-100.

IITA (1985). Cassava in Tropical Africa. A reference manual, International Institute of Tropical Agriculture, Ibadan, Nigeria.

Nweke FI (1996). Cassava: A Cash Crop in Africa. International Institute of Tropical Agriculture. (COSCA Ibadan, Nigeria, Working. P. 4.

UN/ECA (1974). The Changing and Contemporary Role of Women in Africa Development. UNECA, Addis Ababa, Ethiopia.

Enhancing farm profitability by growing wheat for chapatti quality markets in Haryana, India

David R. Coventry[1], Randhir Singh Poswal[2], A. Yadav[3], Ramesh Kumar Gupta[2], B. S. Duggal[4], R. S. Chhokar[2], V. Kumar[5], A. S. Riar[2], Subash Chandra Gill[2], Anuj Kumar[2], Ramesh Kumar Sharma[2], Ramesh Chand[2], Sendhil R[2], S. G. L. Kleemann[1] and J. A. Cummins[6]

[1]School of Agriculture Food and Wine, University of Adelaide, Waite Campus, SA, Australia.
[2]Directorate of Wheat Research, Agarsain Marg, Karnal-132001, Haryana, India.
[3]CSS Haryana Agricultural University, Hisar, Haryana, India.
[4]State Department of Agriculture, Haryana, Panchkula, India.
[5]CSS Haryana Agricultural University, KVK, Bawal, Haryana, India.
[6]Rural Solutions SA, Australia.

In field experiments in six districts of Haryana, best practice management (including zero tillage and 3-way split application of N-fertiliser) provided small (2%) improvements in grain yield when compared with local farmer practice. This best practice management also gave consistent improvement in chapatti quality as measured by grain hardness (11.4%) and chapatti score (3.2%). Where a 'shot-gun' mix of micronutrients and sulphur was included with farmer practice treatment, or with best practice treatment, grain yield was increased by about 3% and there were further improvements in quality. Net returns were calculated for wheat production, the large cost saving associated with not undertaking the 4 cultivations used in local farmer practice, the best practice treatment was always the most profitable (by 9%), and even if no yield improvement was assumed this treatment would still be 6.7% more profitable. Even though it had a 30 to 40% yield penalty, the most profitable practice involved using variety C-306. This is an old wheat variety recognised as the standard for chapatti quality, and C-306 gets at least twice the market value of wheat sold at prices set as a minimum support price. High yields of wheat and its availability is an on-going priority for India, particularly for maintaining food security for the large population. Therefore, any wholesale change in production that moves farmers away from optimizing total grain production is untenable. However, 16 to 20% of farmers and consumers already have some direct marketing/purchasing of the wheat harvest via local markets, and these often involve quality considerations.

Key words: Chapatti quality, profitability, wheat marketing.

INTRODUCTION

Wheat and other crops are grown sequentially in an irrigated, double cropping pattern in Haryana and Punjab states (North-West India) (Coventry et al., 2011c) on about 5.7 M ha area. Among the various cropping rotations, rice-wheat is a major cropping system for sustaining food security in the region and these states play a dominant role in building grain buffers for the whole country. There are millions of farmers and agricultural workers dependent on this system for their livelihoods. Profitability is the key motivation for these farmers and their aim has been to harvest maximum yields. The majority of the wheat grown is consumed as traditional Indian flat bread (chapatti). Chapatti quality is important to consumers and many are willing to pay more for better quality wheat flour. However, farmers do not specifically target quality parameter as their wheat is mostly sold without segregation and premium price (Coventry et al., 2011b).

In South-West Haryana, C-306 and WH-283 varieties are grown to a limited extent with excellent chapatti quality.But these are lower yielding compared with the more widely grown varieties such as PBW-343 and PBW-502 (Coventry et al., 2011b), and it is likely at the present time that farmers will have to manage with these best-quality varieties if they are to obtain premium chapatti quality flour. What is not known is whether the economic returns associated with a practice that does not specifically target optimising grain yield, such as using varieties C-306 and WH-283, will be acceptable to farmers.

In the study reported here, we provide an economic analysis utilising data from extensive field experiments to establish whether an alternative production system that targets chapatti quality is feasible for wheat growers in Haryana. Further, we explore the current processes of wheat and wheat flour purchase based on information from a consumer survey in urban areas of Karnal district (Haryana).

MATERIALS AND METHODS

Field sites

During the winter seasons of 2009-2010 and 2010-2011 a series of experiments were conducted in farmer fields to compare wheat productivity and quality between recommended best practice (Coventry et al., 2011a, b, c) and the dominant local farmers' practice. The studies were conducted in split-plot design field experiments with different wheat-based rotations chosen according to location in Haryana. With rice-wheat rotations there were two experiments during the first year.

The first experiment consisting of two varieties (DBW-17 and PBW-550) in the main plot and four management options in sub-plots (local area farmer practice consisted (FP), FP+ micronutrients (MN), recommended practice (SP), SP+ MN). The FP and SP treatments had application of NP (150:26 kg ha^{-1}) and NPK (150:26:33 kg ha^{-1}) respectively. The fertilizer recommendation for wheat in Haryana is to apply 150 kg N ha^{-1} (2-split schedule: 1/3 at sowing and 2/3 at the first irrigation) and 26 kg P ha^{-1} at sowing (Mishra et al., 2005; Srivastava et al., 2006). In the FP treatment P and some N was applied as basal using di-ammonium phosphate (DAP) and the remaining nitrogen was supplied in two equal splits using urea at the first and second irrigation.

In the SP treatment, P, K and one-third nitrogen were applied as basal and remaining nitrogen in two equal splits through urea at the first and second irrigation. The micronutrients applied were Zn, Fe, Mn, B and Cu and their sources were $ZnSO_4$, $FeSO_4$, $MnSO_4$, Borax and $CuSO_4$, respectively. The concentrations of $ZnSO_4$ (1%), $FeSO_4$ (1%), $MnSO_4$ (0.5%), Borax (0.2%) and $CuSO_4$ (0.2%) were applied as spray solutions. The spraying was done at 55 and 75 days after sowing using a backpack sprayer with a spray volume of 375 L ha^{-1}.

In the FP treatment, three to five ploughings along with a harrowing were undertaken prior to sowing. With the SP treatment the seed was direct seeded. The plot size was 4×15 m and sowing was done using a ferti-seed drill with a seed rate of 100 kg ha^{-1}. The experiment was conducted at 6 locations (Table 1) located in Karnal, Kaithal and Yamunanagar districts of Haryana. In the second experiment conducted during 2009-2010 in the rice-wheat region, seven varieties (C-306, DBW-17, PBW-343, PBW-502, PBW-550, Raj-3765 and WH-283) with two management strategies (FP and SP+MN) were evaluated at the Darar and Ramba, sites. During 2010-2011 four wheat varieties (C-306, HD-2851, WH-1025 and WH-283) and the above described four management options (FP, FP+MN, SP, SP+MN) were tested at the sites situated in village Ramba, Darar and Kalri.

In the non rice-wheat rotations as practiced in West and South-West Haryana (cotton–wheat, pearl millet–wheat and cluster bean–wheat), an experiment was conducted at six sites (Table 1) during 2009-2010 with the seven varieties (C-306, DBW-17, PBW-343, PBW-502, PBW-550, WH-283 and WH-711) and four nutrient management (FP, FP+MN, SP, SP+MN) treatments. Treatments and management were the same as described above except that $ZnSO_4$ was applied as 25 kg ha^{-1} to all treatments. In the second year (2010-2011), the experiment was conducted at 7 locations (un-replicated) with 4 varieties (C-306, HD-2851, WH-1025 and WH-283) and two management options (FP and SP+MN).

The crops were manually harvested from an area of 2×5 m for grain yield and threshed using a small plot thresher. A random grain sample was taken from each treatment for quality analysis (protein, hardness and chapatti score). Protein was estimated using a standard NIR (Near Infra Red) instrument. The Grain Hardness Index (no units) was determined on about 300 kernels using a Perten model SKCS4100 single kernel characterization system. For the evaluation of chapatti quality, various parameters like water absorption, nature and colour of dough (before and after maturation), chapatti appearance, colour, aroma, taste, puffing height, pliability and loss of water were evaluated by giving a score out of 10. All quality analyses were undertaken at the Wheat Quality Laboratory, Directorate of Wheat Research, Karnal. Soil samples (0 to 10 cm depth) were taken at each of the sites in 2009-2010. Soil analyses were undertaken using standard techniques at the soil testing laboratory, State Department of Agriculture, Karnal (Haryana). For available phosphorus in soil the sodium bicarbonate procedure of Olsen et al. (1954) was used. For available potassium in soil the ammonium acetate solution extraction procedure of Richards (1954) was used and for available sulphur the calcium chloride solution extraction procedure of William and Steinbergs (1959) was used. Zinc, iron and manganese were determined using the DTPA test of Lindsay and Norvell (1978). The agronomy and quality data were subjected to analysis of variance using the GENSAT package.

Farmer and consumer survey

A farmer survey involving 823 farmers from 103 villages (7 to 8 farmers per block) in 19 of the 21 districts of Haryana, and from each village there were two marginal (<1 ha), two small (1 to 2 ha), two medium (2 to 4 ha) and two large (>4 ha) farmers surveyed. The survey targeted information on current wheat production practices, including information on operational costs, and also information on farmer demographics and attitudes (including understanding of wheat quality and marketing issues). Another survey targeting 200 consumers (personal interview) was conducted in the urban areas of Karnal district covering Radha Swami Market, Saini Colony, Model Town, Dayan and Colony, Sector 4, Sector 6, Sector 8 Part II, Sector 9, DWR-Karnal, NDRI-Karnal and Sadar Bazaar. The wheat market survey was conducted to explore consumers' sources of grain/flour , their preference for

Table 1. Soil chemical characteristics for the experimental sites (0-15 cm depth) and sufficiency rating for each measure.

Site[1]	Rating	pH (1:2)	EC (1:2) dS/m	OC (%)	P (mg kg^{-1})	K (mg kg^{-1})	S (mg kg^{-1})	Zn (mg kg^{-1})	Fe (mg kg^{-1})	Mn (mg kg^{-1})
	Low	<6.5	<0.8	<0.40	<5.0	<60	<10	<1.0	<4.5	<1.0
	Medium	6.5-8.7	0.8-2.5	0.40-0.75	5-12.5	60-140	10-20	1.0-3.0	4.5-9.0	1.0-2.0
	High	>8.7	>2.5		>12.5	>140	>20	>3.0	>9.0	>2.0
Rice-Wheat Haryana										
Darar (Karnal)		8.3	0.36	0.39	9.23	145.0	36.2	5.3	14.1	14.9
Ramba(Karnal)		8.0	0.48	0.58	8.76	156.0	28.9	7.1	12.7	14.3
Kalri(Karnal)		8.4	0.57	0.43	7.86	143.0	22.3	6.7	32.1	12.7
Sadura (Yamunanagar)		8.5	0.47	0.35	8.16	136.7	32.9	4.5	35.7	10.9
Thana (Kaithal)		8.1	0.37	0.41	9.12	155.1	25.8	7.1	29.3	8.9
Chhapper (Yamunanagar)		8.2	0.78	0.29	7.19	148.7	21.7	5.7	15.9	9.9
South West Haryana										
Kapriwas (Riwari)-1		8.0	0.30	0.16	2.11	118.3	63.0	0.59	2.65	1.08
Tigra (Mahendergarh) 2 & 3		8.3	0.44	0.16	1.39	62.3	35.8	0.50	0.76	1.08
Tigra (Mahendergarh) - 4		8.3	0.38	0.24	1.27	105.8	34.0	1.39	0.29	0.75
Begu Sahapur (Sirsa) - 5		8.2	0.44	0.25	2.35	282.2	67.7	1.74	2.21	2.41
Bhangu (Sirsa) - 6		8.6	0.71	0.22	1.90	170.2	96.7	1.14	3.82	1.97

[1] Village and district (parentheses).

days after sowing using a backpack sprayer with a spray volume different brands, present and future consumption pattern of wheat based products and their sensitivity to price and quality.

Marginal analysis

Data from the field experiments were used for providing comparative data in a marginal analysis that compares the benefits and costs of the different management alternatives used (FP, FP+MN, SP and SP+MN). By examining the incremental change on total revenue and total cost for any change in the output and input used in each alternative, the best economic practice was determined. Operational costs were the only costs considered in establishing the profitability of the selected farming practice. The operational costs include expenses on labour employed to perform different cultural operations and expenses incurred on material inputs *viz.,* seeds, manure, fertilisers, plant protection chemicals, irrigation, micro-nutrients and herbicide (Table 7). These costs were collected from the farm survey conducted in Haryana during 2010. Gross returns (price x yield) and net returns (gross returns minus operational costs) were calculated to establish the margin accrued to the farmer after meeting all working expenses. Benefit-Cost ratio (Gross Income ÷ Operational costs) and net returns per rupee of investment (Net returns ÷ Operational costs) were thus calculated for each farming practice (Table 8).

RESULTS

Site characteristics

The sites chosen for the farm-based experiments were typical of both the rice-wheat (RW) regions of Haryana, and the non rice wheat areas in south-west Haryana (SW) where cotton, pearl millet and cluster bean (but not rice) are included in the rotation (Table 1). The soils of all the sites were alkaline (pH 8.0 to 8.6) in nature with low-medium electrical conductivity (0.29 to 0.78 dSm^{-1}) and low organic matter. The values for phosphorus from all the sites were very low whilst sulphur and potassium were in the high range. Zinc was high in the RW rotation (> 4.5 mg kg^{-1}) but was in the low-medium range for the SW sites (0.5 to 1.7 mg kg^{-1}). Iron and manganese had a similar pattern with the higher measures in the RW sites and low-medium measures in the SW sites.

Table 2. Mean values for grain yield, protein, hardness and chapatti score for 6 sites and 4 management treatments in rice-wheat rotations, 2009-2010 season.

Site	Grain yield (tha^{-1})				Protein (%)				Hardness (%)				Chapatti score (%)			
	FP[1]	FP+MN	SP	SP+MN	FP	FP+MN	SP	SP+MN	FP	FP+MN	SP	SP+MN	FP	FP+MN	SP	SP+MN
Ramba	5.5	6.3	6.2	5.8	11.72	12.02	12.19	12.55	73.8	76.8	79.8	82.2	7.46	7.54	7.61	7.69
Darar	4.6	5.4	4.9	5.4	11.82	11.95	12.18	12.45	75.0	77.5	80.7	82.8	7.43	7.54	7.64	7.69
Kalari	4.1	4.7	4.2	4.6	11.58	11.87	12.05	12.41	71.0	74.7	77.5	79.2	7.42	7.47	7.55	7.64
Thana	3.4	3.7	2.9	3.2	11.94	12.06	12.46	12.77	77.7	80.8	83.8	86.5	7.53	7.59	7.68	7.75
Sadura	5.4	5.7	4.9	5.1	11.23	11.51	11.69	12.05	68.8	72.0	75.8	77.3	7.38	7.44	7.52	7.60
Chapper	3.9	4.3	4.5	5.0	10.83	11.14	11.30	11.64	69.5	72.0	74.7	77.2	7.28	7.37	7.46	7.53
Site	0.3				0.2				1.9				0.07			
Management	0.3				0.2				1.6				0.06			
sitexmgt	0.6				ns				ns							

[1]FP - Local area farmer practice; FP+MN – FP + micronutrients; SP - recommended practice; SP+ MN – SP + micronutrients.

Grain yield and quality data

In the rice-wheat region in the experiment located at six sites (2009-2010) there were positive responses with grain yield to the additional management compared with the recommended farmer practice, although these responses differed in extent between sites (site x management interaction) (Table 2). Four of the six sites had higher grain yield where the SP treatment was used, and at all sites the addition of the micronutrient supplements gave extra yield (mean 8.2%), except one location (Ramba). Three quality measures reported here show a consistent step-wise response pattern at all sites, with both the SP treatment and the addition of micronutrients resulting in extra protein, grain hardness and chapatti score (Table 2). Regarding the quality measures there were no interactions between management and the grains sourced from the different sites. In this experiment, DBW-17 and PBW-550 (mean data shown) varieties were chosen because they represent different protein extremes, with DBW-17 being a high protein variety (Table 3). In the second experiment, wheat varieties, DBW-17, PBW-343, PBW-502, PBW-550 produced similar grain yield, whereas C-306 had less grain yield by about 40% and WH-283 less by 23% compared to PBW-502. In this experiment there was no significant response to the additional management input with grain yield, but this enhanced management package resulted in higher protein, grain hardness and chapatti score with all varieties. C-306 and WH-283 had the highest measures for chapatti quality (grain hardness and chapatti score) while PBW-502, PBW-550 and Raj-3765 showed poorest chapatti quality measures. The variety PBW-550 had the lowest protein and Raj-3765 the highest protein. In the 2010-2011 growing season two other varieties were compared with the two recognised high quality varieties (C-306, WH-283). HD-2851 which is a variety well adapted to late sowing had much higher yield (>46%) than C-306, but the grain yield of WH-1025 was only similar to C-306 in this environment (Table 4). Similarly to the variety experiment in the previous season the quality measures increased in a stepwise manner with the management inputs (Table 4). HD-2851 is also known as a high protein cultivar, but was not comparable to C-306 and WH-283 with respect to chapatti quality parameters.

With the sites used in the west (cotton-wheat) and south-west (pearl millet/cluster bean-wheat) areas of Haryana there were some small increases in grain yield with the improved practice (SP) compared with the farmer practice (FP) and with micronutrients alone. However, where both SP and micronutrients were used together there was a consistent yield increase of about 4 to 5% at all sites (Table 5). Of the two improved management components, micronutrients had the biggest effect on protein, but SP also increased protein and there were similar increases with hardness and chapatti score responses. There was a site x management interaction for these quality measurements, with the cotton-wheat sites being more responsive to the improved management inputs (Table 5). In the same experiment where varieties were compared at these six sites, PBW-343, DBW-17, PBW-502, PBW-550, and WH-711 all had much the same

Table 3. Mean values for grain yield, protein, hardness and chapatti score for 2 sites, 7 varieties and 4 management treatments in rice-wheat rotations, 2009-2010 season.

Site	Grain yield (tha^{-1})				Protein (%)				Hardness (%)				Chapatti score (%)			
	Ramba		Darar		Ramba		Darar		Ramba		Darar		Ramba		Darar	
	FP	SP+MN	FP	SP+MN	FP	SP+MN	FP	SP+MN	FP	SP+MN	FP	SP+MN	FP	SP+MN	FP	SP+MN
C-306	4.17	4.75	3.72	4.18	11.89	11.99	12.07	12.13	92.5	95.5	96.25	98	8.08	8.17	8.15	8.22
DBW-17	6.54	5.56	6.08	5.81	12.00	12.19	12.20	12.35	81.25	85.5	84.5	89.75	7.63	7.74	7.75	7.77
PBW-343	6.22	5.19	6.03	5.38	11.98	11.99	11.93	12.16	79.25	82	82	85.75	7.51	7.64	7.62	7.68
PBW-502	5.74	5.84	6.57	5.52	11.82	11.84	12.03	11.99	73.25	75.75	76.25	80	7.43	7.52	7.52	7.59
PBW-550	6.15	5.59	5.96	5.63	11.23	11.44	11.44	11.87	76.75	79.25	79.5	82.5	7.36	7.45	7.48	7.53
Raj-3765	4.68	5.00	5.33	5.03	12.15	12.43	12.37	12.60	72.75	76	75.75	79.5	7.62	7.74	7.70	7.79
WH-283	4.85	5.02	4.63	5.02	11.59	11.75	11.78	11.91	93.75	96.5	96.75	100.25	8.03	8.09	8.14	8.12
Site	0.21				0.13				1.2				0.03			
Mgt	ns				0.13				1.2				0.03			
Variety	0.39				0.24				2.26				0.05			
Sitexvar	0.55															

Table 4. Mean values for grain yield, protein, hardness and chapatti score for 4 varieties and 4 management treatments in rice-wheat rotations, mean of three sites, 2009-2010 season.

Site	Grain yield (tha^{-1})				Protein (%)				Hardness (%)				Chapatti score (%)			
	FP	FP+MN	SP	SP+MN	FP	FP+MN	SP	SP+MN	FP	FP+MN	SP	SP+MN	FP	FP+MN	SP	SP+MN
C-306	2.99	2.93	3.04	3.06	10.75	11.16	11.65	12.09	96.67	99.89	102.44	106.00	7.97	8.08	8.19	8.28
HD-2851	4.38	4.20	4.45	4.68	11.24	11.71	12.14	12.74	88.56	92.33	93.89	97.11	7.84	7.92	7.99	8.06
WH-1025	2.99	2.89	2.96	2.99	11.29	11.52	11.99	12.34	86.56	91.33	96.00	99.22	7.79	7.90	7.97	8.05
WH-283	3.98	3.87	4.12	4.33	11.34	11.83	12.37	12.69	92.89	96.22	98.99	101.67	7.92	7.98	8.07	8.15
variety	0.15				0.19				1.42				0.03			
Mgt	0.15				0.19				1.42				0.03			
Varxmgt																

yield, and WH-283 had 4% less grain yield and C-306 had 28% less yield compared to PBW-502. There was a significant variety x site interaction. Higher yielding varieties performed much better than C-306 at the higher yielding sites (eg. Site 5). Varieties C-306 and WH-283 had the highest quality measures (Table 6 – only grain hardness shown), with WH-711 next highest, and PBW-502 and PBW-550, the lowest. In 2010-2011, the full management package did not influence grain yield but protein and hardness, in-line with other seasons and locations were increased by 10 to 15% (data not shown). At these western locations, HD-2851 yielded 32% higher than C-306 and WH-1025 the same as C-306.

Marginal analysis

Data from the farm survey undertaken in 2010 was used to establish the operational costs

Table 5. Mean values for grain yield, protein, hardness and chapatti score for 6 sites and 4 management treatments in SW Haryana rotations, 2009-2010 season.

Site	Grain yield (tha^{-1})				Protein (%)				Hardness (%)				Chapatti score (%)			
	FP	FP+mn	SP	SP+mn	FP	FP+mn	SP	SP+mn	FP	FP+mn	SP	SP+mn	FP	FP+mn	SP	SP+mn
Kapriwas(PM-W)	4.21	4.38	4.25	4.40	10.73	11.29	11.10	11.59	77.75	81.89	80.07	84.96	7.55	7.61	7.57	7.65
Tigra 1(PM-W)	4.95	5.02	5.04	5.21	10.52	11.05	10.85	11.38	75.75	79.57	78.61	83.50	7.50	7.55	7.54	7.59
Tigra 1(CB-W)	4.39	4.44	4.43	4.54	10.97	11.48	11.29	11.78	75.50	81.32	82.57	84.43	7.46	7.66	7.53	7.71
Tigra 2(CB-W)	5.15	5.26	5.22	5.36	10.66	11.49	11.31	11.83	76.07	83.96	82.36	87.25	7.49	7.68	7.65	7.71
BeguSahapur(C-W)	4.82	4.90	4.90	5.07	10.72	11.66	11.31	12.02	79.07	84.93	82.46	89.61	7.56	7.78	7.74	7.83
Bhangu (C-W)	4.16	4.25	4.18	4.35	10.97	11.87	11.50	12.23	80.21	86.75	84.39	91.07	7.49	7.71	7.67	7.73
Site	0.08				0.03				0.48				0.01			
Mgt	0.07				0.03				0.39				0.01			
sitexmgt	Ns				0.06				0.52				0.02			

Table 6. Mean values for grain yield and hardness for 6 sites and 7 varieties in SW Haryana rotations, 2009-10 season.

Site[1]	Grain yield ($t\ ha^{-1}$)							Grain hardness (%)						
	C-306	DBW-17	PBW-343	PBW-502	PBW-550	WH-283	WH-711	C-306	DBW-17	PBW-343	PBW-502	PBW-550	WH-283	WH-711
Site 1	3.52	4.64	4.58	4.68	4.19	4.28	4.98	89.8	80.8	76.3	71.6	75.3	89.3	85.3
Site 2	3.76	5.31	5.34	5.46	5.25	5.07	5.33	87.6	79.5	74.4	69.6	73.3	87.6	83.4
Site 3	3.64	4.66	4.70	4.75	4.43	4.48	4.97	88.8	80.5	76.2	73.1	72.6	90.6	84.9
Site 4	3.82	5.47	5.62	5.62	5.50	5.30	5.49	91.1	81.6	77.6	73.0	76.5	90.6	86.6
Site 5	3.40	5.27	5.10	5.30	5.22	5.00	5.64	92.9	83.6	79.1	74.6	78.0	92.2	87.8
Site 6	3.35	4.47	4.33	4.47	4.44	4.26	4.69	94.8	85.4	80.8	76.3	79.5	93.6	88.9
Site	0.083							0.48						
variety	0.089							0.52						
site x var	0.219							1.27						

[1]For site location (Table 5).

(shown in Indian Rupee (INR); 1USD = 48 INR) for the four management practices (FP, FP+MN, SP, SP+ MN) used in the experiments reported here that were undertaken on farmer fields in the rice-wheat areas (RW) and the western and south-western areas (SW) (Table 7). Marginal analysis was done comparing three wheat varieties, PBW-502 representing the currently recommended high yield variety, and, C-306 and WH-283 two varieties that are recognised as 'best quality' for chapattis (Table 8). Based on the data from these experiments, a 2% yield gain was assumed for the SP treatment compared with FP, and a 3% yield gain was assumed where MN was included. A yield of 5.5 tha^{-1} was allocated to PBW-502 for RW and 5.0 tha^{-1} for SW and C-306 was assumed to yield 40 and 30% less than PBW-502 in the RW and SW areas respectively, and WH-283 to yield 25 and 5% less than PBW-502 in these regions. The market prices available at the time for these 3 varieties were INR 11000 and11700 (PBW-502) (Ministry of Agriculture, Government of India, 2012), and around INR 20000 (C-306) and INR 15000 (WH-283) per

Table 7. Operational costs (INR) associated with wheat production, Haryana, India.

Particulars	Haryana			
	FP	FP+MN	SP	SP+MN
Field preparation	3256.50	3256.50	--	--
Seed	1803.87	1803.87	1803.87	1803.87
Sowing	1238.34	1238.34	1238.34	1238.34
Irrigation charges	2350.31	2350.31	2350.31	2350.31
Herbicide and application	1158.03	1158.03	1158.03	1158.03
Fertiliser and application	1772.00	1771.00	2237.00	2237.00
Plant protection chemicals application	1116.92	1116.92	1116.92	1116.92
Micro nutrients	-	2236.30	-	2236.30
Micro nutrients application	-	518.92	-	518.92
Harvesting	3530.78	3530.78	3530.78	3530.78
Threshing	1725.31	1725.31	1725.31	1725.31
Marketing cost	500.00	500.00	500.00	500.00
Total operational costs	**18452.06**	**21206.28**	**15660.56**	**18415.78**

tonne (usually 1.5 to 2 times higher than minimum support price fixed by the Government of India on the recommendation of Commission for Agricultural, Costs and Prices).Of all the combinations shown in Table 8, growing WH-283 in the SW region with the SP treatment was the most profitable operation. The second most profitable outcome was growing C-306 in the SW region with the SP treatment. With all the combinations using the SP treatment was on average 9% more profitable, but where the micronutrient mix was included there was a 2.4% loss obtained (Table 8).

Consumer survey

The average family size of the respondents to the consumer survey was 4.52. About 84% of respondents purchased/retained wheat grain; this was either from the market (58%), directly from farmers (18%) and those who had rural background and retained produce from their own farm (8%). The average amount purchased was 358 kg wheat grain for their annual consumption. The respondents who kept their own produce had 360 kg wheat for annual consumption. The remainder of the respondents (16%) purchased flour either from retailers (14.5%) or mini flour mills (1.5%). They purchased a local brand (*Brand A*) solely (10%), *Brand A* and sometimes a more expensive brand (*Brand B*) (1.5%) and solely *Brand B* (3%) flour. Most of the respondents (70%) who purchased *Brand A* flour made comments about quality concerns. All respondents were satisfied with the quality of *Brand B* flour, and although they look for better quality flour usually the preference is to seek lower price. The respondents purchased flour mainly for making chapatti, although 3.5% of those who purchased flour from the retailers used it for indigenous products like *halwa (a* sweet dish served to the guests or prepared on the occasion of festivals for their own consumption), *puri* (small sized chapatti deep fried in ghee/ vegetable oils) and noodles. The respondents reported that the consumption pattern is changing and the younger generation is increasingly more inclined towards pasta, bread and noodle (wheat based products). Within the farmer survey of farmers throughout the 19 districts surveyed (total 823 farmers; data used in Table 7), 19.8% reported some direct marketing to the consumers of their wheat harvest.

DISCUSSION

Given the intensity of crop production in Haryana (182% annual cropping intensity), particular concern is taken in monitoring the on-going nutrient status of these soils (Singh, 2008). In the field experiments reported here, and in previous experiments (Coventry et al., 2011a), where a mix of micronutrients plus sulphur was applied, there were small (3%) increases in grain yield obtained, and consistent and larger increases in grain protein, hardness and chapatti quality. However the application of this 'shot-gun' mix of nutrients was not economic within this wheat phase of the rotation (Table 8). A recent survey has identified more than 50% of Haryana soils as deficient in Zn (Quinquennial Report, 2005-2006), and this trend isevident in the SW sites used here, but it is not evident in the main rice-wheat districts where the Zn levels have been improved by regular supplementation in the rice phase of the rotation. Nearly all sites used here would be classified as having adequate or high S, although within non-rice districts where S is rarely applied there are some reports of areas of S deficiency (Hegde and Babu, 2007). The lack of S can limit the efficiency of N utilisation in

Table 8. Marginal economic analysis for 3 wheat varieties with different production practices in Haryana, India.

Variety	Particulars	Unit	RW				SW			
			FP	FP_{MN}	SP	SP_{MN}	FP	FP_{MN}	SP	SP_{MN}
PBW-502	Yield	tha^{-1}	5.50	5.67	5.61	5.78	5.00	5.15	5.10	5.25
	Change in yield	%	-	3	2	3	-	3	2	3
	Price of output	INR t^{-1}	11000	11000	11000	11000	11000	11000	11000	11000
	Value of output	INR ha^{-1}	60500	62315	61710	63561	55000	56650	56100	57783
	Operational costs	INR ha^{-1}	18452	21206	15661	18416	18452	21206	15661	18416
	Net income	INR ha^{-1}	42048	41109	46049	45146	36548	35444	40439	39367
	Benefit-Cost ratio (Operational costs)	INR ha^{-1}	3.28	2.94	3.94	3.45	2.98	2.67	3.58	3.14
	Net returns per rupee of investment	INR ha^{-1}	2.28	1.94	2.94	2.45	1.98	1.67	2.58	2.14
	Change in profit over respective categories	%		-2.24	9.52	-1.96		-3.02	10.65	-2.65
C-306	Yield	tha^{-1}	3.30	3.40	3.37	3.47	3.50	3.61	3.57	3.68
	Change in yield	%	-40	3	2	3	-30	3	2	3
	Price of output	INR t^{-1}	20000	20000	20000	20000	20000	20000	20000	20000
	Value of output	INR ha^{-1}	66000	67980	67320	69340	70000	72100	71400	73542
	Operational costs	INR ha^{-1}	18452	21206	15661	18416	18452	21206	15661	18416
	Net income	INR ha^{-1}	47548	46774	51659	50924	51548	50894	55739	55126
	Benefit-Cost ratio (Operational costs)	INR ha^{-1}	3.58	3.21	4.30	3.77	3.79	3.40	4.56	3.99
	Net returns per rupee of investment	INR ha^{-1}	2.58	2.21	3.30	2.77	2.79	2.40	3.56	2.99
	Change in profit over respective categories	%		-1.63	8.65	-1.42		-1.27	8.13	-1.10
WH-283	Yield	tha^{-1}	4.13	4.25	4.21	4.33	4.75	4.89	4.85	4.99
	Change in yield	%	-25	3	2	3	-5	3	2	3
	Price of output	INR t^{-1}	15000	15000	15000	15000	15000	15000	15000	15000
	Value of output	INR ha^{-1}	61875	63731	63113	65006	71250	73388	72675	74855
	Operational costs	INR ha^{-1}	18452	21206	15661	18416	18452	21206	15661	18416
	Net income	INR ha^{-1}	43423	42525	47452	46590	52798	52181	57014	56439
	Benefit-Cost ratio (Operational costs)	INR ha^{-1}	3.35	3.01	4.03	3.53	3.86	3.46	4.64	4.06
	Net returns per rupee of investment	INR ha^{-1}	2.35	2.01	3.03	2.53	2.86	2.46	3.64	3.06
	Change in profit over respective categories	%		-2.07	9.28	-1.82		-1.17	7.99	-1.01

field crops and although with the high yields obtained with the wheat used here there will be a high demand for N, it is still unlikely that there would be any NxS interactions evident given 150 kg Nha^{-1} was applied (Jamal et al., 2010). Despite the sites receiving regular inputs of P fertiliser at recommended rates (26 kg P ha^{-1}) the soil measures for P are in the low range, prompting Coventry et al. (2011a) to suggest that this recommended rate may be too low for soil P maintenance levels. Manganese and iron levels were mostly in the high range from the rice-wheat sites, but the levels at all the non-rice sites are low (Table 1; Singh, 2008).This difference may be due to the continuous submergence and wet tillage during the rice cycle, rather than to any inherent soil differences. The reduced conditions created by wet tillage (puddling) and ponding of water has many benefits such as neutralizing soil pH to improve the availability of plant nutrients (P, K, Ca, Mg, Mn, Fe) and also allow for accumulation of organic matter (Ponnamperuma, 1972; Sahrawat, 2005). Copper is another micronutrient that has been identified as deficient over much of Haryana but was not measured here. Haryana State has already targeted subsidy schemes available for micronutrients and gypsum, and given the intensive nature of the production systems, the use of selected micronutrients and sulphur is an essential requirement. Thus monitoring and targeting specific crop requirements is required as part of best practice management.

Various authors have shown that there is no grain yield penalty, and there may be some yield advantage, where zero till seeding practice is used rather than using multiple cultivations prior to seeding (Coventry et al., 2011a; Erenstein and Laxmi, 2008; Erenstein et al., 2008; Kumar et al., 2006; Malik et al., 2002; Singh et al., 2006; Yadav and Malik, 2005). Indeed already 25% of the wheat crop in Haryana is sown using a zero till practice. Coventry et al. (2011c) have also shown that a 3-way split with the recommended N application rate (150 kg N ha^{-1}) can increase grain yield in the SW region. Thus, we have used both zero till and the 3-way split of N-fertiliser underpinning the best practice (SP) treatment in the 2 years of field trials. Consistent with our previous work we only obtained small (or none) grain yield increases, but always had substantial improvements in all chapatti quality measures irrespective of the wheat variety used. Whilst here we could not discriminate between the component differences in the SP treatment, for the marginal analysis we assumed a 2% yield improvement with this recommended practice, and given the large cost saving associated with not undertaking the 4 cultivations, this SP treatment was always the most profitable (by 9%). Even if we had allowed for no yield improvement, the SP treatment would be 6.7% more profitable.

The most profitable outcome from the examples used here involves taking advantage of the higher price available for grain from either the recognised standard for best quality chapatti flour (C-306) or variety WH-283. In the SW region, the net return per rupee of investment was 3.64 where WH-283 was grown compared with a net return of 2.58 where the high yielding PBW-502 was grown. In the SW region, WH-283 only has a small yield disadvantage (5%) compared with the best yielding standards. In the rice-wheat area the best financial outcome was obtained using variety C-306 (net return 3.30) compared with a return of 2.28 when PBW-502 was used. It should be repeated here that C-306 is an old variety (released in 1966) that produces 30 to 40% less grain than the current recommended wheat varieties, but this disparity can be much higher as this tall variety is susceptible to lodging, and does not have the armoury of genetic protection assembled with cumulative advances in breeding. There is concern that traditional agriculture with wheat as the main winter crop is not a profitable venture due to squeezed margins given the low and mostly static support price given by the government. At present, wheat cultivation earns farmers an annual profit of about INR 20,000 to 24,000 per ha (Kharkar, 2012; CACP Report, 2012), so farmers are looking for options to diversify systems to increase profitability.

The past three seasons have seen record high yields of wheat in Haryana and this outcome is an on-going priority for India, both for raising external income and for ensuring basic food availability for the large population. Thus, the variety improvement programs target high yield and particularly management of yellow rust (*Puccinia striiformis*). Therefore, any wholesale change in wheat production that moves farmers away from optimizing total grain production is untenable. Notwithstanding this reality, some farmers are in a better position to take advantage of the marketing situation as it currently exists. The yield data shows that the current two 'best-quality' varieties (C-306 and WH-286) suffer large yield penalties when grown in the rice-wheat districts, but in the west and south-west areas of Haryana, C-306 yield is relatively not as much affected compared with the RW situation (25% compared with 40%) and WH-283 grain yield is only minimally affected. Therefore at this stage it is likely that little opportunity exists for farmers to be seeking the quality markets in the rice-wheat areas, and that more realistic market opportunities exist in the SW areas where the practice is already better suited for quality. Two additional varieties (HD-2851 and WH-1025) that have good chapatti quality were evaluated here, and HD-2851 performed well with quality and also yield. Whilst quality is routinely evaluated as part of the various wheat improvement programs, there must be awareness for quality within localised markets and special market opportunities. Wheat currently is purchased largely by government agencies based on a minimum support price (MSP), irrespective of any quality traits. From the surveys undertaken and reported here we know that the local marketing of grain is an active market and about 18% of farmers sell their grain outside the MSP. In the SW region, a few villages are already actively undertaking locally based marketing of chapatti quality wheat, and varieties WH-283 and C-306 are being grown by many farmers covering an area of about 10% in Rewari and Mahendergarh districts (Coventry et al., 2011b). Generally in the rice-wheat system, the farmers are more concerned about high wheat yield and few farmers (2%) expressed an interest in growing quality wheat, mainly because they feared that they may not get market recognition for the quality produce.

We have shown that it is possible to improve the chapatti quality of wheat within all the varieties evaluated using best practice management (zero tillage and split applications of N-fertiliser). However, it is unlikely that this quality outcome is likely to be a motivator for farmers without some financial recognition. As India seeks more opportunities in the wheat export markets, the move towards segregation of wheat into targeted markets is developing. Segregation at the large scale will become more plausible as better quality attributes are available in the high yielding varieties. As these specific markets develop, more Indian wheat farmers will have incentives for the production of high quality chapatti wheats.

ACKNOWLEDGMENTS

This study was not possible without the excellent collaboration provided by the farmers at each of the field sites. Specifically the authors acknowledge Sahib Singh, Ramba village; Anil Kumar, Kalri village, Mukhtiar Singh, Darar village; Paramjeet Singh, Sadura village; Tanwar Thana village; Ved Prakash and Parvin Saini, Chhapper village; Dharam Pal, Kapriwas village; Shiv Charan, Jarthal village; Raj Kumar, Rakesh and Surender, Tigra village; Raja Ram, Sahapur Begu village and Sawran Singh and Gurdeep, Baingu village. They also acknowledge the financial assistance provided by the Australian Centre for International Agricultural Research for this study.

REFERENCES

CACP rabi report (2012). http://cacp.dacnet.nic.in/RPP/Rabi_Report 2013-14.

Coventry DR, Poswal RS, Yadav A, Gupta RK, Gill SC, Chhokar RS, Kumar V, Sharma RK, Kumar A, Mehta A, Kleemann SGL, Cummins JA (2011a). Effect of tillage and nutrient management on wheat productivity and quality in Haryana, India. Field Crops Res. 123:234-240.

Coventry DR, Gupta RK, Yadav A, Poswal RS, Chhokar RS, Sharma RK, Kumar V, Gill SC, Kumar A, Mehta A, Kleemann SGL, Bonamano A, Cummins JA (2011b). Wheat quality and productivity as affected by varieties and sowing time in Haryana, India. Field Crops Res.123:214-225.

Coventry DR, Yadav A, Poswal RS, Sharma RK, Gupta RK, Chhokar RS, Gill SC, Kumar V, Kumar A, Mehta A, Kleemann SGL, Cummins JA (2011c). Irrigation and nitrogen scheduling as a requirement for optimising wheat yield and quality in Haryana, India. Field Crops Res. 123:80-88.

Erenstein O, Laxmi V (2008). Zero tillage impacts in India's rice-wheat systems: A review. Soil Tillage Res. 100:1-14.

Erenstein O, Farooq U, Malik RK, Sharif M (2008). On-farm impacts of zero tillage wheat in south Asia's rice-wheat systems. Field Crops Res. 105:240-252.

Hegde DM, Babu SNS (2007). Correcting sulphur deficiencies in soils and crops. Indian J. Fertilizers 3:65-79.

Jamal A, Moon Y, Malik ZA (2010). Sulphur -a general overview and interaction with nitrogen. Aust. J. Crop Sci. 4:523-529.

Kharkar RK (2012). Economics of important rabi crops in Haryana to include in package of practices during 2011-12. Report presented in Agricultural Officers Workshop for rabi crops, April 2012, CCS Haryana Agricultural University, Hisar.

Kumar A, Chand R, Fulzele RM, Singh R (2006). Impact of zero tillage technology intervention under IVLP on wheat cultivation. Ind. J. Extension Sci. 1:19-22.

Lindsay WL, Norvell WA (1978). Development of a DTPA test for zinc, iron, manganese and copper. Soil Sci. Soc. Am. J. 42:421-428.

Malik RK, Yadav A, Singh S, Malik RS, Balyan RS, Saroj Jaipal, Hobbs PR, Gill GS, Gupta RK, Bellinder R (2002). Herbicide resistance management and evolution of zero tillage, A success story. Research Bulletin-2002, CCSHAU, Hisar, 1-43.

Mishra B, Jag S, Chatrath R, Chatrath AK, Sharma RK, Gupta RK, Randhir S, Rane J, Anuj K (2005). Cost Effective and Sustainable Wheat Production Technologies. Directorate of Wheat Research, Karnal 132001. Technical Bulletin, No.8:36.

Olsen SR, Cole CV, Watnabe FS, Dean LA (1954). Estimation of available phosphorus in soils by extraction with sodium bicarbonate. U.S. Department of Agric. Circ, pp. 939.

Ponnamperuma FN (1972). The chemistry of submerged soils. Adv. Agric. 24:29-96.

Quinquennial Report (2005-2006). All India Coordinated Research Project on Micro and Secondary Nutrients and Pollutant Elements in Soils and Plants. Department of Soil Sciences, CCS Haryana Agricultural University, Hisar.

Richards LA (1954). Diagnosis and improvement of saline soils. In: USDA Handbook 60. U.S. Salinity Lab, CA, USA.

Sahrawat KL (2005). Fertility and organic matter in submerged rice soils. Cur. Sci. 88(5):735-739.

Singh R, Kumar A, Kumar S, Chand R, Singh S (2006). Participatory research on effect of tillage options on wheat productivity. Ind. J. Extension Sci. 1(1):106-109.

Singh MV (2008). Micronutrient fertility mapping for Indian soil. Technical Bulletin. AICRP Micronutrients, IISS, Bhopal. 7: 1-60.

Srivastava S, Rao AS, Alivelu K, Singh KN, Raju NS, Rathore A (2006). Evaluation of crop responses to applied fertilizer phosphorus and derivation of optimum recommendations using the Mitscherlich–Bray equation. Commun. Soil Sci. Plant Anal. 37:847–858.

William CH, Steinbergs A (1959). Soil sulphur fractions as chemical indices of available sulphur in some Australian soils. Aust. J. Agric. Res. 10:340-352.

Yadav A, Malik RK (2005). Herbicide resistant *Phalaris minor* in wheat, A sustainability Issue Resource Book. Department of Agronomy and Directorate of Extension Education, CCSHAU, Hisar, India 1:152.

The economic study of production and exportation of selected products from Iran

Mahmoud Bakhshinejad[1] **and Esmail Hosseini**[2]

[1]Department of Economics, Islamic Azad University, Shahrekord Branch, Iran.
[2]Department of Banking, Monetary and Banking Research Institute (MBRI), Central Bank of Iran.

This paper presented an examination of Balassa's 'revealed comparative advantage' (RCA). The results of the research showed that using the RCA should always be attuned in such a way that it becomes symmetric.This paper consisted of three major sections: The theoretical section, the analytical tool and the case studies of comparative advantage' (CA). First, we evaluated the theory and various empirical measures. Then, the analytical tool was applied for the analysis of exported products. The results indicated that Iran had not had CA in the export of walnuts, almonds,hazelnuts, apples andoranges. To increase competitiveness, we suggested several policy recommendations such as increasing agricultural productivity, promoting the development of indigenous technological capabilities, and reducing the cost of doing business. The conclusion was based on a theoretical discussion of the properties of the measure and on convincing empirical evidences according to the Balassa index. The index for 2007 illustrated that Iran had not possessed a CA in the export of walnuts, almonds, hazelnuts, apples and oranges andconsequently had reduced its market share in these products.

Key words: Revealed comparative advantage, iran's agriculture sector, almonds, apples, hazelnuts, walnuts, oranges.

INTRODUCTION

Iran has undertaken a series of economic reforms towards opening up the economy, and the role of agriculture is remarkably significant in this respect since it contributes around 11% to the gross national product and employs one third of labor force. To increase the contribution of agriculture to the GNP in recent years, a great deal of investment has been channeled toward producing items such as almonds, walnuts, apples, oranges and hazelnuts.

Having grasped the issues mentioned above, it seemed necessary to pay due attention to the survey of CA related to these products. This necessity came from the common observation that CA usually were not stable and tend to change with time. Therefore, it appeared to be necessary to estimate the CA of various export products on a regular basis. In the new epoch of "diversified agriculture", a frequently-asked question was, "where was the true place of the CA of Iran's agricultural products?"That is, which agricultural product of Iran was relatively more competitive and had a better chance to thrive in the long run, given the national and world-wide competition? To answer this question, one needed to first evaluate the competition faced by each product to determine their relevant competitiveness. Then, Iran's CA in each product could be assessed by comparing the products' competitiveness to the other products.

Iran appeared to have high CA in products which had relatively high competitiveness.

This paper aimed to survey the CA of Iranian export products cited above and their competitiveness in the world market. This investigation might be regarded as the first one that tried to appraise the competitiveness of the afore–cited products in the global market. More specifically, this paper examined the structure of CA enjoyed by Iran in the global market both on an individual basis and in a comparative framework. Following this, an analysis of the CA is intensity factors for the economy to which it is applied. The pattern of CA is also examined for inter-temporal variation over the period 2002 to 2007. To undertake the study of CA, we made use of the Balassa index of revealed comparative advantages (Balassa, 1965).

Using the law of CA has been regarded a proper way to find answers to the question "Who exports what and why?" Analysis of CA is certainly not a novel idea. Although a plethora of studies have occurred in the past 50 years, few have specifically addressed agricultural trade (Winter, 1995). Some suggested that the conspicuous absence of CA analysis from agriculture may be in part due to the magnitude of distortion brought by government policies (Heley, 1985). In the 1950s, Wassily Leontief argued that"...fluctuation in yield here and abroad, not to speak of government intervention, affect foreign trade in from products to such an extent that the amounts of agriculture commodities exported and imported in one single year can be expected to reflect long-run comparative costs much less than is the case for any other type of good."(Leontief, 1974).Clearly, the strength of this argument has been weakened by recent developments in agricultural trade liberalization. As agricultural producers become more expected to market forces, and as dependence on foreign markets increases, the relevance of CA seems to become increasingly apparent.

This study used the "revealed comparative advantage" (RCA) approach to assess Iran's CA in the selected agricultural products. This assessment might provide systematic information about the direct or indirect competition facing Iran's agricultural products. This could be useful for decision-making regarding sustainable agricultural development policies in Iran.

THEORY AND METHODOLOGY

The idea of CA might be attributed to the work of John Stuart Mill, Adam Smith and David Ricardo. It is largely derived from the proposition on opportunity cost and Labor specialization. Smith and Mill first advanced the concept of absolute advantage claiming that a nation will export an item when it is the lowest cost producer of that item. Ricardo refined the idea of CA by recognizing that a nation tends to allocate its resources to their most productive use. A nation may, therefore, import a good even when it is the lowest cost producer of that good (Leontief, 1956). In the meantime, Eli Heckscher and Bertil Ohlin revolutionized trade theory by emphasizing international differences in resource (or factor) endowments. "Factor abundance theory", or the Heckscher–Ohlin model, predicts that a country will export commodities that are relatively intensive in the factor with which the country is relatively well endowed. Thus, a land-abundant country will export land intensive goods, while a capital-abundant country will export capital intensive goods. The purview of the Heckscher–Ohlin model has been consequently extended through the work of Wassily Leontief, Paul Samuelson, Jaroslav Vanek, and others (Memedovic, 1994).

Empirical tests of CA have often used cost or price information to measure efficiency in production, as well as availability and allocation of scarce resources. Transportation models and linear programming techniques, for example, have determined CA through mark proximity or cost minimizing solution, subjects to resource availability and prices. These forms of analysis, however, have frequently been constrained by a lack of reliable and internationally comparable data. Even when survey methods were used to overcome data scarcity, the estimation of exchange rates, purchasing power and valuation of local land, labor and capital could be problematic. Further complications also arise when taking into account these so-called "milieu factors" such as government policy, history and other likely sources of CA that did not easily lend themselves to quantification.

Ideally, measures of CA should reflect regional or cross-country differences in a hypothetical pre-trade environment known as autarky. Autarky is the condition where equilibrium prices are unaffected by influences external to an economy (Hook, 1992). In reality, since all countries engage in some level of international trade, "true" CA in autarky cannot be directly observed. Bela-Balassa (1965) introduced the notion of "Revealed Comparative Advantage" (RCA) as a way to approximate CA in Autarky. According to Balassa, "the concept of RCA pertains to the relative trade performances of individual countries in particular commodities. As the commodity pattern of trade reflects inter-country differences in relative costs as well as in non-price factors, this is assumed to reveal the CA of traditional countries" (Balassa, 1977). If trade performance is determined CA, then direct observations of trade performance should be "revealed" CA. Barring production or export subsidies, the greater the CA in the production of that commodity. The plausibility of this condition has almost certainly been strengthened by recent trade liberalization.

CA reflects relative competitiveness. Competitiveness can be measured by market shares (Mahanta, 2005). The larger Iran's market-share of a product gets, the more competitive it will become relative to other suppliers in the market. For example, if Iran's market share of product A is greater than its market share of product B, then it is relatively more competitive in product A than in product B. In other words. It has stronger CA in product A than in product B. Based on the RCA approach, we collected the data from FAOSTAT and measured Iran's CA in agricultural products by the following index:

$$RCAj = (Sj / s)^{1} \qquad (1)$$

The RCA index thus defined compares Iran's market share of each agricultural product (Sj) to its average market share for all agricultural products under comparison. Iran has above-average CA in products whose RCA scores are greater than 1. Iran has stronger CA in products with higher RCA scores.

Competitiveness and CA are unlikely to be constant over time because of the change in consumers' preference, production cost, transportation costs, regulations, etc. Variation of competitiveness can be measured by the changes in market shares. An increase (or decrease) in the market shares indicates the competitiveness gain (or loss). However, it is usually not appropriate to directly use the change of RCA scores for measuring the variation of CA. The

[1] - Sj denotes each agricultural product.

Table 1. Comparative advantage of Iran's walnuts, almonds, hazelnuts, apples and oranges.

Product	2007	2007 vs, 2002
	RCA	RCAV
Walnuts	0.18	-3.72
Almonds	0.0	-5.63
Hazelnuts	0.005	-1.89
Apples	0.018	5.01
Oranges	0.03	6.04

The calculated RCA index was based on wholesale value of export from FAOSTAT. A measure of variation of comparative advantage computed based on the RCAV equation.

following formula provides a more precise measure of CA:

$$RCAV_i = C_{i,t+1}(RCA_{j,t+1} - \beta RCA_{j,t}) \times 100,^2 \quad (2)$$
$$\beta = (1+g)(1+\sum c_{j,t} g_j)^{-1}$$

The sign and value of RCAV scores indicate the direction and magnitude of CA variation over time. For example, an Iran's agriculture product RCAV score of (+)5 indicates that Iran has gained comparative in this product, and if without the change in CA, sales of this Iran agriculture product would have been 5% lower than its actual level. The implication for negative RCAV scores would be the opposite. Note that the sum of RCAV scores for all the products is equal to Zero. This captures the shift of CA among products. When Iran becomes relatively more competitive in some products, it automatically becomes relatively less competitive in other products.

The method of analysis outlined above has been applied in case studies of the agriculture and manufacturing in Pakistan, Kenya, India, Mali, etc. In this section we provided only a short summary of the conclusions in one study in Kenya. In the study of Kenya, recently collected data were used together with data from the mid-2000s in many cases from the same firms. This paved the way to examine how Kenya's policy reforms of the late 2000 and early 2009 had affected the country's agricultural sector.

Interestingly, the sector seemed somehow to have gained in comparative advantage but lost in terms of competitiveness. This unexpected result was explained by the fact that several major distortions acted as obstacles to competitiveness, in particular the very substantial interest-rate distortion caused to a large extent by government policy.

RESULTS

The data were analyzed in 2007 (the last year for which complete trade data were available) for both Iran and the rest of the world. In order to acquire better appreciation of trade dynamics, the data of the year 2002 were also examined. RCA was evaluated for walnuts, which will be discussed in its own place.

The index for 2007 showed that Iran does not have a comparative advantage in these products. Table 1 shows the RCA scores of the agriculture in 2007 and their RCAV scores between 2002 and 2007. These results could be suitable to two main forces namely declining competitiveness and structural change. We used market shares to measure Iran's competitiveness in five selected agriculture products from 2002 to 2007. Iran reduced its market share in these products (Table 2). We used the RCAV index to measure variation of Iran's comparative advantage in the five products between 2002 and 2007. During this period, Iran decreased its comparative advantage in these agricultural products.

DISCUSSION

This paper contended that the analysis of comparative advantage (CA) for agricultural commodities not only has become relevant but also might in fact be an important tool for understanding the future of the world agriculture. Although autarky precludes direct observation of CA, measures of RCA can provide useful approximation of CA.

While Balassa's export-based RCA index overlooked some aspects of domestic consumption and value-added processing, it was nevertheless a meaningful gauge for measuring the relative strength or weakness of agricultural exports. In the case of walnuts, almonds, hazelnuts, apples and oranges, the measurement of RCA showed that CA was a dynamic condition, not a static one. The dynamics of CA became increasingly apparent as agricultural markets became less insulated by government trade and support policies. The fall in the RCA index across a number of sectors presumably reflected the fact that many indigenous and traditionally labor-intensive sectors have found it increasingly difficult to compete. This was caused by quite high rates of wage inflation in Iran over the last decade and greater competition from international markets. As far as

---

[2]- C_j denotes the ratio of Iran's supply of each agricultural product in its total supply of all agricultural products; t and t+1 are time subscripts. g_j andg denote, respectively, the growth rate of each individual agricultural product and the average rate for all agricultural products. i denotes the country.

Table 2. Iran's Competitiveness in walnuts, Almonds, Hazelnuts, Apples and Oranges.

Product	Walnuts		Apples		Almonds		Oranges		Hazelnuts	
	Supply	Market share	Supply	Market share	Supply	Market share	Supply	Market share	Supply	Market share
	1000$	%	1000$	%	1000$	%	1000$	%	1000$	%
Africa	0	0	0	0	0	0	0	0	0	0
America	19	0	23	0	0	0	0	0	0	0
Asia	2198	0.002	103656	0.005	111732	0.03	0	0	0	0
Europe	7.5	0	340	0	1628	0	0	0	0	0
Oceania	0	0	0	0	0.1	0	0	0	0	0

Wholesale value of export from FAOSTAT.

structural change was concerned, the fall in RCA reflected the fact that CA was dynamic rather than static.

Conclusion

Due to data constraints, the comparative advantage assessment in this article compares the wholesale value of Iran's agricultural supplies to the market with those of other countries irrespective of the final destination of these commodities. A more refined assessment in the future should compare the agricultural production of Iran and that of other countries.

REFERENCES

Balassa B (1965). Trade Liberalization and Revealed Comparative Advantage. Manchester School 33:99-123.

Balassa B (1977). A Stage Approach to Comparative Advantage. World Bank Staff Working Paper, P. 256.

Heley SL (1985). The Theory of Agricultural Comparative Advantage. Unpub, Ph. D. diss, purdue University, West Lafayette.

Hook JP (1992). The Comparative Advantage of Agricultural Economics. Am. J. Agric. Econ. pp. 1059-1065.

Leontief W (1974). Domestic production and Foreign Trade: The American Capital position Re-Examined. Economic International, P. 7.

Leontief W (1956). Factor Proportions and the Structure of American Trade: Further Theoretical and Empirical Analysis. Review of Economic Statistics, P. 38.

Memedovic O (1994). On the Theory and Measurement of Comparative Advantage: An Empirical Analysis of Yugoslav Trade in Manufactures with the OECD Countries. pp. 1970-1986. Amsterdam: Thesis Publishers.

Mahanta AK (2005). The Comparative Advantage of Agricultural Economics. Am. J. Agric. Econ. P. 156-165.

United Nations Food and Agriculture Organization (1999). FAOSTAT database .Online Available http://www.fao.org January 15, 1999.

Winter NA (1995). Measuring the comparative advantage of agricultural activities: domestic resource costs and the social cost-benefit ratio. Am. J. Agric. Econ. pp. 243-250.

An analysis of the determinants of household level production and marketing of beans in Eastern Congo

Eliud Birachi[1], Rachel Zozo[2], Bernard Vanlauwe[3] and Jonas Chianu[4] and W. Chiuri[1]

[1]International Center for Tropical Agriculture-Africa (CIAT), Box 1269 Kigali, Rwanda.
[2]Consortium for Improvement of Agricultural-based Livelihoods in Central Africa (CIALCA), CIAT-TSBF, Bukavu, Democratic Republic of Congo.
[3]International Center for Tropical Agriculture-Africa (CIAT-TSBF), Nairobi, Kenya.
[4]African Development Bank.

Legumes and especially beans provide some of the most widely grown crops in the East and Central Africa region. The question of whether certain crops, specifically beans, are more beneficial to women than to men in relation to cash crops and other cereals, has been raised from time to time. In the central Africa region and in particular in the Democratic Republic of Congo (DRC), bean production has been promoted in the recent years. Beans contribute to better nutrition and soil fertility improvement in the region. The economic benefits and costs associated with beans production has not received adequate attention at the household level while taking into consideration male and female farm actors. Based on beans, this paper examines the differentiated profitabilities and costs of production for beans at the household level, among male and female farmers in the South Kivu Province of the DRC. The paper uses results obtained from a survey of 160 farmers in 10 villages in the South Kivu Province of DRC. Data was collected on household characteristics, costs, prices and quantities associated with bean production. Regression analysis was used to assess factors that influence the observedbean gross margins and production costs among the households. The results indicate the average age of bean farmers is relatively low and that there is no significant difference between the profits received by men and women farmers in spite of men incurring relatively higher production costs than women farmers. the results also indicate that larger surfaces put under beans do not necessarily lead to higher gross margins. Market access had factors such as physical location of the farmers vis a vis markets influenced to raise the costs incurred in bean production for those farmers located much farther away from major markets. Farmers located in remote areas suffer higher production costs in acquiring inputs, usually at higher costs than those located in optimal locations. Interventions that would enable such farmers to affordably access productive inputs are likely to impact positively on the farmers' economic welfare.

Key words: Bean production and marketing, South Kivu, gender, gross margins.

INTRODUCTION

Despite their immense contribution to society (ILO, 2009), women's productive, domestic and community-related activities seem to be underestimated; consequently they are misunderstood and are rendered invisible. The division of farming activities between genders varies according to the enterprise, the farming system, the technology used, and the wealth of the household. The Control over the benefits of production also varies between gender regarding labor input, use of produce for consumption or for sale, cultural norms regarding their enterprises, the dominance of men as the household heads and, their entitlements to the most important resources like land and credit. This makes estimations of the contribution of women to agriculture more complex (Marinda, 2006).

In addition to working in the home and on the farm, rural women are engaged in a diverse range of off-farm reproductive and community activities. There are few enterprises in which women dominate both the workloads and the control of the benefits. But there are several enterprises in which women and men share both the workloads and the benefits. However, in most cases, men are the key players in crop and livestock production, and control income generated from produce sales. In many countries that produce legumes, beans are one of the major pulses produced both for home consumption and for sale. Though men and women share the workload in bean production, the right to access the benefits is very limited for women (Lemlem et al., 2010).

Gender issues are important when addressing the factors influencing economic development of countries, as any national economy will benefit if the gender roles in communities are equal, and prejudices and inequality suffered by women can only downsize any development efforts, as well as poverty, trade and debt-related issues. In Africa, gender system is much more complex, and while women have some land use and other rights, which come into play with changes in agricultural production with women farming in their own right as heads of household (Boserup, 1970).

In South Kivu, most rural households depend on commercialisation of agricultural products where there is surplus, for livelihood. For the last decade, there has been declines in crop yields which do not favour most households to sell their commodities and thus a major proportion of production is used for home consumption at household level. Many development agencies which operate in Eastern DRC and South Kivu in particular have not paid much attention to legume production as a source of livelihood improvement. Most have focussed on emergency aid rather than long term poverty alleviation, food security and income generation strategies that can move most people from a crisis situation. One of the initiatives targeted at improving economic and nutritonal welfare in the South Kivu province is the Consortium for Improved Agricultural Based Livelivehoods in Central Africa (CIALCA) operating in Bas-Congo,South Kivu, North Kivu provinces of the DRC and in Rwanda and Burundisince 2005.One of the main focus of the consortium is on legumes because legumes make some of the best crops to introduce to communities in crisis or just getting out of crisis. They grow fast (60-90 days) but also improve soil fertility. In humid tropics where most nutrients are in the vegetation form war tends to clear all the vegetation and hence the need to introduce nitrogen fixing crops as communities go back to their homes after the war. The war has increased reliance on resource extraction leading to the destruction of traditional livelihoods and alternative sources of protein when communities are forced to flee from their gardens (Hart and Mwinyihali, 2001). These challenges have led to poor food security status and soil and natural resource depletion and degradation that has in turn affected productivity further. Legumes can add substantial amounts of nitrogen from the atmosphere to the soil and still provide food (grains) and animal feed (haulms). Legume-based protein is an important source of protein in South Kivu because many households lost their livestock during the war. In addition legumes have attracted good market values and are being traded locally and regionally.

This paper provides a detailed analysis of the relative costs and profit abilities of bean production in the South Kivu province of the Democratic Republic of Congo. Various studies show that women are incrreasingly playing a crtitical role in agriculture (Ezumah and Di Domenico, 1995) and that there is a growing trend towards women headed households in many countries (Enete and Amusa, 2010), a trend that may affect the productivity of the households.This paper investigates factors influencing returns on bean production,as well as factors that influence the costs faced by male and female bean farmersin order to gain more understanding of household production and marketing dynamics among smallholder subsistence oriented farmers. The paper makes a contribution to new research area of knowledge with respect to livelihoods improvement through legume production and sales especially for crops that have a relatively higher women participation in production and marketing. An analysis of costs and profitability is necessary to provide information on the attractiveness of the legumes as a source of incomes. This will ensure that the benefits of legumes in fixing nitrogen to improve soil fertility is assured, and in addition, a good source of protein for better health is assured for the millions of people in Africa, and especially in conflict areas such as South Kivu.

Household production and marketing of beans

The paper is based on the household decision making theory (De Janvry and Sadoulet, 2003). A farmer makes decisions about what to grow, how much to grow, and the number of necessary inputs in response to market prices of inputs and outputs and the available fixed environmental inputs available on their land. The farmer makes two decisions: Production and consumption, both of which are made simultaneously and this decision has an impact on household utility. The decision farmers make to grow a particular crop is basically to bring more benefits than others and the idea behind the decision is to increase utility. Therefore, the benefits can be represented by the amount of produce consumed and the amount sold. If the amount of produce is not enough for consumption purposes, farmers are likely to buy from the market to maintain their lifestyle.

Economic theory suggests that different factors affect farmers' benefits derived from a crop. Existing literature reveals that the possible factors that may affect the benefits accruing to the farmer include heterogeneity of farmers' resources such as land, resource endowment like education, labor and wealth, a multiplicity of farmers' household needs among which include food, fodder, trade, access to markets, and income motives among others. These factors can be incorporated into a basic household model, iteratively. Economic models for analyzing benefits from an enterprise can be modelled using agricultural household models (Singh et al., 1986). The analytical framework presented here models farmers' benefits from a legume enterprise (beans) as a function of farmers' utility maximization, constrained by resources endowment. Assuming a well-behaved utility function,and based on Van Dusen (2000) a farmer's utility maximization can be set as:

$$\max U= U (G_f, G_p /\varpi_0) \quad (1)$$

Where G_f and G_p refer to consumption of aggregate on-farm produced and aggregate purchased goods, respectively and ϖ_0 refers to the vector of household level exogeneous factors including inputs endowment and heterogeneity (land and labor), human capital (age, sex, and education), asset endowmnet (livestock, land holding and wealth) and market access factors (distance or time to market places) and extension, all conditionning utility. However, utility maximization is subject to certain constraints such as land, cash, labor and other inputs that are available at the disposal of the household and the competing requirements for the resources.

The other most important constraint that the household faces is the production technology constraint which is conditionnal on choices of crop enterprises and varieties. It may be assumed that, in deciding whether to plant beans or not, the producer weights up the expected utility of wealth from the expected benefits represented as $U^*_A(\pi)$ and the expected utility of wealth from planting alternative crops represented as $U^*_N(\pi)$ and a producer will select an enterprise only if $U^*_A(\pi) > U^*_N(\pi)$ (Abdulai and Binder, 2006). This is under the assumption that farmers are risk neutral and that net farm returns (π) represent wealth. The parameters of this decision can be represented by the costs and returns associated with the enterprises and can be captured in monetary terms. Utility of bean enterprise can be related to a set of explanatory variables, Z as follow:

$$U(\pi)= \delta' Z_1+\varepsilon_1 \quad (2)$$

Where δ is a vector of parameters and ε is an error term with mean zero and variance σ^2_ε. The error term includes measurement and factors unobserved by the research but known to the farmer. Variables in Z include all the variables described previously. Following this logical model, when costs are optimised, an empirical model for both costs and profits in this paper can be presented as follows:

$$\mathbf{Y}= \alpha_0 + \alpha_1X_1 + \alpha_2X_2+ \alpha_3X_3+ \alpha_4X_4 \quad (3)$$

Where Y = Gross profit on beans; X_1 = Sex; X_2 = Age of farmer; X_3 = market access indicator (Time to market); X_4 = Bean hactares; X_5 = Education level; X_6 = Land ownership; ε = other factors outside the control of the farmer Similarly, determinants of cost of bean production can be evaluated as follows:

$$\mathbf{Y}= \alpha_0 + \alpha_1X_1 + \alpha_2X_2+ \alpha_3X_3+ \alpha_4X_4+\varepsilon \quad (4)$$

Where: Y= Cost of bean production and marketing; X_1= Sex; X_2= market accessindicator(location of farmer) ; X_3= Marital status; X_4= Bean hactares; ε = other factors outside the control of the farmer.

Using the empirical models both gross margins and costs of producing and marketing beans were analyzed using ordinary least squares regression to evaluate factors influencing the bean gross profit margins and the observed costs.

The gross profit margin is suitable for this study because when using the net profit margin, the costs incurred by farmers are very high and do not express any efficiency on land use and profitability.Gross profit margin is useful for comparing product lines because it allows farmers to make decisions about whether a product is worthwhile, and to keep track of how items perform over time while net profit margin shows the health of the business overall because it indicates how well the activities are doing at keeping expenses low and sales high.

MATERIALS AND METHODS

The paper is based on a A survey was conducted in the eastern DRC, South Kivu province, in 2007. Random sampling procedures were employed to get a representative sample of farmers that were included in the survey. Two sites (Lurhala and Kabamba) were randomly selected from among four CIALCA project intevention areas and an equal number of farmers selected from each of them.A sampling frame was prepared from all villages in these two sites and 5 villages were selected (Buganda, Isimbu, Nyalushoze, Bukalye, Karambi in Lurhala; and Kaboneke, Chifinjo, Ceyal, Lukayo, Kahanga in Kabamba). Sixteen farmers were randomly selected for inclusion in the interviews from each of the villages resulting in a total of 160 farmer respondents. The population of the study consisted of farmers who produce beans regardless of whether or not they produce other crops since 93% of the

Table 1. Definition of variables.

Variable	Definition	Measure
Sex	Sex of farmer	
Location	Market access indicator	1=poor access (Lurhala), 2=good access (Kabamba)
Age of farmer	Number of years since birth	Years
Time to market	Market access indicator (Time it takes to reach main market)	Minutes
Education level	Education levels	1=no formal education 2=primary 3=post primary education
Marital status	Marital status	1= single 2=married-monogamous 3=polygamous 4= widow
Land tenure	Type of land ownership	1=purchasedor rented, 2 = inherited or by donation
Land size	Size of land under beans	HactaresHectares
Quantity produced	Quantity of beans produced	Kgkg/ha
Selling price	Sale prices of the beans	$USD/kg
Cost of production	Cost from land preparation to post-harvest activities	$USD/ha
Gross margins	Margins derived from bean sales	$USD/ha

1 USD = 500 CF (Congo Francs) on 30th July 2007.

population rely on agriculture as the main livelihood activity. A general classification of the two sites indicated that Kabamba had a generally better market access compared to Lurhala.

Data description

Data was collected on landownership and size, age and sex of farmer household heads, labour and capital and other factors that contribute to output producedsuch as experience of the farmer, access to market in terms of time to reach market points as well as bean prices. Data was presented using statistics such as frequencies, mean comparisons, and regression analysis. The variables of interest are described in Table 1. Sales prices were computed after converting the local unit measures into US dollars per kilogram. Costs of production were computed as the sum of all cost incurred (land preparation, planting, weeding, harvesting and post-harvestcosts) per hectare that were incurred by the farmers. Gross margins were computed as the difference between production costs and the sales value of the beans (sales value = quantity sold*price per kg/unit).

RESULTS AND DISCUSSION

Summary statisitcs show that the majority of socio-economic characteristics

The sampled respondents comprised 57% women farmers and 43% men. The mean age of the both male and female farmers was 36 years of age. Most of them were married. A large number (68%) did not have any formal education as presented in Table 2.

Access to productive assets (land, labor, credit, inputs and seeds)

The land tenure systemcan be characterized by a mix of systems: some farmers used land purchased by themeselves, other rented-in land. Other modes include inheritance and donations. A majority of the farmers used land inherited from the families as well as land donated to them (68%) while about 32% used purchased or rented-in land. In this study, land tenure was represented by land either purchased/rented in and land that was inherited or acquired under deonation. These two classifications captured the incentives to invest in the land to improve productivity.

Market access

Accessibility to market can determine whether or not farmers have access to inputs and output markets and whether this has an impact on transaction cost and it is likely to affect crop production and marketing as farmers tend to sell at a lower price due to poor market access. Market access can be measured by time to reach markets. Market access may also be measured by assessing whether a location has been determined to be a good access or poor access location for the farmers involved. Each of the two regions had an equal number of farmers included in the survey.The two regions where the survey was carried could be classified as follows: Kabamba generally was a good market access region while Lurhala was more remote relative to major roads and urban areas (Table 4).

The average time to reach the markets was about 70 min (Table 5). Most of the farmers in the province were more likely to use human backor headto transport products to markets (on foot). The average land under beans per farmer was 0.34 ha, this may be a factor of wealth endownment or show the ability to put up more land under beans. The mean selling price of beans was

Table 2. Responents' marital status, marriage types and education levels.

Marital status	Percent	Education level	Percent	Gender	Percent
Single	9.4	No formal education	68.1	Male	43.1
Monogamy	66.9	Primary school	22.5	Female	56.9
Polygamy	7.5	Secondary school	9.4		
Widow	16.3	Total	100.0		
Total	100.0				

N=160.

Table 3. Land tenure.

Land tenure	Percent
Purchase/rented	31.9
Heritage/donated	68.1
Total	100.0

N=160.

Table 4. Distribution of sampled farmers by location.

Location indicator	Percentage of sample
Lurhala	50
Kabamba	50
Total	100

about 340 FC (USD 0.68) per kilogram. The costs of production were defined as costs related to household labour use, hired labour and other costs such as the input purchases, land preparation, planting, weeding, harvesting, andpost harvest activities. The other costs were the costs of transport and different taxes (both official and informal). The average cost of production per farmer was about 21,000FC (USD 42) while the average gross margin was about 140,000 FC (USD 280).

Factors influencing household level gross margins and production costs

Regression results on the factors influencing bean gross margins are summarized in Table 6. The regression resultsshow that the size of land under bean production was inversley related to the amount of gross margin achieved by the farmers. Farmers with smaller land holdings, most of whom were women, were more likely to achieve higher gross margins from their beans. Due to their relatively poor economic status, women in this study grew beans in land that was not purchased or rented, they did not hire labor and transport equipment and services as their male counterparts did. Thus, they operate relatively smaller plots than men. It would also imply that farmers that had larger land holdings used more inputs, hired labor and equipments which increased the costs of production and transporting beans to markets, thus reducing their overall returns. Conversely, smaller farmers(women) used very low levels of production inputs resulting in relatively less costs.

The time taken to reach markets also had some miminal and negative influence on the grossprofitmargins reported. Longer times to markets appeared to reduce the amount of gross profit margins achievable by the farmers as this increased the costs of accessing markets or prevented farmers from accessing more profitable markets,ending up selling at lower prices. The longer distances hampered the purchase and transportation of inputs necessary to achieve higher production due to various reasons includinglack of regular, reliable motorized transport system. This also affected farmers'capacity to transport their produce to the market and bring back the required farm inputs. Most often famers carried their commodities on heads or backs to reach market places. Due to very poor infrastructure, farmers may not afford to pay for tranport.

The land tenure was distributed between purchase and heritage land and some land rented and donated (Table 3). Related to form of land holdings, farmers who used inherited or donated land were more likely to receive higher gross margins relative to those who rented or purchased land that they used. The paper classified land ownership into purchase/rent and heritage/donation to reflect the influence on the gross margins based on the two major classes of ownership. Both male and female farmers were not willing to invest long time in land that they do not have full usage rights thus leading to lower returns. Such farmers are less likely to apply soil fertility enhancements such as fertilizers, manure and/or soil ammendments that would be benefit beyond one season of renting. Temporary tenure systems, common in DRC are not favourable for increasing land productivity. Those who used purchased land were also more likely to receive lower retuns due to having purchased least productive parcels which requires more inputs and management for maximum production. In addition, DRC land tenure policies and rules are not clear, making the purchasers of land not sure how much they should invest in it.

"Under the formal law, the state owns all the DRC's natural resources (land, water, forests, and minerals);

Table 5. Summary statistics for factors influencing costs and margins of beans.

Variable	N	Mean	Std. deviation
Age of farmer	160	36.3 years	31.1 years
Time to market	57	69.5 min	39.6 min
Bean area in ha	72	0.34 ha	0.63 ha
Selling price	111	342.3 (CF)	56.50 (CF)
Total cost of bean production per hectare	144	21230.5 (CF)	31625.80 (CF)
Gross margin on beans per hectare	53	141769.0 (CF)	175261.30 (CF)

1 USD = 500 CF (Congo Francs) on 30th July 2007.

Table 6. Regression results for factors influencing gross margins achieved in bean production.

Variable	Standardized coefficients	t-value	Sig.
	Beta		
(Constant		-0.542-0.749	0.465
Bean area in ha	-.560	-3.124***	0.007
Education levelTime to market	0.174-0.325	0.874-1.757	0.098
FormAge of farmer	0.576-0.072	2.925***-.381	0.708
Gender	0.109-0.129	0.571-0.714	0.486
Age of farmerLand tenure	0.564	.8963.047***	0.008

R^2=56.7% (adjusted 43%), F=4.2**; ***,**significant at 1 and 5%, respectively; N=160.

Table 7. Regression results for determinants of cost of producing beans in south Kivu.

Variable	Standardized Coefficients	t	Sig.
	Beta		
(Constant)		1.250	0.216
Age of farmer	-0.278	-2.876***	0.005
Sex of farmer (1=male, 2=female)	-0.313	-3.187***	0.002
Location of farmer (1=kabamba, 2=lurhala)Market access (1=Kabamba=good access, 2=Lurhala=poor access)	0.474	5.198***	0.000
Bean area in ha	0.435	4.912***	0.000
Marital status	0.089	.918	0.362

R^2=56% (adjusted 53%); F=16.3; ***, **, significant at 1 and 5%, respectively; N=160.

people can obtain various types of use and exploitation rights under an evolving set of laws and regulations. In practice, customary law endures, and natural resource rights are subject to parallel, incomplete, and often contradictory systems of formal and customary law. Land rights are often ambiguous, usually undocumented, and tenuous. Agricultural land is subject to seizure and land-grabbing. Formal and customary institutions are often ill-equipped to resolve land disputes" (USAID, 2011). Thus, farmers in DRC are not willing to risk investing in land that they cannot be assured of in the long run. It is no wonder then that our results in the survey areas show that there was very minimal use of inputs irrespective of land tenure. Table 7 presents summary results of factors that contribute to the observed bean production and marketing costs. Model statistics indicate that these factors explain over 50% of the observed variation in costs of production and marketing for beans. The overall influence of the variables is significant at the 1% level of significance.

Land size under bean production also influenced amount of costs faced by the bean farmers as expected. This is expected since a large portion of costs faced by smallholder farmers are direct costs related to managing

Table 8. Gross margins between men and women.

S/N	Men	Women
1	Lease, purchase, gift, inherit land	Gift, inherit land mostly
2	Hire labor, purchase inputs, rent equipments	Own and /or family labor, no purchase of inputs or equipments
3	Hire transport equipments	Women transport produce on their heads or backs

the land (labor) and input application. In this regard, Kabamba was expected to incur more production costs since it has relatively larger acreage under beans than Lurhala.

Results of the age of the farmer indicate that younger farmers are more likely to face higher costs of producing and marketing beans than older farmers. This can be attributed to experience where experienced farmers are likely to benefit from lower costs due to social capital and wide knowledge of sources of resources associated with experience in farming.

With regard to the sex of the farmer, results show that male farmers are more likely to experience higher costs in producing beans relative to female farmers (Table 8). Though they may have an advantage in accessing resources such as credit or services more easily compared to the female farmers. They are also likely to have better collaterals and/or alternative sources of resources to support farming activities compared to their female counterparts. They gain knowledge and skills from training organized by NGOs, orientation from the private sector, field visits and knowledge sharing plus other informal sources. Women usually have almost no access to farm inputs yet according to the Huger Project (Weisfeld-Adams, 2008), if women were able to access the same inputs as men they are likely to increase their yields by 22% which would be more beneficial for them and thus increase the household food security.

Market access was measured by the location of the farmers where certain locations are considered to have good or poor market access. Being located in Lurhala was more likely to lead to higher costs relative to being located in Kabamba. Kabamba is characterized as having fair access to market (located next to the main road and market centre) while Lurhala has poor market access (located more than 5 km away from the main road and market centere). Lurhala is situated much further away from the main road and main markets for beans. Thus Lurhala is likely to incur more costs to keep up production as well as in accessing both input and output markets. Longer distances are exacerbated by intractable roads that are not motorable and /or farmers are not being able to afford transportation fees. Thus the poor access to markets led to higher costs of production for some of the regions.

All these may attract higher costs compared to women who rely on informal sources alone for acquiring knowledge and skills that may not have access to the resources. This may explain why female farmers incur lower cost than their male counterparts. In addition, female farmers are more likely to perform some production activities on their own rather than hiring labor due to shortage of resources. For example, while men may hire a wheelbarrow or a bicycle for transport, women rely sorely on putting the loads on their heads and /or backs to and from the market will thus appear to incur lower costs. But this is simply because they do not pay themselves for labor and time spent in food production and marketing.

Implications for technology interventions and uptake

Interventions that would enable such farmers to affordably access productive inputs are likely to impact positively on the farmers' economic welfare. This is necessary for self empowerment and self sustenance in food and incomes that beans can signficantly contribute to in the farming communities.

Due to inefficiency in landuse affecting both profitability and costs, there is need of extension knowledge to increase efficiency of landuse by adopting ISFM technologies. There is need of interventions that would enable farmers to affordably access productive inputs that are likely to impact positively on the farmer's livelihood.

Male farmers access more land to put under beans confirming the gender differentials. Women's production "makes more profit" for various reasons including family labor whose cost is not computed in the production cost. They manage small units which ratyes then more efficient compared to men's bigger beans areas.

In SSA increase in commodity traded is from increased acreage not improved production per unit area. SSA agriculture is known for low use of inputs in general and for bean production in particular.

Rural infrastructure is a challenge and from other studies in the region. Farmers and scholars have listed road infrastructure as one of the major facing small scale agriculture. It confirms studies that show the social cost of infrastructure which is paid by women who end up transporting most of the farm produce to the market. (Blackben and Bhanu, 1999; Birachi and Wanjiku, 2010) show that over 50% of the commodity transport in the region is on human body. Weisfeld-Adams (2008) estimates that they contribute two third of the total rural transport.

Development agents in the region need to lobby for

improved infrasrtucture in the MDHs and CAADP goals are to be met in DRC. As well as ensuring security which also hinders farmers from investing when not sure if they will harvest. There is also needfor relief agents to rethink their intervention strategies and more then from emergency, response to more sustainable long term assistance since relative peace is preventing.

REFERENCES

Abdulai A, Binder CR (2006). Slash and burn cultivation practice and agricultural input demand and output supply. Environ. Develop. Econ. 11(2):201-220.

Arnfred S (2004). Questions of Power: Women's movement, Feminist Theory and Development Aid" in SIDA, Discussing Women's Empowerment, Theory and Practice.Sida Stockholm. Pp.73-86

Birachi E, Wanjiku C (2010). The Trader Baseline Survey Report: Lake Kivu Pilot Learning Site of the Sub-Saharan Challenge Program. Unpublished FARA. http://www.fara-africa.org/media/uploads/library/docs/ssacp/MARKET_SURVEY_BASELINE_REPORT_-_LKPLS.pdf

Blackben M, Bhanu C (1999). Gender Growth and Poverty Reduction: Special Program of Assistance for Africa. Washington, DC: The World Bank.

Boserup E (1970). reprinted 1997). Women's Role in Economic Development. London: Earthscan. ISBN 1-85383-040-2

De Janvry A, Sadoulet E (2003). Progress in the modeling of rural households' behavior under market Failures, in Alain de Janvry and Ravi Kanbur (Eds) Poverty, Inequality and Development, Essays in Honor of Erik Thorbecke, Kluwer publishing.

Enete AA, Amusa TA (2010a). Contribution of Men and Women to Farming Decisions in Cocoa based Agroforestry Households of EkitiState, Nigeria. Tropicultura 28:77-83

Ezumah N, Di Domenico CM (1995).Enhancing the Role of Women in Crop Production: A Case Study of Igbo Women in Nigeria. World Develop. 23:1731-1744.

Hart T, Mwinyihali R (2001). Armed conflict and biodiversity in sub-Saharan Africa: the case of the Democratic Republic of Congo (DRC). Washington, DC, Biodiversity Support Program.

ILO (2009). Global Employment Trends for Women, Geneva, 2009. ISBN: 9789221220466

Lemlem A, Bishop-Sambrook C, Puskur R, Ephrem T (2010). Opportunities for promoting gender equality in rural Ethiopia through thecommercialization of agriculture. IPMS (Improving Productivity and Market Success) of Ethiopian Farmers Project Working Paper 18. ILRI (International Livestock Research Institute), Nairobi, Kenya. P. 84.

Marinda P (2006). Effects of Gender Inequality in Resource Ownership and Access on Household Welfare and Food Security in Kenya. A Case Study of West Pokot District. Development Economics and Policy - Volume 51. Peter Lang International Academic Publishers.

Singh I, Squire L, Strauss J (1986). Agricultural household models: Extensions, applications and policy. The Johns Hopkins University Press, Baltimore.

USAID Country Profile (2011).Property Rights and Resource Governance Democratic Republic of Congo.http://usaidlandtenure.net/usaidltprproducts/country-profiles/democratic-republic-of-congo/democratic-republic-of-congo-country-profile(23/11/2011)

Van Dusen ME (2000). In SituConservation of Crop Genetic Resources in the Mexican Milpa System.PhD, dissertation, University of California, Davis, California.

Estimating farmers patronage level of informal credit in Imo State: Implication for efficient formal and informal financial linkage and rural development

N. C. Ehirim and N. N. O. Oguoma

Department of Agricultural Economics, Federal University of Technology Owerri, Imo State, Nigeria.

Low agricultural productivity in Imo State is blamed on poor grassroot financial development. Farmers are caught between maginalization, clumsy, costly and grossly inefficient credit delivery system. An efficient informal credit patronage that engenders informal-formal credit linkage system in Imo State was investigated. Data were collected from 59 arable crop farmers and 15 Informal Financial Agencies (IFAs) in Imo State and analysed using both descriptive ststistics and econometric tools such as frequency/means and Ordinary Least Square regression analysis (OLS) respectively. The result showed that 93.2% of the farmers must have accessed credit from town union/age grade while 55.5 and 33.9% were with Esusu and cooperatives respectively. The mean informal credit patronage ratio of 0.712, ranging from 0.64 (town unions) to 0.97 (esusu groups) was recorded. Average loan of N12,756.61 (town unions) to N12,959.79 (esusu) was obtained. Though there were disbursement with more members in the town unions yet the level of credit patronage was highest (97.0%) with esusu group of farmers. Increase in education, farming experience, household size and social capital each has a significant and positive effect on informal credit patronage while interest rate reduces it. It is recommended that the formal credit sectors should cash on this high informal credit patronage to market loan facilities to informal credit agencies at a lower cost for an informal-formal credit linkage.

Key words: Patronage, coperative, esusu, town union, formal and informal linkage and credit delivery.

INTRODUCTION

Resource poor farmers in Nigeria are faced with low productivity due to lack of access to improved farm inputs as credit facilities to acquire them were beyond the reach of rural farmers. Resource poor farmers were marginalized by most conventional financial houses. Anyanwu (2004) noted that, the poor socio-economic disposition of the farmers do not allow easy facilitation of credits from most conversational banks, hence acquisition of the necessary improved farm implements for large scale farming is difficult. The consequence is the use of any available crude farm implements, which hardly increase productivity. Commercial banks traditionally, lend to medium and large enterprises that they judged to be creditworthy. It was difficult to secure credit facilities from most formal financial institutions due to high cost of credit (high transaction cost), lack of collateral security, high level illiteracy, poor loan quality and wide spread and default, and financial management as deficiencies (Oni, 1972; Ijose and Abaelu, 1973; Oshunotgun, 1973). They avoided doing business with the poor and their micro enterprises because of the associated cost and risks (Anyanwu, 2004) and this led to a review of rural credit scheme in 1975.

The scheme was a move by the federal government to ensure a geographical dispersal of commercial banks branches in the rural areas to facilitate credit needs of the resource poor farmers (Anwanwu et. al., 1997). The scheme, which actually started in 1977 and ended its third phase in 1989, recorded a total of 756 out of the intended 766 branches across the nation. The reformation was followed by a liberalization policy of Structural Adjustment Programme (SAP), in 1986. The reforms, rather than improving the grass root credit transactions, expand technology and increase food production, exacerbated the poverty (Olomola, 2006). The loanable funds from government sources dropped considerably and the cost of borrowing escalated substantially, living the resource poor farmer to no other choice than the subsistence farming with crude farm implements. Despite the expanded branches across Nigeria and impressive performance recorded by the banks, rural farmers especially the low-income earners were still poor, informal and formal financial linkage expected to facilitate rural development dropped as farmers patronized mostly the traditional financial agencies.

The unwillingness or inability of the formal financial institutions to perform its role at the grass root and unsustainable support from the government credit schemes, have led to the growth of private sector financial services at the grass root in Nigeria. The group provides savings and credit services to their members.

The informal finance arrangements operate under different names across the country: it is called 'esusu' among the Yorubas of Western Nigeria, 'etoto' by the Igbos in the East and 'adashi' in the North by the Hausas (CBN, 2000). The key features of these informal schemes are savings and credit components to the rural populace, informality in operations (though at higher interest rates) allows a large proportion of farmers to participate in the services and they work together for the mutual benefits of their members. Prior to the emergence of formal microfinance (FMF) institutions, it is important to note that, informal financial agencies (IFAs) activities had already flourished all over the country. They however, offer little capital for agricultural development.

Otu and Anyanwu (2003) and CBN (2005), noted that, a large number of rural dwellers rely on the informal sector for their financial services on the account of its relatively low information and transaction cost, easy access to low income groups, timeliness of operation, simplicity and flexibility in financial procedures. The non-traditional, formalized microfinance institutions (NMEIs) are operating side by side with the informal services. They include the NGOs and foreign agencies such as World Bank assistance to the farmers. The private sector seemed to be bracing up to the challenges of grass-root credit needs to resource poor farmers (Olomola, 2006). Past literatures indicated that, these lenders are capable of eliminating the problems of high transaction and information cost, as well as management deficiencies that surrounds extension of credit to small-scale farmers.

Although it was alleged that the fear of no legal and fiscal regulatory and prudential frame work of the monetary and financial authority will increase the risk of engaging in a reasonable financial net work with the agencies, the existence of social capital acquired through social integrations between the lenders and the borrowers can enhance trust and confidence in microfinance. The rural environment in Nigeria provides such social mobility for informal financial transactions to thrive on.

The need for rural development through self help initiative and productivity requires a linkage. It becomes very important to know the level of IFAs credit services to farmers and how it can be linked to micro financial institutions in the area for a sustainable rural development in Imo State. The relative size of the participants and the factors that accounted for the high patronage of informal credit in Nigeria and Imo State in particular has not been determined to facilitate a financial linkage in the state. This study therefore wants to estimate the volume of credit transactions by farmers with these informal financial institutions and the factors that accounted for farmers' patronage. The study will help to enhance the volume of credit given to farmers if the factors that enhance patronage of informal credit are incorporated into their management strategies while collaborating with the informal credit agencies.

METHODOLOGY

The study was carried out in Imo State. The sate is located in the rain forest zone of Nigeria with rainfall and temperature distribution of between 1500 to 2500 mm and 21 to 33°C, respectively annually. There is bimodal climate distribution supports double arable crop production per cropping season. Arable crop production is performed by many small-scale farmers widely distributed in the rural area (Imo ADP, 2004).

Imo state is divided into three zones for administrative purposes and they are Owerri, Orlu and Okigwe agricultural zones (ADP, 2996) and the three zones were used for this study. The study drew its sample from both small scale arable crop farmers and IFAs drawn from six Local Government areas with two from each zones. The areas are Aboh-Mbaise and Ngor-Okpala in Owerri zone, Ideato North and Oguta in Orlu zone and Onu-Imo and Isiala Mbano in Okigwe zone were purposefully selected for the study because of their involvement in arable crop production in the state.

Two different research instruments were designed for the study; one is for the farmers and it elicited information on their socio-economic features, volume of loan obtained and repaid, membership association (social capital) and amount of capital needed for crop production. The other instrument elicited information concerning the number of members each agency has, the total amount disbursed, the repayment capacities, and the type of securities demanded etc. Fifteen farmers who patronized one or more of the IFAs and 3 IFAs were randomly selected from each Local Government Area. About 90 and 18 questionnaires were distributed to farmers and IFAs, respectively but only 59 and 15 questionnaires were found useful for data analysis.

Data analysis was done using simple ratios and or descriptive statistics such as mean and relative frequencies as well as econometric tool such as ordinary least square regression estimate (OLS). The socio-economics characteristics of the farmers such as age, educational levels, farming experience, level of social capital volume of loan obtained, volume of loan disbursed by each agency etc., were described using this method. The patronage level was estimated using a modified simple patronage ratio or index approach as developed by Okereke (1978), and adopted by Benchendo (2005). The index is expressed as:

$$P_{patronage\ index} = \frac{X}{Y} \quad (1)$$

Where, X = Aggregate volume of loan an ith farmer obtains from all the agencies the farmers belongs, and Y = Aggregate volume of the loan available to the farmer from all the credit agencies which the farmer belongs or at least patronized in the area

The patronage estimate obtained and the socio-economic data of the farmers are fitted into four functional forms to isolate the determinants of farmer's' patronage of informal credit in Imo State. The model is implicitly expressed as:

$$P_{P1} = f\,(X_1, X_2, X_3, X_4, X_5, X_7, X_8, e)\ldots \quad (2)$$

Where, P_{P1} = Patronage ratio, X_1 = Farmers age level measure in years, X_2 = Number of years spent in acquiring formal education measured in years. X_3 = Farming experience measured in year, X_4 = House hold size (number of people under one catering arrangement), X_5 = Interest on borrowed capital measured in naira, X_7 = Social capital which is number of association each farmer belongs, (Olomola, 2006), e = Stochastic (random error) variable.

RESULTS AND DISCUSSION

The socio-economic features of the farmers shown in Table 1 shows that, farmers mean age is 39 years. More than 28% of the farmers are between 31 and 40 years of age while about less than 2% is relatively old of above 60 years. The age distribution showed that, majority of the farmers is young and active and can participate in informal financial services, accept new innovations and improved farming techniques (Ejiogu et. al., 2002; Anebunwa, 2008). It therefore implies that, the respondents are capable of taking any risk associated with such improved packages and financial services. The result also showed that, over 52% of the farmers are females and over 42% of the total respondents have acquired secondary education. Less than 6% had tertiary education and about 13.6%. It implies that, majority of the farmers have acquired at least 12 years of formal education and so can withstand the shocks of informal financial risk. The result further showed that, 32.2% of the farmers had about 1 to 10 years of farming experience while about 39% of them had about 11 to 20 years of farming experience.

The mean household size of the distribution is 7 persons per household. About 44.5% of the respondents have about 4 to 7 persons per house hold while only 25.4% of them have above 8 persons per house hold but less than 11 persons per household. The result also showed that, 15.3% of the total respondents had 12 to 15 persons per household. Majority of the farmers (67.8) have farming as their major occupation. 81.4% of them plant cassava while about 59.3% of them plant maize and 44.1% of them planted beans.

The distribution of respondents according to their level of involvement in informal financial services showed that, majority of the farmers (93.2%) belongs to age grade or town union category offers a lot of social mobility and social capital to rural farmers than other agencies because the farmers are reorganized even from their local setting or by the village heads. The ability of these farmers is easily picked due to close interactions they share among themselves. Another reason for more involvement in age grade than other informal agencies by the rural farmers could be due to small volume of interest and collaterals required before the loan is disbursed. The group does not require much of these items as the farmers can even be traced down to their origins before such money is released to them. In case of default, it is easy to recover money from farmers by the agency as the Union can easily confiscate or seize the property of the farmers. This action guided by customary law and the age grade or town union applies it in case of self help actions on rural development.

Table 2 showed the average volume of loan disbursed by the IFAs. Apart from having the highest number of members than other agencies, the town union has the largest mean of disbursement and the highest total of disbursable credit to farmers. The result revealed a mean volume of loan of N26, 175.93 was disbursed by all the agencies in the state. On individual disbursement, the town unions and age grades disbursed a mean amount of N26,709 ranging from 9,000 to N60,000 while esusu disbursed a mean value of N25,727 ranging from 9,000 to N60,000 and cooperative society disbursed a mean amount of N25,450. This value ranges from a minimum amount of 9,000 to N59,000. The total amount of loan disbursed by the cooperative thrift is N509,000 while the Esusu disbursed a total of N849,000. The town unions disbursed a total of N1,469,000.00. This implies that, the town union/age grade disbursed about 52% of the total credit available to farmers for arable crop production through IFAs. The town union is capable of offering this volume of credit services to farmers due to the available social capital it enjoys with its members.

The result further revealed a patronage index of 0.712. This implies that, an aggregate patronage of 71.2% is recorded for informal credit in the state. It implies that, more than 71% of the farmers are interested in informal credit acquisition. The high rate of demand for informal credit despite its high interest rate could be relevant to micro finance if it is properly coordinated. The formal and informal financial linkage in the state should work towards incorporating this group of farmers for convenient and efficient grass root banking to facilitate rural development in the state. The volume of loan disbursed will increase while the number of farmers increased with the flexible credit services offered by the informal financial services. This is in addition to the informal qualities of loan administration adopted by the town unions that can enhance the repayment performance of the farmers in the

Table 1. Socio-economic characteristics of the farmers in Imo State, Nigeria.

Variable	Frequency	Relative (% age) frequency
Age (years)		
20 – 30	12	20.3
31 – 40	17	28.8
41 – 50	24	14.7
51 – 60	5	8.6
61 – 70	1	1.7
Total	59	100
Mean = 39 years		
Educational level		
Non-formal education	8	13.6
Adult education	2	3.5
Primary education	21	35.6
Secondary education	25	42.2
Tertiary education	3	5.2
Total	59	100.0
Gender (Sex)		
Male	28	47.5
Female	31	52.5
Total	59	100.0
Farming experiences (years)		
1 – 10	19	32.2
11 – 20	23	39.0
21 – 30	11	18.6
31 – 40	6	10.2
total	59	100.0
Mean = 16 Years		
House hold size		
1 – 3	9	15.3
4 – 7	26	44.5
8 – 11	15	25.4
12 – 15	9	15.3
Total	59	100.0
Mean = 7		
Occupation		
Farming as a major occupation	40	67.8
Farming as a major occupation	19	32.2
Total	59	100
Types of crop planted		
Cassava	48	81.4
Maize	35	59.3
Beans	26	44.1
Total	109^{+}	
Informal financial groups		
Esusu	33	55.9
Cooperative	20	33.9

Table 1. Contd.

Age grade/town union/club	55	93.2
Total	108+	

Source: field survey 2007. (+) = multiple responses.

Table 2. Level of membership and volume of informal credit disbursement in imo state.

Informal agency	number of farmers	total credit	mean credit	min. credit	max. credit	std. deviation
Esusu	33 (55.5)	849.000	25,727	9000	60,000	12,959,79
Cooperative	20 (33.9)	509,000	25,450	9000	50,000	12,959.79
Town union	44 (93.2)	1,469,000	25,709	9000	60,000	12,756.61
Total	108+	2,827,000	26,175.93			

Source: field survey 2007, (+) multiple response

Table 3. Patronage index of informal credit agencies in Imo State, Nigeria.

Informal agency	Total credit	Total cash	Patronage ratio	Patronage index
Esusu	849,000	875257.73	0.97	
Cooperative	509,000	783076.92	0.65	0.712
Town union	1,469,000	2,309.748.43	0.636	
Total	2,827,00	3,968,083.08		

Source: field survey 2007.

state. Table 3 shows the rate at which farmers patronize informal credit for initial capital. The patronage ratio is the ratio of total credit disbursed to the farmers and the total cash available for loan in each informal credit agency in the state.

The result showed a mean patronage level of 71.2% from the farmers in all the agencies. This result is consistent with that of Ojo (1991) findings that between 70 to 90%, credit needs of farmers are obtained from informal agencies and this may be a pointer to slow status of production system as these agencies hardly disburse enough and quality credit to meet the farmers' capital need in the rural areas. However, the patronage ratio of the farmers in the state ranges from 63.6% as in the case of town unions to 97% as it is with esusu. Despite the high membership of the town union, Esusu group recorded the highest patronage index. This is because 97% of their total cash for disbursement are released to farmers as farmers form the bulk of there membership. As for the town unions, only 63.6% of their credit is made available to the farmer. The remaining 36.4% of their total cash are kept for other community development projects. This implies that, there is a credit rationing behaviour with most town unions in the area so that, an integrated rural development can be achieved. The cooperative thrift only disbursed about 65% of their total credit for farmers while 35% of their total cash were used for the growth of the cooperative in the area.

Table 4 showed the determinants of farmers patronage of informal credit. The patronage level of individual farmers and their socio-economic factors as well as social capital and the amount contributed by each farmer were fitted into four functional forms. The result showed that, the linear model is the best fit. It has the highest coefficient of multiple determinations, the highest number of significant explanatory variables and highest F-statistics. The F-statistic is significant at 1% and the coefficient of multiple determinations is 0.527. This implies that, the null hypothesis which includes explanatory variables such as social capitals, socio-economic factors and the amount contributed by each farmer as well as the interest charged have significant effect on farmers rate of patronizing the informal credit services and also about 52.7% explanation to the variation in the rate at which informal credit services is patronized, traditional farmers can be explained by these included variables.

Increasing educational level, farming experience, household size and social capital of the farmers by a unit will give a direct increase in the rate of patronage by a more than proportionate unit of 0.40, 4.9 0.32 and 0.615%, respectively. Again reducing the interest rate charges on a unit of credit by the agency will lead to a more than proportionate increase in rate of patronage of informal credit by 2.1%. Amount contributed by farmers to the agency has an insignificant and inverse effect on

Table 4. Determinants of farmers patronage of informal financial services in Imo State, Nigeria.

Variable	Constant C	Age (X_1)	Edulevel (X_2)	FarmExp. (X3)	Household size (X4)	Interest rate (X5)	Social capital (X6)	Amount contributed (X7)	R^2	Adj. R^2	F-stat.
Linear model coeff.	15.689	0.11	0.405	4.901	0.321	-2.12	0.615	-1.527			
Std. error	4.269***	0.133	0.105***	1.115***	0.133***	0.998***	0.189***	2.421	0.527	0.423	3.092
/t-value/	3.675	0.838	3.82	4.396	2.414	2.122	3.25	0.631			
Semi-log model coeff.	34.598	-5.523	5.251	0.929	2.391	-2.102	2.124	1.657			
Std. error	21.956	4.797	1.755**	0.513*	1.178**	1.072*	2.480	2.556	0.504	0.407	2.086
/t-value/	1.579	1.151	2.997	1.811	2.030	1.960	0.856	0.648			
Cob-douglas coeff.	3.793	-0.369	0.210	5.06×10^{-6}	8.04×10^{-2}	6.78×10^{-2}	0.170	-0.154			
Std. error	1.325**	0.289	0.106*	0.0109***	0.0131***	0.150	0.164	0.155	0.446	0.378	1.522
/t-value/	2.863	1.276	1.996	4.64	6.137	0.453	1.036	0.997			
Exponential coeff.	2.914	7.12×10^{-2}	1.17×10^{-2}	3.48×10^{-2}	2.04×10^{-2}	-1.1710^{-2}	5.69×10^{-2}	-9.74×10^{-2}			
Std. error	0.269***	0.008	0.007*	0.097	0.0071**	0.0569	0.027**	0.153	0.423	0.302	1.388
/t-value/	10.830	0.849	1.763	0.481	2.873	0.206	2.10	0.638			

Source: Result analysis 2007; *,**, ***significant at 10, 5 and 1% respectively; coeff., coefficient.

rate of patronage. This means that, less emphasis is placed on level of contribution by the farmers but on social capital acquisition for credit disbursement. The formal credit agencies or the micro finance houses should avail themselves of this important feature in credit transactions with the traditional society as this can be a major linking factor in formal and informal credit transaction and rural development in the state.

CONCLUSION AND POLICY RECOMMENDATION

The level of credit marginalization that characterized the grass root financial services in the rural areas of Nigeria makes formal and informal financial linkage imperative. Showcasing a harsh conventional banking system on the traditional society and unrecorded credit transactions that characterized seemingly effective formal financial services for grass root rural development cannot solve the problems of declining food production in Nigeria. There is need to show a direct credit linkage between the informal and formal credit agencies in Nigeria Data were obtained from both the IFAs and arable crop farmers and analyzed using both descriptive and econometric tools.

It was discovered in this study that, the Town unions has both the largest membership and the highest credit transaction but recorded the lowest patronage. The study further to showed that, increasing the educational level, farming experience and social capital, which include social mobility and social organization membership, increases rural farmers credit patronage. Again incorporating and synergizing the activities of informal-formal credit systems can facilitate quick and increasing deliveries to rural farmers in Nigeria. The study therefore suggests that, micro finance institutions should cash on these advantages to increase farmer's credit patronage in the rural areas.

The study recommended that, farmers' level of education be increased through extension education to enhance grass root financial performance. Secondly, the micro finance services should work with the IFAs to serve the interest of the farmers or they should prefer to work with farmers as a group rather than individuals if the grass root financial status is to be raised in the area.

REFERENCES

Anebunwa FO (2008). Cost and Returns Analysis of Yam

Marketing in Southern Ebonyi State of Nigeria. In the proceedings of the 42nd Annual Conference of the Agricultural society of Nigeria, Ebonyi State University Abakaliki, Nigeria.

Anyanwu JC, Oyefusi A, Oaikhenan H, Dimowo FA (1997). The structure of Nigerian Economy (1960-1997) Joanee Educational Publishers LTD. Onitsha Anambra State, Nigeria, pp. 131-188.

Anyanwu CM (2004). Microfinance Institutions in Nigeria: Policy, Practice and Potential. Central Bank of Nigeria paper Presented at the G24 Workshop on "Constraints to Growth in Sub Saharan Africa", Pretoria, South Africa, November 29-30, 2004.

Central Bank of Nigeria (2000). The Changing Structure of the Nigerian Economy and Implications for Development.

Central Bank of Nigeria (2005). Microfinance Policy, Regulatory and Supervisory Framework for Nigeria. Publication Guidelines on Microfinance Policy in Nigeria. Central Bank of Nigeria, Abuja, December 2005.

Ejiogu AO, Eboh EC, Okorji EC (2002). Determinants of Membership of Traditional Credit and Savings Association among Rice Farmers in Imo State. In the proceedings of the 36th Annual Conference of the Agricultural society of Nigeria, Federal University of Technology, Owerri Imo State.

Ijose A, Abaelu JN (1973). Institutional Credit for Smallholder Farmers: A Case Study of Western Nigeria Agricultural credit Corporation", AID, Spring Review of Small Farmer Credit, VI.

Imo State Agricultural Development Programme (2004). Quarterly Report on Imo Sustainable Food Production Reports, Imo State.

Olomola AS (2006). Social Capital, Microfinance Group Performance and Poverty Implications in Nigeria by Nigerian Institute of Social and Economic Research (NISER) Ibadan.

Oni SA (1972). Credit in Rural Development: An Appraisal of the Supervised Credit Schemefor the settlers in Western Nigeria. Proceedings of the Nigerian Economic Society.

Oshunotgun A, Oludimu O (1981). "Extending Agricultural Credit Through Public Institutions in Nigeria: A Comparative Study of the Ondo State Agricultural Corporation. Oxford Agrarian Stud. 10:85-97.

Otu MC, Anyanwu CM (2003). Informal Credit Market and Monetary Management in Nigeria in Central Bank of Nigeria Research 'Department. An Occasional. P. 29.

29

Does low yield heterosis limit commercial hybrids in wheat?

Rajiv Sharma

CIMMYT-Afghanistan, Kabul, Afghanistan.

Heterosis has contributed to productivity gains in several crops like maize, rice, sorghum, cotton etc. Wheat breeders have largely been unsuccessful to take advantage from this technology at commercial level. Lack of commercial level yield heterosis is regarded as a major reason for this failure as compared to other technical barriers like difficult pollination control and seed production. The allopolyploidy nature of wheat endows even wheat purelines with a fixed intergenomic heterosis which perhaps is the foremost reason for lack of classical yield heterosis in wheat. The coming together of three diverse but functionally similar genomes causes differential gene expression among several other outcomes and leads to a diploid behaving self-sustaining intergenomic hybrid. A long history of highly successful pureline breeding and shortage of nicking parents are other two reasons responsible for failure to realize commercial level heterosis in wheat. Molecular biology tools now make it possible to dissect the phenomenon of heterosis into detectable Mendelian factors to tailor nicking parents to develop commercially sustainable wheat hybrids. This review probes the reasons for the absence of commercial-scale heterosis in wheat.

Key words: Wheat, *Triticum*, hybrid, heterosis, yield.

INTRODUCTION

Commercial exploitation of hybridity has been one of the greatest achievements of crop improvement research. Heterosis increases commercially utilizable yields in farmers' fields, leading to higher productivity and income levels. Farmers and breeders, especially for cross-pollinated crops like maize, have shifted emphasis from open-pollinated varieties to hybrids to gain from the technology. Shull (1952) defined heterosis as "the interpretation of increased vigour, size, fruitfulness, speed of development, resistance to disease and to insect pests, or to climatic rigors of any kind manifested by crossbred organisms as compared with corresponding inbreds, as the specific results of genetic unlikeness in the constitution of the uniting parental gametes." Heterosis has been studied extensively in maize because of (1) a large expression in terms of yield, (2) extensive exploitation in maize, and (3) ease of pollination control. Wheat, on the other hand, employs cleistogamy in a relatively small flower, making pollination control extremely difficult. Single-plant and small-plot estimates have reported a wide range of yield heterosis in wheat (Singh et al., 2010), however, economically sustainable hybrid exploitation at farmer field level is still limited.

WHAT EXPLAINS HETEROSIS?

Shull (1908, 1909) observed that when maize plants were selfed, their vigor and grain yield declined, but when two inbred lines were crossed, both vigor and grain yield increased and even exceeded the mean of the two parents. These reports established the concept of heterosis breeding, and efforts have been made since then to use the phenomenon to benefit all crop plants.

The International Maize and Wheat Improvement Centre (CIMMYT) convened an international congress in 1998 on heterosis breeding to take stock of the application of hybrid technology to different crops, including wheat. One essential characteristic of heterosis is that it happens only when hybridity brings together different alleles originating from different parents into one genotype. There have been several explanations for the phenomenon of heterosis:

a) Dominance hypothesis: Heterosis is the joint action of multiple loci, with favorable allele being either partially or completely dominant (Bruce, 1910; Keeble and Pellew, 1910; Jones, 1917; Collins, 1921).

b) Overdominance hypothesis: Heterosis occurs due to overdominant gene action at many loci (East, 1936; Hull, 1945; Crow, 1948).

c) Epistasis hypothesis: Heterosis arises because of epistatic interactions between non-allelic genes (Richey, 1942; Schnell and Cockerham, 1992).

d) Quantitative genetic explanation: Researchers have employed both first-degree and second-degree statistical estimates to explain heterosis (Hallauer and Miranda, 1981) and both provided a clear answer about the relative importance of various types of gene actions involved in expression of heterosis. Falconer and Mackay (1996) gave a quantitative genetic expression for heterosis in terms of gene effects. Heterosis for a cross between two populations was equal to "hy^2", where "h" is the genotypic value of the heterozygote cross and "y" is the difference in gene frequencies of two parental populations. According to this concept, the first critical component determining extent of heterosis is the genotypic value of a heterozygote at any given locus, and it would be dependent on interallelic or within-locus interaction at a locus, namely, dominance or overdominance. Also, when "hy^2" is summed over all loci affecting the trait under consideration, non-allelic interactions would also be accounted for. Therefore, the quantitative expression draws its strength from all three above hypotheses.

The second crucial component is the difference in gene frequency, meaning how diverse two parental populations are or how diverse the parents are. Therefore, two "Ds" viz., -diversity, reflected in differences in gene frequency, and desirability, measured as "h" summed over all loci contributing to the trait under study will decide whether the trait will manifest heterosis or not. A similar analogy was put forward by Fu and Dooner (2002) on the basis of DNA sequencing data in maize. They suggested that different maize lines often lack different functional genes. When a nicking pair of lines is crossed, the lines complemented each other's weaknesses, bringing in the two "Ds," and heterosis results.

e) Physiological and molecular diagnosis: Developments in molecular biology enabled molecular dissection of heterosis. Quantitative trait locus (QTL) analyses have begun to further our understanding of heterosis by breaking it down into Mendelian factors and studying their modes of inheritance (Yuan et al., 2012; Hua et al., 2003; Li et al., 2001). However, the relative importance of different genetic phenomena explaining heterosis varies from system to system and trait to trait. One important outcome of several such studies is that heterosis can be defined by a limited number of Mendelian factors. Though an association between QTLs and heterosis has been shown, cloning and further utilization are yet to be accomplished (Lippman and Zamir, 2007). Lack of genetic colinearity at gene level has also been proposed as a reason for heterosis. Gene deletions causing lack of colinearity are also said to be functionally compensated by duplicate copies elsewhere in the genome (Fu and Dooner, 2002). However, a hemizygous complementation of these gene deletions by the other genome leads to heterosis in a hybrid.

Altered gene expression in a hybrid, as a result of two genomes coming together, has been proposed as another molecular regulatory mechanism causing heterosis. Song et al. (2009) proposed differential protein expression as a cause of heterosis in wheat. A total of 49 of approximately 900 protein spots expressed differentially in seedling leaves in a hybrid between common wheat and spelt wheat. Of the 49 differentially expressed proteins, 30 were involved in metabolism, signal transduction, energy, cell growth and division, disease defense mechanism and secondary metabolism. Song et al. (2009) postulated that these protein differences were involved in diverse physiological pathways that might be responsible for observed heterosis. Wang et al. (2006), observed differential expression in roots for 27.52% of 990 fragments among hybrids and their parents at the jointing stage and concluded that this differential expression was important for heterosis in root system traits. However, Bottley et al. (2006) observed differential transcriptional silencing of up to 27% of genes in leaves and about 26% of those in roots, in which one (and rarely two) members of a set of three homoeoalleles were not present in the cDNA (complementary DNA) of either roots or leaves. They failed to detect a trend in their study of 236 single-copy genes, each of which mapped to one locus of the three homoeologous chromosomes, but the phenomenon of differential transcriptional silencing definitely hints at some method of genome/ allele compensation. On the other hand, Feldman et al. (2012) discovered that non random intergenomic silencing occurred only for about 10% of the genes and that for most other genes, the extra copies were making a favorable effect through positive intergenomic interactions. Gu et al. (2004) compared a sequence of a 307 kb physical contig covering the high molecular weight (HMW) glutenin locus from the A genome of durum wheat (*Triticum turgidum* [*T. durum*] AABB) with the orthologous regions from the B genome of the same wheat and the D genome of the diploid wheat *Aegilops tauschii*. Based on sequence comparison, they concluded that hexaploid wheat might have more than one tetraploid ancestor, further supporting allelic diversity contributing to intergenomic heterosis.

HYBRID WHEAT

Heterosis was first reported in wheat by Freeman (1919) for plant height. Since then, there have been several reports on heterosis in wheat since then. The whole subject of hybrid wheat was reviewed by Pickett and Galwey in 1997 and more recently by Singh et al. (2010). A few successful hybrids have been reported by the private sector, for example in Europe and India (Saaten, 2013; Mahyco, 2013); however, even the hybrids currently available in market offer the best economic advantage only under less than optimum growing conditions. This has been observed in several other studies (Sharma and Tandon, 1995; Solomon et al., 2007).

The question remains why wheat breeders are unable to bring together best of the two "Ds" to develop profitable commercial hybrids. Wheat is the third most important cereal crop in the world, and perhaps the number one cereal consumed directly as food by humans. A number of reasons have been suggested for the lack of a commercial hybrid, including the absence of economical, sustainable hybrid seed production and the absence of a high enough level of heterosis to compensate for the costs involved in seed production and thus become a profitable option for farmers and seed companies. This paper focuses on the reasons for the absence of commercial-scale heterosis.

What limits yield heterosis in wheat?

Pickett and Galwey (1997) concluded that the most serious technological barrier to the development of a successful commercial wheat hybrid is the absence of adequate heterosis. They argued that if the benefits of hybrid varieties were sufficiently attractive, other difficulties could be overcome by researchers, as is evident from several promising recent reports on male sterility and seed production systems (Song et al., 2013; Titan and Meglic, 2011). Three key factors limit yield heterosis in wheat: multiple genomes, the long history of wheat breeding, and the lack of nicking parents.

Multiple genomes

Polyploids are organisms that contain two or more sets of basic chromosomes. Allopolyploids contain doubled interspecific genomes, meaning that hybrid vigor and heterozygosity are permanently fixed like in wheat, *Brassica napus,* and *Geum urbanum* (Vandepitte et al., 2011). Polyploids often not only exceed the diploid progenitors' phenotypes but also exhibit phenotypes not existing in diploid progenitors. Some of these traits either contribute to heterosis or confer adaptation to new ecological niches. The advent of molecular tools (Aversano et al., 2012) offers opportunities to explore the molecular effects of polyploidization. These allopolyploids allow heterosis-causing interactions between homoeologous genes on different genomes, much like between different alleles of one gene, causing heterosis in heterozygote diploid genotypes (Figure 1). In allopolyploids, such interactions occur even in genotypes (Figure 1a), and thus they are supposed to have fixed heterosis.

Polyploids are in fact very successful as about 70% (Masterson, 1994) of flowering plants are polyploids, and about 75% of those are allopolyploids (Grant, 1981; Brochmann et al., 2004). Allopolyploids are formed spontaneously either by crossing unreduced gametes of the participating species or by chromosomal doubling of the interspecific hybrid. Some allopolyploids, such as Brassicas (UN, 1935), evolve through multiple origins involving reciprocal crosses, whereas others, like wheat (Salamini et al., 2002), are formed by a single or a few hybridization events. It is interesting to note, however, that some allopolyploids, like cotton, are still able to have highly successful commercial hybrids over and above their allopolyploid character, while we are still struggling to achieve full scale commercial hybrids in wheat. Having understood the role of desirable diversity in realizing commercial heterosis, it is critical to assess whether there is sufficient desirable allelic diversity in wheat or its inadequacy itself has become a limiting factor.

Wheat, being an amphiploid, has three genomes and is therefore considered to be a natural hybrid. This multiple-genome ancestry perhaps increases wheat's ecological amplitude and evolutionary success (Meimberg et al., 2009) and gives it a very wide agro-climatic adaptability.

The allopolyploidization in wheat causes genome restructuring, including sequence elimination from the parental genomes, which appears to be dictated by parental genomes, ploidy level, and sequence type (repetitive, low-copy, or coding) and affects preferentially the larger parental genome (Bento et al., 2011). This elimination phenomenon was found to be non-random and reproducible and augmented the differentiation of homoeologous chromosomes, providing the physical basis for a diploid-like meiotic behaviour (Ozkan et al., 2002).

This process also prevented intergenomic recombination, ensuring full fertility and permanent heterosis between alleles of different genomes (homoeoalleles). However, polyploidization causes some degree of intergenomic homoeologous gene silencing and differential gene expression and perhaps contributes

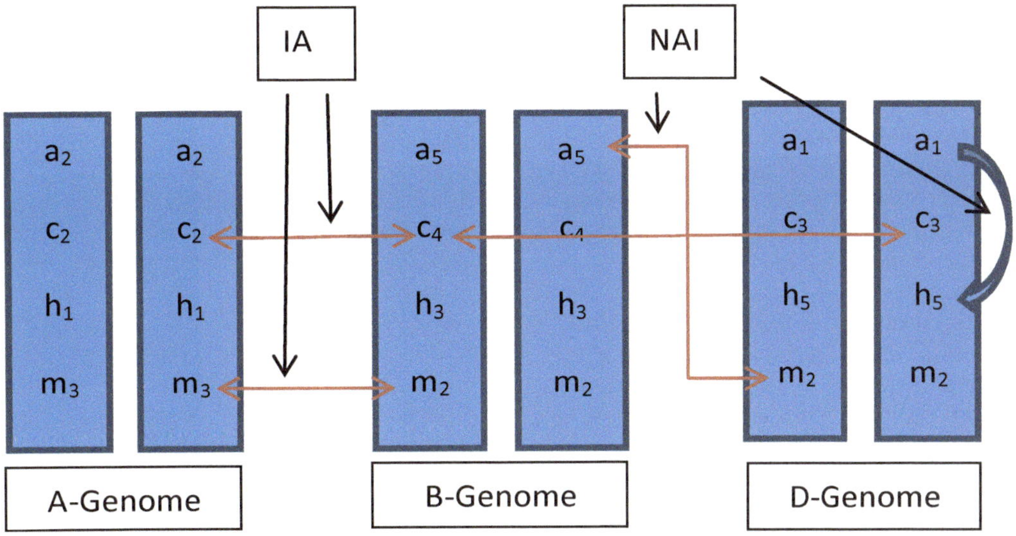

Figure 1a. Fixed interallelic (IA) and non-allelic (NAI) interactions among three wheat genomes in a wheat pureline leading to fixed intergenomic heterosis.

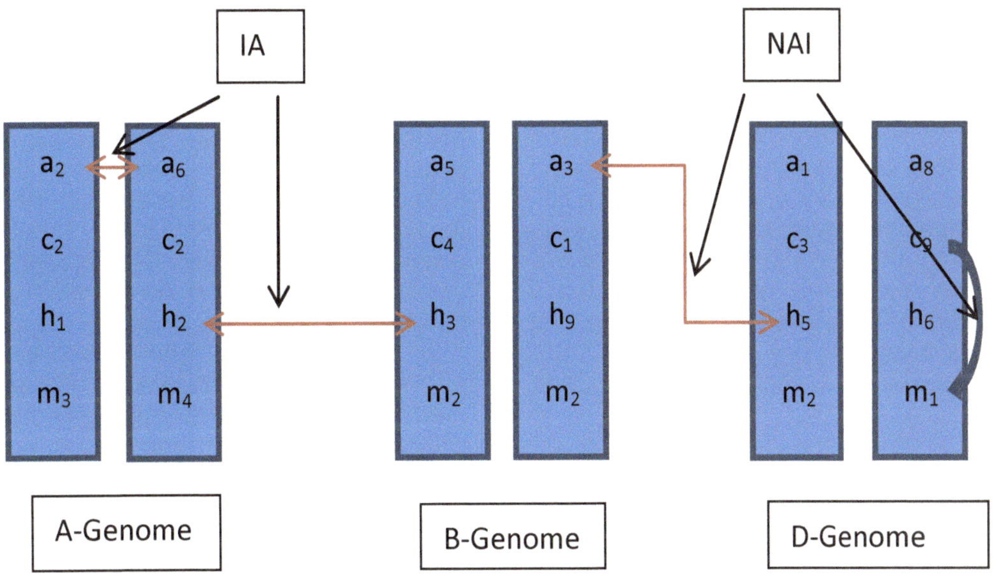

Figure 1b. Heterozygosity conferred interallelic (IA) and non-allelic (NAI) interactions among three wheat genomes in a conventional F_1 wheat hybrid.

to intergenomic heterosis (Mochida et al., 2003). It possesses fixed heterosis resulting from homoeologous alleles on its three homoeologous chromosomes, which in fact is even better than classical heterosis, as the latter has only two genes at each locus (Figure 1a). When two such hexaploid wheat lines are crossed, the fixed intergenomic heterosis is considered to be one of the reasons for low classical heterosis.

The presence of three homoeologous alleles (homoeoalleles) in any pure line already imparts enough hybridity to benefit from the phenomenon of heterosis, and a classical cross, which brings in six homoeoalleles at each locus, perhaps fails to bring in enough desirable hybridity across all loci to endow the resultant hybrid with a significantly higher commercial-level heterosis over and above the fixed intergenomic one in best commercial cultivars (Figure 1b). Developments in molecular biology are now allowing investigations into fixed heterosis employing specially constructed amphiploid populations using their diploid progenitors (Abel et al., 2005;

Bansal et al., 2012).

Long history of breeding

It can be argued that centuries of wheat breeding have already accumulated three diverse desirable alleles at most loci and that a gradual improvement in this direction is under way in all conventional breeding programs around the world. A collection of 9,000 gene-associated single nucleotide polymorphisms were used to study a worldwide sample of 2,994 hexaploid wheat landraces and modern cultivars to detect regions of wheat genome subject to selection during improvement (Colin et al., 2013). The study revealed that ancestral variation has been used extensively, selection likely acting on multiple functionally equivalent alleles. Can we really conclude that this long-term selection accumulated different but functionally similar alleles over decades, further enhancing intergenomic heterosis within wheat?

While maize has a heterozygous balance, wheat has a homozygous balance, meaning there are no deleterious alleles in pure lines. However, this does not rule out the dominance model as an explanation of heterosis at some loci in wheat. The dominance model (Bruce, 1910; Jones, 1917) postulates that inbred parents contain deleterious or inferior alleles at several loci, inhibiting overall good performance, and that these are dominated by superior dominant alleles from the other parent in a hybrid, resulting in a superior performance. Wheat can be a little different in that there might not be deleterious alleles in a pure line (because there is no inbreeding depression), but there could be inferior alleles (decreasers in case of quantitative traits) in one parent and superior alleles (increasers in case of quantitative traits) in the other parent, which, when brought together, result in a superior performing hybrid.

For example, for a thoroughly investigated locus like HMW glutenin subunits (HMW-GS), Ribeiro et al. (2011) reported a total of 56 patterns in a set of 134 hexaploid wheat accessions in Portugal. For the three loci Glu-A1, Glu-A2, and Glu-A3 of HMW-GS, they discovered 4, 10, and 6 alleles, respectively. This type of not-yet-investigated allelic variability for other economically important traits in different genomes has been utilized by breeders over centuries and has been accumulated in pure lines conferring a high level of intergenomic heterosis to present-day wheat cultivars. This also implies that it is theoretically possible to accumulate all possible desirable alleles to get pure lines as high-yielding as a hybrid. This theoretical possibility might have been turned into a reality by centuries of wheat improvement work throughout the world, and even if this has been partially achieved, it would explain lack of frequent commercial-level heterosis in wheat (Figure 1). This was somewhat supported by Cui et al. (2002) when they found only 11.14% heterosis in 20 intervarietal hybrids compared to 111.39% in interspecific hybrids involving spelt wheat, highlighting the importance of interspecific gene transfer to broaden the common wheat gene pool for further yield improvement, including heterosis exploitation. A potential useful introgression from *T. tauschii* has been suggested (Mohammadi et al., 1999), and this possibility was demonstrated by Snape and Parker (1985) when they successfully combined yield-determining alleles dispersed between two parents in one inbred line.

Lack of nicking parents

It is understood that yield is a complex trait-the sum of several qualitative and quantitative traits-and yield heterosis cannot be fully explained by the oversimplified hypotheses that have been put forward by several researchers. One certain requirement for any commercial heterosis is the presence of genetic differences in the two parental lines (East, 1936); however, diverse parents will not always yield heterosis (Solomon et al., 2007). Choosing the right parental combination is central to achieving heterosis. Overdominance (Shull, 1908; East, 1936; Crow, 1948) and pseudo-overdominance (Semel et al., 2006) models can definitely explain heterosis happening at several loci, even in a crop with homozygous balance like wheat. Overdominance postulates interallelic interaction as the cause of heterosis at some loci, whereas pseudo-overdominance results from complementation of two or more linked dominant and recessive alleles in repulsion, in which the dominant and recessive alleles are located on opposite homologues of the two genes acting as overdominance.

Since heterosis is a genome-wide expression, and one model explaining heterosis for one locus does not preclude other models explaining heterosis for other loci, different models might explain different locus-specific heterosis expressions. These different genetic models, therefore, only explain different temporal and/or spatial changes in gene regulatory network caused by hybridity (Omholt et al., 2000). However, it needs to be realized that the unique intergenomic hybridity of bread wheat also fixes all forms of non additive heterosis available in the three constituent genomes (Figure 1a). A simple estimate of diversity does not necessarily mean that parents will be heterotic. The allelic constitutions of the three constituent genomes of both the parents need to nick well to give a commercial heterosis over and above the heterosis already fixed in the two parents (Figure 1b). Most studies estimating diversity using molecular markers (El-Zanaty et al., 2013; Chen et al., 2010; Solomon et al., 2007; Zhan et al., 2006; Dreisigacker et al., 2005; Zhang et al., 2003; Corbellini et al., 2002; Xu et al., 2002) concluded that this simple diversity estimate was unable to predict the heterotic potential of the identified parents.

CONCLUSION

Heterosis is a powerful tool benefitting several important crops with yield advantages ranging from 15 to 50%. Wheat breeders have naturally been interested in exploiting this option. A molecular deciphering of the phenomenon of heterosis will definitely help explain the lack of commercial-level heterosis in wheat. Any phenotype, especially those involved in commercial heterosis, involves multiple quantitative traits expressed sequentially or simultaneously during development. A molecular dissection will relate genome-wide tempo-spatial gene expression data with the relevant phenotypes.

Zachary et al. (2006) proposed a “phenomics” platform to measure multiple traits in the context of high-resolution QTL mapping as a tool to dissect heterosis into its fundamental components. QTL mapping in both rice (Hua et al., 2002, 2003; Li et al., 2001; Luo et al., 2001; Xiao et al., 1995; Yu et al., 1997) and maize (Stuber et al., 1992) have broken down heterosis into Mendelian factors displaying inheritance as hypothesized in classical models of dominance, overdominance, and epistasis. The central idea remains to identify the responsible markers/QTLs/phenomena and then select the right type of parental combinations to maximize heterosis. QTL mapping/cloning approach can be effective in identification of heterosis-causing Mendelian factors and heterotic parental combinations in a multigenomic amphiploid like wheat. What is therefore needed is substantial basic research to determine causes of yield heterosis in wheat and then identify right type of parental combinations.

REFERENCES

Abel S, Mollers C, Becker HC (2005). Development of synthetic Brassica napus lines for the analysis of “fixed heterosis” in allopolyploid plants. Euphytica 146:157-163.

Aversano R, Ercolano MR, Caruso I, Fasano C, Rosellini D, Carputo D (2012). Molecular tools for exploring polyploid genomes in plants. Intl' J. Mol. Sci. 13:10316-10335.

Bansal P, Banga S, Banga SS (2012). Heterosis as investigated in terms of polyploidy and genetic diversity using designed Brassica juncea amphiploid and its progenitor diplod species. PLoS ONE 7: e29607.

Bento M, Gustafson JP, Viegas W, Silva M (2011). Size matters in Triticeae polyploids: larger genomes have higher remodelling. Genome 54:175-183.

Bottley A, Xia GM, Koebner RMD (2006). Homoeologous gene silencing in hexaploid wheat. Pl. J. 47:897-906.

Brochmann C, Brysting AK, Alsos IG, Borgen L, Grundt HH, Scheen AC, Elven R (2004). Polyploidy in arctic plants. Biol. J. Linn. Soc. 82:521-536.

Bruce AB (1910). The mendelian theory of heredity and the augmentation of vigor. Science 32:627-628.

Colin RC, Shiaoman C, Shichen W, Bevan EHuang, Stuart S, Seifollah K, Kerrie F, Cyrille S, Gina LB-G, Alina A, Deven S, Guihua B, Michael P, Luxmi T, Debbie W, Stephan K, Matthew R, Marta LS, Harold B, Luther T, James AA, Susanne D, Stephen B, Arron C, Viktor K, Peter LMorrell, Jorge D, Matthew KM, Mark ESorrells, Matthew JHayden, Eduard A (2013). Genome wide comparative diversity uncovers multiple targets of selection for improvement in hexaploid wheat landraces and cultivars. Proceed. Nat'l Acad. Sci. USA 110:8057-8062.

Chen XD, Sun DF, Rong DF, Sun GL, Peng JH (2010). Relationship of genetic distance and hybrid performance in hybrids derived from a new photoperiod-thermo sensitive male sterile wheat line 337S. Euphytica 175:365-371.

Collins GN (1921). Dominance and vigour of first generation hybrids. Am. Nat. 55:116-133.

Corbellini M, Perenzin M, Accerbi M, Vaccino P, Borghi B (2002). Genetic diversity in bread wheat as revealed by coefficient of parentage and molecular markers and its relationship to hybrid performance. Euphytica 123:273-285.

Crow JF (1948). Alternative hypothesis of hybrid vigour. Genetics 33:477-487.

Cui GH, Ni ZF, Wu LM, Li YQ, Sun QX (2002). Study on wheat heterotic groups. V. Relationship between microsatellite marker based genetic distance and heterosis between *Triticum aestivum* and *T. spelta*. J. Triticeae crops 22:5-9.

Dreisigacker S, Melchinger AE, Zhang P, Ammar K, Flachenecker C, Hoisington D, Warburton ML (2005). Hybrid performance and heterosis in spring bread wheat and their relation to SSR based genetic distances and coefficient of parentage. Euphytica 144:51-59.

East EM (1936). Heterosis. Genetics 21:375-397.

El-Zanaty AFM, El-Hadary MH, Ismail M, El-Gammal AA (2013). Genetic diversity of wheat genotypes based on RAPD relative to F1 hybrid performance. Int'l J. Agron. Pl. Prod. 4:1098-1107.

Falconer DS, Mackay TFC (1996).Introduction to quantitative genetics, 4th edition. Longman, Essex, England.

Feldman M, Levy AA, Fahima T, Korol A (2012). Genomic asymmetry in allopolyploid plants: wheat as a model. J. Ex'l Bot. 63:5045-5059.

Freeman GF (1919). Heredity of quantitative characters in wheat. Genetics 4:1-9.

Fu H, Dooner HK (2002). Intraspecific violation of genetic colinearity and its implications in maize. Proc. Natl. Acad. Sci. USA 99:9573-9578.

Grant V (1981). Plant Speciation, Columbia University Press. Columbia.

Gu YQ, Colman-Derr D, Kong X, Anderson OD (2004). Rapid genome evolution revealed by comparative sequence analysis of orthologous regions from four Triticeae genomes. Pl. Physiol. 135:459-470.

Hallauer AR, Miranda JB (1981). Quantitative genetics in maize breeding, Iowa State University Press, Ames USA.

Hua J, Xing Y, Wu W, Xu C, Sun X, Yu S, Zhang Q (2003). Single locus heterotic effects and dominance by dominance interactions can adequately explain the genetic basis of heterosis in an elite rice hybrid. Proc. Natl. Acad. Sci. U.S.A. 100:2574-2579.

Hua JP, Xing YZ, Xu CG, Sun XL, Yu SB, Zhang Q (2002). Genetic dissection of an elite rice hybrid revealed that heterozygotes are not always advantageous for performance. Genetics 162:1885-1895.

Hull FH (1945). Recurrent selection for specific combining ability in corn. J. Am. Soc. Agron. 37:134-145.

Jones DF (1917). Dominance of linked factors as a means of accounting for heterosis. Genetics 2:466-479.

Keeble F, Pellew C (1910). The mode of inheritance of stature and of time of flowering in peas (*Pisum sativum*). J. Genet. 1:47-56.

Li ZK, Luo LJ, Mei HW, Wang DL, Shu QY, Tabien R, Zhong DB, Ying CS, Standel JW, Khush GS, Paterson AH (2001). Overdominant epistatic loci are the primary genetic basis of inbreeding depression and heterosis in rice. I. Biomass and grain yield. Genetics 158:1737-1753.

Lippman ZB, Zamir D (2007). Heterosis: Revisiting the magic. Trends Genet. 23:60-66.

Luo LJ, Li ZK, Mei HW, Shu QY, Tabien R, Zhong DB, Ying CS, Stansel JW, Khush GS, Paterson AH (2001). Overdominant epistatic loci are the primary genetic basis of inbreeding depression and heterosis in rice. II. Grain yield components, Genetics 158:1755-1771.

Mahyco (2013). http://www.mahyco.com/field_crops.html; Accessed October, 2013.

Masterson J (1994). Stomatal size in fossil plants: evidence for polyploidy in majority of angiosperms. Science 264:421-424.

Meimberg H, Rice KJ, Milan NF, Njoku CC, McKay JK (2009). Multiple origins promote the ecological amplitude of allopolyploid *Aegilops*

(Poaceae). American J. Bot., 96:1262-1273.
Mochida K, Yamazaki Y, Ogihara Y (2003). Discrimination of homoeologous gene expression in hexaploid wheat by SNP analysis of contigs grouped from a large number of expressed sequence tags. Mol. Genet. Genom. 270:371-377.
Mohammadi V, Ahmadian-Tehrani P, Abd Mishani S, Ghannadha MR (1999). Study on genetic diversity in wild diploid wheat (Triticum tauschii) using seed storage protein electrophoresis. Iran. J. Agril' Sci. 30:55-64.
Omholt SW, Plahte E, Oyehaug L, Xiang K (2000). Gene regulatory networks generating the phenomena of additivity, dominance and epistasis. Genetics 155:969-980.
Ozkan H, Levy AA, Feldman M (2002). Rapid differentiation of homeologous chromosomes in newly formed allopolyploid wheat. Israel J. Pl. Sci. 50:S65-S76.
Pickett AA, Galwey NW (1997). A further evaluation of hybrid wheat. Plant Cultivars Seeds 10:15-32.
Ribeiro M, Carvalho C, Carnide V, Guedes-Pinto H, Igrejas G (2011). Towardsa alleleic diversity in the storage proteins of old and currently growing tetraploid and hexaploid wheats in Portugal. Gen. Resour. Crop Evol. 58:1051-1073.
Richey FD (1942). Mock-dominance and hybrid vigour. Science 96:280-281.
Saaten (2013). http://www.saaten-union.com/index.cfm/nav/682.html; Accessed October, 2013.
Salamini F, Ozkan H, Brandolini A, Schafer-Pregl R, Martin W (2002). Genetics and geography of wild cereal domestication in the near east. Nat. Rev. Genet. 3:429-441.
Schnell FW, Cockerhan CC (1992). Multiplicative vs. arbitrary gene action in heterosis. Genetics 131:461-469.
Semel Y, Nissenbaum J, Menda N, Zinder M, Krieger U, Issman N, Pleban T, Lippman Z, Gur A, Zamir D (2006). Overdominant quantitative trait loci for yield and fitness in tomato. Proc. Natl. Acad. Sci. U.S.A. 103:12981-12986.
Sharma RK, Tandon JP (1995) Effect of heat stress on variation in parental lines and their hybrids in wheat: a comparison. Indian J. Genet. Pl. Breed. 55:455-456.
Shull GH (1908). The composition of a field of maize. Am. Breed. Assoc. Rep. 4:296-301.
Shull GH (1909). A pure line method f corn breeding. Am. Breed. Assoc. Rep. 5:51-59.
Shull GH (1952). Beginings of the heterosis concept, In: JW Gowen (ed) Heterosis. Iowa State College Press, Ames, IA. Pp 14-48.
Singh SK, Ravish Chatrath, Mishra B (2010). Perspective of hybrid wheat research: A review. Indian J. Agric. Sci. 80:1013-1027.
Snape JW, Parker BB (1985). The genetical basis of heterosis for yield in wheat. Annual Rep. Pl. Breeding Institute, Cambridge 1984:73-74.
Solomon KF, Labuschagne MT, Viljoen CD (2007). Estimates of heterosis and association of genetic distance with heterosis in durum wheat under different moisture regimes. J. Agric. Sci. 145:239-248.
Song X, Ni ZF, Yao YY, Zhang YH, Sun QX (2009). Identification of differentially expressed proteins between hybrid and parents in wheat (*Triticum aestivum* L.) seedling leaves. Theor. Appl. Genet., 118:213-225.
Song X, Zhang L, Zeng J, Qian H, Li H, He B (2013). Development of thermo sensitive cytoplasmic male sterile (TCMS) lines of wheat characterized by complete male sterility at lower temperatures and partially restored fertility at higher temperatures. Euphytica 192:393-399.
Stuber CW, Lincoln SE, Wolff DW, Helentjaris T, Lander ES (1992). Identification of genetic factors contributing to heterosis in a hybrid from two elite maize inbred lines using molecular markers. Genetics 132:823-839.
Titan P, Meglic V (2011). Several approaches for heterosis exploitation in common wheat (*Triticum aestivum* L.). Acta Agric. Slov. 97:137-144.
UN (1935). Genome analysis in Brassica with special reference to the experimental formation of B. napus and peculiar mode of fertilization. Jpn. J. Genet. 7:389-452.
Vandepitte K, Jacquemyn H, Roldan-Ruiz I, Honnay O (2011). The consequences of mating over a range of parental genetic similarity in a selfing allopolyploidd species. J. Evol. Biol. 24:2750-2758.
Wang Z, Ni Z, Wu H, Nie X, Sun Q (2006). Heterosis in root development and differential gene expression between hybrids and their parental inbreds in wheat (*Triticum aestivum* L.). Theor. Appl. Genet. 113:1283-1294.
Xiao J, Li J, Yuan L, Tanksley SD (1995). Dominance is the major genetic basis of heterosis in rice as revealed by QTL analysis using molecular markers. Genetics 140:745-754.
Xu Z, Zhang A, Liu D, Guo X, Zhang S (2002). The heterosis for kernel weight of main spike and its relation to genetic diversity revealed by RAPD markers in wheat (*Triticum aestivum* L.). J. Genet. Breed. 56:107-111.
Yu SB, Li JX, Xu CG, Tan YF, Gao YJ, Li XH, Zhang Q, Saghai MA (1997). Importance of epistasis as the genetic basis of heterosis in an elite rice hybrid. Proc. Natl. Acad. Sci. U.S.A. 94:9226-9231.
Yuan Q, Deng Z, Peng T, Tian J (2012). QTL based analysis of heterosis for number of grains per spike in wheat using DH and immortalized F2 populations. Euphytica 188:387-395.
Zachary BL, Dani Z (2006). Heterosis: revisiting the magic. Trends Genet. 23:60-66.
Zhan KH, Sun H, Gao X, Wang LH, Song YH, Wu SF (2006). Relationship between the genetic distances based on molecular markers and heterosis in hybrid wheat. J. Triticeae Crops 26:27-31.
Zhang ZQ, Zheng YL, Wei YM, Zhou YH, Liu DC, Lan XJ, Liu H (2003). Relationship between heterosis and genetic distance based on simple sequence repeat markers in wheat (*Triticum aestivum*). J. Schuan Agric. Univ. 21:6-9.

Adoption of improved production technology of Mandarin in Rajasthan, India: A review

B. C. Yadav, Ravish Choudhary and P. L. Saran

Indian Agricultural Research Institute Regional Station, Pusa, Samastipur (Bihar)-848 125, India.

Fruits are of great importance in the field of human nutrition. Citrus is one of the most important fruit crop in India. In India common citrus fruits are mandarin, sweet orange and lime. India also earns foreign exchange by exporting orange to Iran, Bahrin, Singapore etc. The demand for mandarin fruit is very high because of its nutritive value. Citrus fruits have a special place in Indian diet as they are rich source of vitamin C, vitamin A and B and fruit sugars. In some part of Rajasthan such as Jhalawar, Kota, Sikar and Ganganagar districts, citrus is the leading fruit crop. Nagpuri mandarin is quite important and produced commercially at large scale of the various types of citrus. Majority of the farmers (69.00%) were found to have medium level of knowledge about improved production technology. The study shows farmers possessed maximum knowledge about "varieties" and "plantation" while minimum knowledge was reported in "manure and fertilizer" and "plant protection measure". Majority of farmers (64.00%) were found to have medium level of adoption of mandarin production technology. The study indicates that farmers had high adoption level for the practices viz., varieties and plantation and low adoption practices viz., manure and fertilizer and plant protection measures. The knowledge of the farmers about improved production technology of mandarin cultivation was positively and significantly associated with their caste, occupation, educational level, social participation, size land holding while their market distance and irrigation potentiality were negatively and non significantly co-related with their knowledge level.

Key words: Citrus, mandarin, improved production technology, farmers.

INTRODUCTION

Fruits are of great importance in human diet. India is the second largest producer of fruits in the world. Its share in world's fruit production is 11%. The major fruits grown in India are mango, banana, citrus, guava, pineapple, grape and papaya in tropics and subtropics and apple in the temperate region. Apart from these, sapota, aonla ber, pomegranate, litchi, peach, pear plum and walnut are grown on a sizable area. A number of other fruits such as jack-fruit, lasoda, phalsa, mulberry, beal, fig, datepalm etc, are also grown in different regions. Major fruit growing states are Uttar Pradesh, Andhra Pradesh, Bihar, Karnataka, Tamil Nadu, Maharashtra, Kerla and West Bengal. The other states which have substantial area are Gujarat, Assam, Madhya Pradesh and Orissa. The per capita availability of fruits in the country is only 46 g per day against to the requirement of 92 g as prescribed by the Indian Council of Medical Research, this may be due to very low productivity of the large unproductive old orchard, small size of land holding and poor management of the orchards, especially in the

Figure 1. Fruit bearing tree and fruits of mandarin.

rainfed area which constitute area over 60% of the total areas under fruits. Besides this, lack of quality of planting material of improved varieties and several production constraints also contribute to this low productivity (Anonymous, 2004).

Citrus (Linnaeus) is one of the most economically important fruit crops of India, belonging to the subtribe Citrinae, tribe Citreae, subfamily Aurantioideae of the family Rutaceae. It is widely distributed throughout the tropical and subtropical regions of the world and believed to be originated in Southeast Asia, particularly, in the regions extending from North-east India eastward through the Malay archipelago; North into China and Japan, and South to Australia (Swingle and Reece, 1967; Moore, 2001). Mandarin (*Citrus reticulata* Blanco) is one of the premier commercial citrus grown in central India. It is grown in 92.2 thousand hectare with a total production of 11.2 lack tonnes and an average productivity 11.5 tonnes per hectare, Nagpuri mandarin was introduced in 1894 by late Shri Rahuji Raje Bhonsale in Central India (Shrivastva and Singh, 2003). Citrus is world's leading fruit crop. It is also known as fancy fruits and is eaten as desert fruit. It is originated in all probability in Cochin, China and has been cultivated in China and Japan over long period. Citrus plant is characterized by the following diagnostic characters: shrubs or small to large trees; stem usually angular and spiny, leaves unifoliolate, usually with narrow to broadly winged petiole and pellucid dots (oil glands) on leaf surface, hesperidium fruit (a type of berry with a leathery adherent pericarp), spongy or fibrous mesocarp and fleshy endocarp filled with pulp vesicles containing sour or sweet juice. *Citrus* species have diploid chromosome number, 2n = 18, but some species show polyploidy, that is, 27 and 36 chromosomes (Cameron and Frost, 1968; Guerra, 1984).

Thus there is a great scope of increasing the fruit production by increasing the productivity besides bringing more area under fruits crops. According to the national commission on agriculture, the total production in the country should be increased to 52 million tonnes to provide enough fruits for a population of about one billion people by the turn of the century. Another significant factor which contributes to low availability of fruits to the people is the considerably high post-harvest losses; almost 20 to 30% of the fruits are lost at the post harvest stages which could be saved by adoption of improved harvesting, packaging, handling and storage techniques. Development of processing industry should result in sustained fruit production by ensuring strong sink for disposal. However, at present, only about 0.3% of the fruits are used for processing against a demand of about 0.5%, there is considerably scope of increasing the supply of fruits in processed form.

Utilization of fruits has been inherent in the Indian way of life from ancient times, medicinal uses of fruits like aonla, beal, citrus, ber, karronda, lemon, lime, fig, jamun, mango, orange, banana, pomegranate, almond etc. have been mentioned in charak samhita and sushrut samhita, mango, banana, beal, aonla and coconut have been associated with the festival and ritual in India. The several fruits like mango, banana, fig, grape and date palm were the most favorite. The mandarin group includes all type of loose jacket oranges commonly called by the Indian name santra such as Nagpuri santra, khasi orange and desi type orange.

India has an enormous diversity of citrus genetic resources, both cultivated and wild. Citrus occupies the second position in terms of area (5, 02,800 ha) and third position in terms of production, that is, 4,396,700 million tonnes (mt) of fruit crops in India (NHB database, 2012). Being the second most important fruit crop of the world trade for fresh fruits, more than fifty countries are growing citrus fruits commercially in different agroclimatic conditions for its diversified use and increasing demand worldwide. The rich genetic diversity particularly in mandarins, sweet oranges, lemons and limes immensely contributed to improvement of citrus cultivation and industry in India.

At present mandarin are commercially grown in India. The tree is small, spiny with a dense top (Figure 1). The leaves are long, lanceolate, petiole slightly, margined, articulated, flower white, solitarily, perfect. The size of fruit is medium to large, flattened or depressed globose. The colour of fruit is yellow, reddish orange (Figure 1). The core is hollow. The rind is thin, rind and segments are easily separated. The number of segment is 10 to 14. The pulp quality is exceptionally fine. The seeds are small beaked, cotyledon green, highly polyembryonic.

The mandarin is locally knows as "Nagpuri santra" has reputation and it is the best of its kind grown in India. The regular excessive bearing in the shallow soil exhaust "Nagpuri mandarin" plants, earlier than those in the deep soil. If not managed properly and timely, the plant starts declining their vigour and yield potential whereas, the mandarin plants respond well to regular and timely supply of fertilizers with proper irrigation schedule in shallow soils. The demands for mandarin are very high because of its nutritive value. It is very rich in vitamin C (ascorbic acid), fruit sugars and in addition to this it also contains vitamin A and B. Juice content is abundant, yellow in colour with excellent flavour, sweet in taste. The number of seeds per fruit is 6 to 7. It provides vitamin P which keeps the small blood vessels in healthy condition in body. Mandarin is the most important of citrus fruits alone occupy 40% of the land under citrus cultivation. Citrus cultivation is of major economic importance. India earns foreign exchange by exporting oranges to Iran, Bahrin, Singapore, Nepal, Switzerland etc.

In Rajasthan, kinnow occupies 5522 ha area under cultivation (Anonymous, 2008, 2009). The mandarin is being grown in different districts of which important are Jhalawar, Kota, Sikar, Baran, Bhilwara and Rajsamand. Jhalawar district has highest area (5693.12 ha) and production (4784.66 qtl.) (Revenue office, Jhalawar). Citrus fruits are good sources of citric acid, flavonoids, phenolics, pectins, limonoids, ascorbic acid, etc. (Dugo and DiGiacomo, 2002). The dietary, nutritional, medicinal and cosmetic properties of citrus fruits are well documented (Swingle and Reece, 1967; Dugo and Digiacomo, 2002). Recent pharmacological studies support the traditional medical applications of citrus fruits in scurvy, obesity, cancer, HIV/AIDS, contraception, cough, and reducing blood pressure (Mabberley, 2004). Oranges are used for treatment of cold, fever, liver disorders, gall bladder problems, acne, and ringworm (Imbesi and Pasquale, 2002; Mabberley, 2004; Arias and Laca, 2005). During last few years, the urgent need for undertaking the basic and strategic research work has been done in the country on various aspects such as varieties, propagation, irrigation and pruning for stabilizing and increasing the production, productivity and quality of mandarin fruits. The production of poor quality of mandarin fruit is due to lack of knowledge in the mandarin growers about mandarin production technology.

Despite the availability of improved varieties, improved package of practices, scientific technology and poor management of the mandarin orchard, it is a large gap between the yield of mandarin orchard recorded at the research farm and its production at farmer's field. The low production of mandarin fruits may be due to the non adoption or poor adoption of recommended technology of mandarin by the farmers which may be due to their being unaware of about latest recommended technology and they may be facing some problems in its adoption at their own farm and there may be certain factors which may be affect adoption of recommended technology of mandarin cultivation.

Purpose statement

Mandarin is most suitable fruit crop for India, but when we look to the productivity of mandarin, it stands very low productivity as compared to other countries. This is a challenging task for the scientists and the farmers. The scientists have evolved latest technologies but the farmers have not adopted latest technologies due to the fact of low knowledge and awareness, about new technologies. Under such conditions it is quite imperative that reasons for the technological gap in mandarin should be identified and studied critically in order to face the existing challenge of low adoption of improved technology. It is true that improved practices are not only the things, which cause adoption or rejection, but certainly other factors are also directly or indirectly responsible for technological gap.

Some enthusiastic farmers in Jhalawar district planted a few saplings of loose skinned range after importing from Nagpur in fifties. These plants started bearing in 1955. The growers found it similar in quality and better in yield than the original place of it. Which was motivated the farmers to extend area under orange plantation. At the end of 1960, its production was for marketing. The first attempt to orange plantation was made in Jhalawar district during 1972 with the financial assistance of NABARD. This area is known as "Nagpur of Rajasthan" for orange production. Jhalarapatan, pirawa and pachpahad are the main basis in Jhalawar district where orange (mandarin) is produced on large scale.

IMPROVED PRODUCTION TECHNOLOGY

Adoption of improved production technology of mandarin cultivation is directly or indirectly related to knowledge of mandarin growers. Hence, it was considered necessary to assess the knowledge of the farmers about improved production technology of mandarin cultivation while the knowledge about the technology had influence on the decision making about its adoption with this view in mind the knowledge test was applied on the farmers to test their knowledge about improved production technology of mandarin cultivation.

Phuse et al. (2007) found that the progressive farmers of the village were found to be the most credible source of information for the orange growers. The soil type and use of planting; recommended rootstock use; application of fertilizer doses and irrigation practice; total no of irrigation, method of application for manures were the practices partially adopted by more than half of the respondents. However, it could further be seen that majority of the respondents did not adopt the important improved cultivation practices such as; proper harvesting time for fruit (T.S.S: Acidity ratio) was (81.50%) and up to 44% respondent did not adopt the proper time of harvesting. It is due to they cannot take the risk in market, as well as lack of awareness but they are totally dependent up on mediator. It is also seen that (54.50%) respondent's did not follow soil testing at the time of planting as well as at bearing stage. So, there is a need to adopt the full-improved recommended package of practices for increasing the productivity with quality in mandarin orange.

Improved production technology has revolutionized the fruits cultivation and India has a reputed position in the field of fruit production. The increasing demand of fruits clearly indicates that the production is an urgent need to boost up the production of mandarin fruit to cater the demands of growing population. Some of the factors responsible for non adoption of latest technology by farmers may be lack of awareness, low technical know-how and less knowledge regarding resources of information factors such as credit, storage and marketing etc. The knowledge level of farmers about improved production technology of mandarin cultivation was measured by a knowledge test developed by the Mohammad (2000). The adoption level of farmers about improved production technology of mandarin cultivation was measured by scale developed by Mohammad (2000) with slight modification. Whereas, the constraints faced by the farmers in adoption of improved production technology of mandarin cultivation as measured by schedule prepared by the investigator in light of suggestion of the experts, the independent variables like caste, occupation, educational level, social participation, size of land holding, market distance and irrigation potentiality were measured by modified scale of Trivedi (1963). The source of information was measured by scale of Ramchandarn (1974) with slight modification as per need of the study.

The majority of kinnow growers belonged to medium adoption category for the recommended kinnow production technology such as 'application of N', 'application of potash', 'planting pit size' and 'filling the pit'. The least adoption was for the practices such as 'plant protection measure' and 'application of growth regulators' and high adoption for the practices such as 'plant population', 'irrigation intervals', 'application of FYM', 'plantation depth', 'recommended method of propagation' and 'application of phosphorus'. The knowledge level, education level, size of land holding and occupation were positively and significantly associated where as their market distance was negatively and significantly associated, while social participation and irrigation potentiality were positively and non-significantly related with the adoption level of farmers about recommended kinnow production technology (Choudhary et al., 2011).

Nainawat (1990) found that majority of farmers have moderate knowledge about recommended ber production technology. Nimje et al. (1991) observed that there is a need to transfer the ber cultivation techniques to the ber growers. About 25% ber growers did not have sufficient knowledge of various aspects of ber cultivation. Very simple aspects like digging pits, selection of bud grafts, selection of buds, grafting methods etc. were not known to the ber growers. Mohammad and Punjabi (1997) reported that majority of mandarin growers (80%) were having medium level of knowledge of mandarin production technology while 20% of them had low level of knowledge. In case of non mandarin growers, 70% medium level of knowledge and about 30% possessed low level of knowledge. Sharma and Bangarwa (1997) concluded that majority of the respondents (about 60%) had medium knowledge level about tree plantation, followed by 26.66% who had moderate knowledge level. Only 8 and 6% of respondent had low and high knowledge level about tree plantation respectively.

Deshmukh et al. (1998) found that 93.33% of the respondent had knowledge about the varieties and time of harvesting of custard apple. As much 92.90% of the respondents had knowledge of recommended soil type for custard apple cultivation. Similarly the knowledge about the importance of digging pit and size of pit was possessed by 81.66 and 65.85% respondent, respectively. Devi and Agarwal (2000) reported that majority of the farmers had medium level of knowledge about improved technology of pea cultivation. Dayama (2000) concluded that majority of farmers has medium knowledge level (68%) about improved practices of aonla plantation. Mohammad (2000) observed that guava orchard owners had maximum knowledge about fruit varieties (100%). It was followed by planting (67.22%), cultural practices (66.66%), harvesting and marketing (65.33%) and propagation (55.55%). Comparatively less knowledge was reported in plant protection measures (47.91%) among the guava orchard owners. Jangid (2001) reported that 68% farmers were categorized in medium knowledge level, 22% farmers were in low knowledge level and 10% farmers were in high knowledge about adoption of recommended cultivation practices of chilli.

Meena (2001) found that there exists a significant knowledge difference with regards to all the practices of onion production technology between trained and untrained respondents. Pandaria et al. (2002) revealed that the potential of crop regulation technology in orchard growers but also highlighted the useful implications.

Table 1. Distribution of farmers under different knowledge levels about improved production technology of mandarin cultivation (N = 100).

S/N	Categories of knowledge level	No of farmers	Percentage of farmers
1	Low knowledge (Score upto 26.10)	11	11
2	Medium knowledge (score from 26.11 to 34.08)	69	69
3	High knowledge (score above 34.08)	20	20
	Total	**100**	**100**

X = 30.09 σ3.99.

Firstly, participating in trials, growers gained knowledge about the technology which they can adopt and disseminate among fallow growers. Secondly, the participatory trials at different locations provided opportunity for the growers to develop their capability for on farm experimentation and also build up their confidence in designing, implementing and evaluating the on farm experimentation along with scientists. Poonia (2002) found that majority of respondents (65.00%) had medium level of knowledge of improved kinnow cultivation practices. However, respondents found in high and low knowledge category were (18.33%) and (16.67%), respectively. Srivastava et al. (2002) it was concluded that majority of chilli growers were having over all medium level of knowledge regarding S-49 chilli cultivation technology. In case of practice wise level of knowledge, nearly 14.00 to 24.00% of chilli growers had high level of knowledge while in case of low possession of knowledge of chilli growers varies from nearly 14.00 to 22.00% regarding different agronomic practices of chilli cultivation technology.

Arneja and Khangura (2003) it was concluded that the pea growers were also categorized into three different categories according to their overall performance of knowledge score that 40% of farmers had low, while 32% of respondents had medium level of knowledge and only 28% of the pea growers had a high level of knowledge regarding various pea cultivation practices. Thus, it can be concluded (72.00%) had low to medium level of knowledge regarding various pea cultivation practices. Meena (2004) observed that 60.00% farmers were categories in medium knowledge level, 21.70% farmers were in high knowledge level and 18.30% farmers were in low knowledge level about improved practices of guava plantation. Jhajharia (2005) concluded that the study shows farmers possessed maximum knowledge about soil and soil preparation and spacing while minimum knowledge reported in plant protection and seed treatment about improved ber production technology.

The knowledge level of farmers about improved production technology of mandarin cultivation was measured with the help of knowledge test developed by Mohammad (2000) which was used with slight modification by the researcher for the study purpose. Eleven technologies/practices having 25 questions were included in the schedule after discussion with Horticulture scientists and extension personnel's to test the knowledge level of the respondents regarding improved production technology of mandarin cultivation. The range of knowledge score were from zero to 50. The knowledge score were assigned according to the performance of respondents in the knowledge test. One mark was given to every right answer and zero mark for every wrong answer. This range was divided into three categories based on the mean (30.09) and standard deviation (3.99) as presented below.

(1) The farmers who obtained knowledge score upto 26.10 were categorized as having low knowledge level.
(2) The farmers who obtained knowledge score from 26.11 to 34.08 were categorized under medium knowledge level.
(3) The farmers who obtained knowledge score above 34.08 were categorized under high knowledge level.

The statistical data regarding the knowledge level of farmers about improved production technology of mandarin cultivation have been presented in Table 1. The data in Table 1 concluded that the 69.00% of mandarin growers were having medium knowledge level, 20.00% of farmers having high knowledge level and 11.00% farmers having low knowledge level about improved production technology of mandarin cultivation (Yadav, 2006).

It was also evident that all the farmers were having knowledge about "varieties", this might be due to the fact that majority of the (33%) farmers were literate to highly educated in the study area and hence may know the latest varieties by reading the related literature and also they remain in contact with neighbors, friends, progressive farmers etc. Hence, they may discuss about the latest varieties with the fallow farmers. Hence they had good knowledge about the varieties of mandarin. The farmers had low knowledge about "manure and fertilizers" and "plant protection measures" this might be due to the reason that the farmers might have not understood the instruction written on the container of chemicals because of its complex language they might also not get proper technical guidance about these aspects. It might also due to fewer contacts of farmers with plant protection specialists. The findings of the study are in conformity of

Table 2. Distribution of farmers under different adoption categories towards improved production technology of mandarin cultivation.

S/N	Categories of adoption level	No of farmers	Percent of farmers
1	Low adoption (Score upto 28.51)	15	15
2	Medium adoption (Score from 28.52 to 34.59)	64	64
3	High adoption(Score above 34.59)	21	21
	Total	**100**	**100**

X 31.55, σ = 3.04.

Agarwal (2000), Jangid (2001), Meena (2004), Mohammad (2000) and Jhajharia (2005).

EXTENT OF ADOPTION OF IMPROVED PRODUCTION TECHNOLOGY OF MANDARIN CULTIVATION

Nainawat (1990) reported that 57.84, 83.43, 50.00, 58.43, 80.62, 53.31 and 4.12% of the respondents adopted the recommended improved variety, spacing, soil treatment, manure and fertilizer application, irrigation, intercultural operations and plant protection measures, respectively. Urede et al. (1991) observed that majority of respondents (87.80%) planted mango orchard followed by Indian jujuba (44.66%) on their farm. Only 4.00% farmers have planted pomegranate orchard. Chiprikar and Khuspe (1992) reported that a majority of grape growers were medium to low adopters of grape production technology.

Sen (1995) found that the maximum percentage of adoption of production technology of guava was about planting (70.37%). Manjula (1996) found that over all adoption of farm women about recommended mulberry cultivation practices revealed that 39.58% of the farm women had medium level of adoption with respect of recommended mulberry cultivation practices. Thakare et al. (1996) stated that the majority of orange growers had medium level of adoption of recommended orange cultivation practices (60.50%), while 25.50% of the respondents were in low level of adoption and only 14% of respondents were having high level of adoption. Bhople et al. (1996) found that 62% of the orange growers had adopted recommended cultivation practices of orange to medium extent. Only 16% of them were found to be higher adopters. It was also noted that 22% of orange growers were found to be low adopters. Mohammad and Punjabi (1997) found that maximum adoption gap was reported in propagation (35.56%) this was followed by plant protection measures (22.23%), planting (11.86%) and harvesting and marketing (8.19%).

Agarwal (2000) reported that adoption of improved cultivation of pea by the farmers was not upto mark in soil treatment (13.50%), fertilizer application (19.57%), weed management (31.66%) and plant protection measures (33.20%), whereas, they had high adoption regarding recommended seed rate (88.00%) and recommended spacing (66.93%). Mohammad (2000) found that maximum adoption of guava production technology was reported in varieties (100%) pit digging of recommended size plant sampling in orchard (89.44%), irrigation interval as per the suggested irrigation schedule (88.88%) and scientific method of irrigation for higher water use efficiency (83.33%). On other hand, comparatively less adoption was found in the use of plant protection measures as per the advice of experts (33.83%) grading of fruits after harvesting (36.11%) and use of suitable method of propagation (10.55%) among the guava growers. Meena (2001) observed that a significant difference was reported in adoption between trained and untrained respondents regarding all the practices of onion production technology. Meena (2002) reported that 64% tomato growers were in the medium adoption group and 21% farmers were in the high adoption group, while only 15% farmers were in the group of low adoption. Poonia (2002) found that 60% respondents had medium level of adoption and 22.50% respondent having low and 17.50% having high level of adoption of recommended kinnow production technology. Meena (2004) was found that 88.83% of guava growers found to be medium adopters and 25.00% farmers were high adopter and 16.67% of the respondents were low adopters. Jhajharia (2005) concluded that farmers had high adoption level for the practices viz., harvesting, soil and soil preparation and low adoption practices are plant protection and weed management in improved ber production technology.

Based on adoption score obtained by the farmers, the mean 31.55 and standard deviation (3.04) were computed for the purpose of classifiying the adoption level in to three categories namely low level, medium level and high level adopters. In this way the farmers were categorized into three groups as follows:

(i) Farmers who obtained adoption score upto 28.51 were categorized as low adopters.
(ii) Farmers who obtained adoption score from 28.52 to 34.59 were categorized as medium adopters.
(iii) Farmers who obtained adoption score more than 34.59 were categorized as high adopters.

The statistical data regarding the extent of adoption categories towards improved production technology of mandarin cultivation by the farmers have been presented in Table 2. About 64.00% of farmers were found to be

Table 3. Extent of adoption of improved production technology of mandarin cultivation by the farmers.

S/N	Package of practices	MPS	Rank
1	Soil and soil preparation	80.33	III
2	Varieties	88.67	I
3	Propagation	55.33	VI
4	Plantation	83.44	II
5	Hoeing and weeding management	56.33	V
6	Irrigation management	41.50	IX
7	Inter-cropping	45.67	VIII
8	Manure and fertilizers	36.33	X
9	Training and pruning	51.00	VII
10	Plant protection measures	19.67	XI
11	Harvesting and marketing	76.55	IV
	Over all	**57.43**	

medium adopter, while 15.00% farmers were low adopter and 21.00% of farmers were high adopters (Table 3). The average extent of adoption of improved production technology of mandarin was 57.43%. Out of eleven selected practices, the extent of adoption of "varieties" was 88.67 MPs which were higher than the other adopted practices of mandarin cultivation. The adoption of "plantation" (83.44 MPS) and "soil and soil preparation" (80.33 MPS) recorded second and third position, respectively after adoption of the "varieties" (Table 3). Hence, it may be inferred from the above results that the respondents were very conscious about the adoption of "varieties" of mandarin whereas, they were least bothered about the adoption of "manure and fertilizers" and "plant protection measures" due to lack of knowledge in plant protection and manure and fertilizers.

It was found that majority of the respondents 64.00% were medium adopter, 15.00% were low adopters and 21.00% of the respondents were high adopter of improved production technology of mandarin cultivation. It might be due to the fact that various extension activities like demonstration, training etc. are frequently organized by the extension field functionaries in the villages, which may have helped in convincing the farmers about the improved production technology of mandarin cultivation, which have resulted in increasing the adoption of improved production technology of mandarin cultivation, but still there is an increasing recognition of the need to convert the medium adopters into high adopters and low adopters into medium adopters, therefore, all the essential supplies and services for transfer of technology through extension activities should be made available to the farmers and intensive efforts by all concerned to convince the farmers about improved production technology of mandarin cultivation is needed. In case of extent of adoption of improved production technology of mandarin cultivation, the farmers had appropriately adopted the recommended "varieties", "plantation", "soil and soil preparation" and "harvesting and marketing". Such practices were adopted by farmers as they had good return due to adoption of these technologies/practices. Also these technologies/practices neither require any extra investment, nor complicated in using the probable reason for medium adoption of "hoeing and weeding management", "propagation", "training and pruning", "inter cropping", "irrigation management" may be due to lack of knowledge about these technologies/practices and lack of proper technical guidance provided to them by the state department of Agriculture/Horticulture.

The probable reason for low adoption of "manure and fertilizer" and "plant protection measures" may be due to high cost of plant protection chemical and plant protection equipment, fertilizers and weedicides. The untimely availability of fertilizers, lack of credit facility, and lack of knowledge about irrigation management may also be the other reasons, for low adoption of these practices. The low adoption of plant protection measures might also be because of the non availabilility of latest plant protection equipment at farmer's field, their hazardous effect on human beings and animals, harmful residual effect and lack of any subsidy provision. The findings of the present study are in line with the findings of Manjula (1996), Meena (2004) and Jhajharia (2005).

MAJOR CONSTRAINTS

Lokhande and Wangikar (1991) observed that about 79.16% respondents expressed the problem about unavailability of cuttings of improved varieties in time, 76.66% of the respondents did not receive know how about improved varieties of grape, while 70.83% respondents expressed the problems about less resistance of varieties to diseases. Urade et al. (1991) reported difficulties in successful plantation for want of technical guidence, untimely supply of information and difficulty in procurement of bank loan, non-availability of protective irrigation and trampling by stray cattle. Difficulty in selection of dry land fruit crop, lack of knowledge of pests and their control measure were also reported as problems by them. Pandey (1993) reported that in cultivation of mango, guava, pomegranate, aonla and other fruit plants, there was a great problem of unavailability of reliable plant material. Plants were purchased indiscriminately from nearby states, from private nurseries that have no pedigree records of their mother plants. He further added that many of the plants die during transit period.

Singh (1993) reported that in case of papaya, mosaic disease had restricted the farmers for cultivation of this profitable fruit crop having tremendous potentiality. Non availability of seeds of reliable varieties to the common cultivator was a great problem. Bhople et al. (1996) found that orange growers faced marketing constraints like costly packaging material (80.00%), non availability of processing units (76.67%) and high transportation charges for transport of fruit from orchards to nearby market (54.90%). Sharma (1997) reported that provision should be made for adequate and timely supply of essential inputs such as grafted mango plants of regular bearing varieties, fertilizers etc.

He further added that timely finance and credit facilities should be provided to mango growers at low interest rate. Sutar et al. (1997) observed that the constraints in adoption of selected scientific technologies in grape cultivation are the high cost of cuttings of required varieties and fertilizers as well as non availability of improved varieties, fertilizers and credit to overcome these constraints in the adoption of modern grape cultivation.

Handiganur et al. (1998) revealed that 90% farmers expressed the problem of scarcity of water and non-availability of labour were the main problems viewed by 67.50% of the farmers expressed the problems of severity of pests and diseases. Shrestha et al. (1998) found that the lack of technical know-how was the main constraint for trying the technology followed by low production and low economic status for cellar store in mandarin orange fruit. Gomase and Patil (1998) revealed that major constraints perceived by the kagzi lime growers that inadequacy of irrigation water, irregular power supply, non availability of labour, lack of knowledge of insect pest and disease and its control measures and high wages of labour.

Devi and Monoharan (1999) revealed that low price obtained for the produce in the market, lack of quick transport facilities, lack of storage facilities, non availability of middle men where the major constraints faced by the guava cultivators that high cost of fertilizers, soil and water problems, non availability of credits and lack of input availability are the area where guava farmer requires the external agency assistance. Mohammad (2000) concluded that high intensity constraints regarding aspects of orchard development with their relative position in the hierarchy as expressed by the orchard owners were lack of technical knowhow, long juvenile period, perishable nature of commodity, a high cost of establishment of orchards, lack of storage facility in the area, malpractices of the middle men, problematic soil, erratic rainfall, sub division and fragmentation of land and lack of need based training programme.

Poonia (2002) found that lack of technical knowhow, irregular water supply from the canal, perishable nature of commodity, high cost of insecticides and pesticides, lack of storage and preservation industry in the area and unfavourable weather condition, were the constraints in the adoption of kinnow orchard by the owners. Rajput et al. (2002) reported that the major constraints felt by the most of the respondents includes non availability of soil turning plough, lack of soil treatment, im-balanced use of fertilizer, infestation of several insect and diseases and lack of good quality of pesticides. Sale of cotton (produce) through mandi may be encouraged to measure producers share in consumer rupee. Meena (2004) reported that majority of cabbage growers were experienced the constraints such as lack of knowledge, high cost of inputs and lack of skill regarding the adoption of improved cabbage cultivation practices.

Waman and Girase (2003) it was found that the present study implied that sent percent pomegranate growers had adopted grading and packing of fruits in card board boxes of standard sizes. Majority of the growers preferred to sale fruits in other states that selling of fruits in local market to fetch attractive prices. They faced many problems in marketing, it is recommended that the government may establish markets at nearly place and to create infrastructure facilities to preserve and process the fruits.

This may enable them to fetch remunerative prices, stabilize rates, to reduce transformation charges and to assure timely receipt of payments. Jhajharia (2005) it was revealed that among all the categories of constraints, technical constraints were perceived with highest intensity by the ber growers.

The constraints “no preservation industry” (97.33 MPS) was the most perceived constraints among all the constraints faced by the farmers which was responsible for the adoption of improved production technology of mandarin cultivation. Hence it was awarded first rank. The constraints “reluctance for fruit growing” (31.00 MPS) was the least perceived constraint by the farmers and was ranked at last position (Table 4). The problem “no preservation industry” might be due to the fact that the unawareness of people, no proper guidance and avoidness policy of govt. and therefore, it was not possible to establish the industry without awareness of people and economic help by Govt. and agriculture department in their operational area that’s why these constraints were reported.

Table 4. Constraints perceived by the farmers in adoption of improved production technology of mandarin cultivation in the study area, N = 100 (Multiple response).

S/N	Constraints	MPS	Rank
A	**Technical constraints**		
1	Long vegetative period	36.33	V
2	Lack of regular water supply	34.33	VI
3	Poor quality of water	33.00	VII
4	Lack of technical know how	59.67	I
5	Unsuitable soil for orcharding	38.33	IV
6	High mortality of plant during initial stages	57.00	II
7	Disease sensitive	45.33	III
B	**Economical constraints**		
1	Perishable commodity results in economic losses	95.67	I
2	Labour intensive affairs	45.33	VII
3	High cost of transportation of fruit and plant material	49.67	VI
4	Lack of credit facility in the area	52.00	V
5	High cost of establishment at the initial stage	60.33	IV
6	High cost of insecticide and pesticides	94.67	II
7	Costly plant material	92.33	III
C	Storage and marketing constraints		
1	Lack of storage facility	96.33	II
2	No preservation industry	97.33	I
3	Lack of co-operative marketing system	53.00	V
4	Lack of proper market	42.67	VI
5	Low purchasing power of the consumer	39.00	VII
6	General unawareness about mandarin in product and their nutritional importance	69.67	IV
7	High fluctuation in market prices	95.67	III
D	**General constraints**		
1	Small land holding	32.67	V
2	Undulated land	32.67	V
3	Threat from wild and stray animal	59.33	II
4	Chances of theft	61.33	I
5	No proper road	50.33	III
6	Jealously of neighbours	40.33	IV
7	Reluctance for fruit growing	31.00	VI

CONCLUSION

Majority of the farmers had medium knowledge about the improved production technology of mandarin cultivation. Among the various aspects of different improved production technology, all the farmers had knowledge about plantation, "soil and soil preparation", "harvesting and marketing", "hoeing and weeding management", and propagation". Majority of the farmer had medium adoption about the improved production technology of mandarin cultivation. The knowledge of the farmers about improved production technology of mandarin cultivation was positively and significantly associated with their caste, occupation, educational level, social participation, size land holding while their market distance and irrigation potentiality were negatively and non significantly co-related with their knowledge level about mandarin cultivation. The caste, occupation, educational level, source of information utilized, social participation and size of land holding were the important variables for predicting the knowledge level of farmers about improved production technology of mandarin cultivation. The extent of adoption of improved production technology of mandarin cultivation by the farmers was positively and significantly related with caste, occupation, educational level, irrigation potentiality, source of information utilized and knowledge about mandarin cultivation while market distance was negatively and significantly co-related and social participation and size of land holding were non significantly co-related with their extent of adoption of improved production technology of mandarin cultivation. The caste, occupation, educational level, market distance, information source utilized and knowledge about mandarin cultivation were the important variables for predicting the extent of adoption of improved production technology of mandarin cultivation.

REFERENCES

Agarwal JK (2000). "Knowledge and adoption of improved pea cultivation practices in Jaipur district of Rajasthan". M.Sc. (Ag.) Thesis, RAU, Bikaner, Campus-Jobner.

Anonymous (2004). "Vital Horticulture Statistics of Rajasthan". Dir. Agric. Jaipur.

Anonymous (2008). "Vital Horticulture Statistics of Rajasthan. Directorate of Agriculture, Jaipur.

Anonymous (2009). "Vital Horticulture Statistics of Rajasthan. Directorate of Agriculture, Jaipur.

Arias BA, Laca RL (2005). Pharmacological properties of *Citrus* and their ancient and medieval uses in the mediterrnean region. J. Ethnopharmacol. 97(1):89-95.

Arneja CS, Khangura RS (2003). "A study of the knowledge of pea growers and its association with their selected socio personal characteristics". Rural India 66:166-168.

Bhople RS, Shinde PS, Nimje VR (1996). "Production and marketing constraints faced by orange growers". Mah. J. Ext. Edu., 15:57-62.

Cameron JW, Frost HB (1968). Gentics, breeding and nucellar embryony. *In:* The citrus industry II. W. Reuther, L. D. Batchelor and H. J. Webber California Univ. Press, Berkeley, pp. 325-370.

Chiprikar BP, Khuspe TS (1992). "A study of grapes grower with special reference to transfer of grape technology". Mah. J. Ext. Edu. 11: 314.

Choudhary H, Bangarva GS, Singh N (2011). "ADOPTION of recommended kinnow production technology. Raj. J. Ext. Edu. 19:190-193.

Dayama MC (2000). "Factors affecting adoption of improved technologies of aonla plantation in chomu tehsil of district Jaipur (Rajasthan)". M.Sc. (Ag) Thesis, RAU, Bikaner, Campus-Jobner.

Deshmukh PR, Wangikar SD, Wakle PK (1998). "Knowledge and adoption of recommended cultivation practice of custard apple". Mah. J. Ext. Edu. 17:136.

Devi MN, Manoharan M (1999). "Strategies for enhanced guava production". Agric. Ext. Rev. 2(2):11-12.

Dugo G, DiGiacomo A (2002). Citrus . The Genus *Citrus*: P. 642.

Gomase AS, Patil RL (1998). "Adoption behaviour of kagzi lime growers". Thesis abstract 24(1):201-202.

Guerra M (1984). Cytogenetics of Rutaceae. II. Nuclear DNA content. Caryologia 37:219-226.

Handiganur SS, Sastry KNR, Vishweshwar SP (1998). "Problems conforting the procedure about production and processing grape". Rural India 61:137-139.

Imbesi A, Pascuale A. de (2002). *Citrus* species and their essential oils in traditional medicine. *In:* Giovanni, D. and Giacomo, A. de. *Citrus*: the genus citrus. CRC Press ISBN 0-415-28491-0: 577.

Jangid NL (2001). "Analysis of constraints in the adoption of recommended cultivation practices of chilli cultivation by the farmers of Panchayat Samiti Sambhar Lake of Jaipur district (Rajasthan)". M.Sc. (Ag.) Thesis, RAU, Bikaner, Campus-Jobner.

Jhajharia P (2005). "Knowledge and adoption of improved production technology of ber cultivation by the farmers in Chomu tehsil, Jaipur district of Rajasthan. "M.Sc. (Ag.) Thesis, RAU, Bikaner, Campus-Jobner.

Lokhande VD, Wangikar SD (1991). "Constraints in the adoption of selected scientific grape technologies". Mah. J. Ext. Edu. 10(2):138.

Mabberley DJ (2004). Citrus (Rutaceae): A review of recent advances in etymology, systematics and medical applications. Blumea 49:481-498.

Manjula N (1996). "Adoption of mulberry cultivation practices of farm woman". Rural India, pp. 257-259.

Meena ML (2001). "Impact of National Horticulture Research and Development foundation in adoption of onion production technology among the farmers of Kishangarh Bas Panchayat Samiti of Alwar district, Rajasthan". M.Sc. (Ag.) Thesis, MPUAT, Udaipur.

Meena RK (2002). "Adoption of improved cultivation practices of tomato by the farmers of Bassi Panchayat Samiti of Jaipur district (Rajasthan)." M.Sc. (Ag.) Thesis, RAU, Bikaner, Campus-Jobner.

Meena V (2004). "Knowledge and adoption of improved technology of guava plantation in Sawai Madhopur district of Rajasthan". M.Sc. (Ag.) Thesis, RAU, Bikaner, Campus-Jobner.

Mohammad A (2000). "Problems and prospects of orchard development in Hadauti region of Rajasthan state". Ph. D. Thesis MPUAT, Udaipur.

Mohammad A, Punjabi NK (1997). "A study on knowledge and adoption of improved cultivation practices of mandarin among farmers in Jhalawar district of Rajasthan". M.Sc. (Ag.) Thesis, Thesis, RAU, Bikaner, Campus- Udaipur.

Moore GA (2001). Oranges and Lemons: Clues to the taxonomy of *Citrus* from molecular markers. Trends Genet. 17:536-540.

Nainawat RK (1990). "Factors affecting adoption of improved technology of Ber plantation in Jaipur district of Rajasthan". M.Sc. (Ag.) Thesis, RAU, Bikaner, Campus-Jobner.

NHB database (2012). National Horticulture Board (NHB) Government of India – an autonomous society under the Societies Registration Act 1860. Institutional Area, Sector 18, Gurgaon (Haryana).

Nimje NR, Kulkarni VR, Choudhary DP (1991). "Knowledge and skill about ber cultivation practices among farmers". Mah. J. Ext. Edu. *10*(2):108.

Pandaria RN, Singh G, Singh AK (2002). "Validation and refinement of technology for crop regulation in guava". Ind. J. Ext. Edu.38(3,4):117-124.

Pandey RR (1993). "Planting material". Indian Horticulture, pp. 14-17.

Phuse AP, Vitonde AK, Thipse CD (2007). "Adoption of recommended mandarin orange production practices. Ind. Res. J. Ext. Edu. 7 (2&3):98-100.

Poonia A (2002). "Technological gap among the Kinnow (*Citrus deliciosa)* orchard owners in Sriganganagar district of Rajasthan". M.Sc. (Ag.) MPUAT, Udaipur.

Rajput AM, Saxena KK, Singh RL (2002). "Production performance of cotton varieties and economic constraints in adoption of cotton production practices on tribal farms in Madhya Pradesh". Ind. Res. J. Ext. Edu. 1:72-75.

Ramchandarn PK (1974). "A multivariate study on information sources utilization of big medium and small farmers". Ph.D. Thesis, Div. of Agril. Ext. IARI, New Delhi.

Sen R (1995). "Problems and prospects of orchard development in Jhadol Panchayat Samiti of district Udaipur (Rajasthan)". M.Sc. (Ag.) Thesis, MPUAT, Udaipur.

Sharma DD (1997). "Constraints in adoption of recommended mango cultivation practices by growers". Mah. J. Ext. Edu. 16:362.

Sharma SC, Bangarwa GS (1997). "Level of knowledge and attitude of farmers towards tree plantation". Ind. J. Soc. Res. *38(*7-9):163-166.

Shrestha B, Subedi PP, Thapa JJ (1998). "Socio-economic factors in adoption of cellar store as a post harvest technology for mandarin orange in western hills of Nepal". Lumle Agric. Res. Centr. 98:34-54.

Shrivastva, Singh (2003). "Establishing soil constraints-free mandarin orchards". Indian Horticulture, pp. 48(1)10-13.

Singh SP (1993). "Integrated pest management in horticulture crops". Indian Horticulture, pp. 25-28.

Srivastava KK, Trivedi MS, Lakhera ML (2002). "Knowledge and adoption behaviour of chilli growers". Agricul. Ext. Rev. 14:22-25.

Sutar AM, Deotale AB, Ingle LA, Awn AH, Joshi DA (1997). "Constraint in adoption of related scientific grape technologies". J. Soil Crops 7(1):59-62.

Swingle WT, Reece PC (1967). The botany of *Citrus* and its wild relatives. *In*: The Citrus Industry, Vol. 1. (eds Reuther, W., Webber, H. J. and Batchelor, L. D..), Univ. of Calif., Berkeley, pp. 190-430.

Thakare PV, Chikale NJ, Deshmukh PV (1996). "Factor affecting adoption of orange technology by growers". Mah. J. Ext. Edu., 15:176-178.

Trivedi G (1963). "Measurement and analysis of socio-economic status of rural families". Ph.D. Thesis, Division of Agric. Extn., IARI, New Delhi.

Urade PN, Bhople RS, Choudhary DP (1991). "Adoption of dry land horticulture technology". Mah. J. Ext. Edu. 2:108.

Waman GK, Girase KA (2003). "Constraints analysis in marketing of pomegranate". Rural India. 66:43-45.
Yadav BC (2006). A study on knowledge and adoption of improved production technology of mandarin by the farmers in Jhalarpatan panchayat samiti of Jhalawar district of Rajasthan. M.Sc. (Ag.) Thesis, S.K.N. College of Agriculture, Jobner.

An econometric analysis of agriculture-growth nexus in the Gambia: 1966 to 2009, it is more people per acre and fewer yields per acre, and the reverse should have been the case

Bukhari Sillah

Department of Economics, College of Business Administration, King Saud University, Saudi Arabia.

Agriculture is a major engine for the economic well-being of the Gambia. The successive governments from the colonial periods to the present have all recognized this importance of the agricultural sector. But most of their policies could not bear any fruit, because they were working against natural laws. They, in many instances encouraged slash and burn technology and putting increasingly more people onto the land instead of increasing the yields per acre. More people than before do remain still on the land, but they are now poorer than their forefathers. Their forefathers had one person to an acre producing on average 1100 kg per acre. Today, there are two persons to an acre producing the same 1100 kg per acre. Using autoregressive and vector error correction techniques to examine the growth-agriculture relation in the Gambia for the period from 1966 to 2009, it is found that the capital per worker is a significant and relevant factor input for the economic growth. The agricultural labor per acre is found to be irrelevant in both the short run and the long run analyses. The agricultural productivity measured as crop yields per acre is the most important variable for the economic growth in the Gambia. It boosts both the economic growth and the capital formation in the country. The agricultural policies should be focused on increasing the crop yields per acre not having more people back to the land. The processing, services and small manufacturing sectors should be developed and built into the agricultural policies in order to create redeployments for the agricultural labor surpluses.

Key words: Gambia, colonial era, agricultural policies, land, labor, acre, crop yields.

INTRODUCTION

The 'governments' of the Gambia, in the colonial era, in the 'first republic' and in the 'second republic', have all endeavored relentlessly to improve the livings of the poor agricultural farmers of the Gambia, who constitute more than 70% of the population. The situation for more than a century now is far from desirable. Poverty still ravages more than 60% of the agricultural farmers. In each instance, an intervention was introduced – and the intervention was often an initial government investment, or a donor-driven investment or both – the situation improved and promised better living for the agricultural farmers; and when the intervention was removed, by demanding the project to be self-sustaining, the situation deteriorated and became worse than pre-intervention era, causing havoc in the social fabrics and economic foundations of the agricultural farmers. The result is massive rural urban migrations and increasing number of illegal migration of able-bodied young men to Europe and other parts of the world. The few agricultural farmers left behind do subsistence farming to complement the remittances to sustain women, children and elderly. Those who have surplus remittances would erect structures and engage in distributive trade. If no further intervention is mounted, the agricultural industry will not take off, and assume that the remittances are not sustained increasingly - which could be done by increasing the number of new and affluent migrants to replace the old migrants, who have become emptied out of able-bodiness or got assimilated into new found lands - many of the farming communities will become ghost villages, the country may become one city reeling with unemployment and poverty.

This study believes that agricultural industry in the Gambia is just like that of Japan and Europe is sacrosanct and untouchable, therefore it should be supported as long as the people of the Gambia continue to eat and drink. This study is organized as follows: subsequently, the background is given; thereafter, reviews on some relevant literature; then it discusses the methods of estimation; analyzes the findings; and derives the conclusions and policy recommendations. The study employs 'error correction method' to examine both long run and short run relations between the agricultural variables and per capita GDP; it employs the vector error correction technique to estimate the long run co-integrating vectors. It examines the Granger causality effects between the agricultural variables and the per capita GDP. Agriculture has been held to be the mainstay of the Gambian economy, unfortunately no scientific research has been undertaken in order to formulate evidence based policies. This study expects to contribute significantly to filling in this gap in the agricultural intervention projects, which are widely awash in the country.

BACKGROUNDS

Here, we discuss some relevant efforts and interventions introduced by different governments in order to try to improve the living of the farming communities. We achieve this by examining agriculture in the pre-independence, the First Republic and the Second Republic.

Agriculture in the pre-independence

Trading in agricultural produce with international world was introduced by the British in 1820[1] in response to the high demand in North America and Europe for the groundnut oil, which was used for cooking oil, industrial oil, lubricants and production of soap and candles. The nut itself was introduced by the Portuguese in the sixteenth century. The farmers responded to the policy in order to obtain the colonial money to pay taxes and purchase the imports. By 1904, the welfare of the country was solely dependent on this groundnut, a situation that raised a concern of the government as they state in the 1904 Report, "It is a difficult matter in a small colony like the Gambia to deal year after year with question of agriculture. The groundnut is the main product, and the export has so often been described that there is nothing fresh to add beyond once again stating that the welfare of the colony depends entirely on this crop. The people devote almost their whole attention to it, only growing small quantities of rice, millet, Koos and corn for food".

The agricultural technology was simply slash and burn, and land was not of a constraint, but labor. The population of the country was just 196101[2] in 1911. Labor shortage was addressed by boosting population growth through the strange farmers, who would come from neighboring countries to produce groundnut in the Gambia during the rain season. Their number reached 32, 220 in 1915[3], and they contributed annually on average one third of the groundnut export. The national population rapidly increased to 210530 in 1921 indicating an increase of 44% from that of 1911. The school expansion was restricted in the rural regions to further increase labor on the land. The agriculture industry was formalized with the establishment of Department of Agriculture in 1924. Its main function was to demonstrate species of plants for suitability to the Gambia. The groundnut trade flourished inducing further population growth as the farmers multiplied themselves, and the more strange farmers preferred to stay to avoid the frequent to and fro trips to their homelands. To improve land fertility, farmers practiced two techniques, one is fallow system, whereby the land would be left uncultivated for 15 to 30 years, and the other was to interchange crops in the same land. These techniques were dependent on the availability of abundant arable lands. But as the population kept rising, the fallow duration diminished, and lands were annually cultivated with increasing number of labor. This allowed the diminishing returns to quickly set in; the groundnut yields dwindled and hence the farm incomes which are also used to finance the food purchases, the farmers were not producing adequate food cereals because the groundnut sales were then sufficient to purchase food.

The government response was to subsidize food imports particularly and mechanize the agriculture that was carried out 100% with human labor. They brought animal-powered moldboard ploughs, Emcot ridges and ox carts in 1940. The food insecurity continued to haunt farmers and the government, and food import subsidy became a costly burden on the government budget. This forced the government to try import substitution of rice, which became a staple diet for the average Gambian. The Colonial development Corporation enclosed an area of 9600 ha for the rice irrigation project, which involved canalization of leveled fields to be fed by the tidal flow of the river Gambia. Mechanical traction was introduced to rice production. Due to poor canalization system, less than 10% of the hectares planted survived the flooding. The project also converted rice farmers into low wage laborers, a situation which led many to shun it. It later changed to sharecropping, which landed the farmers into accumulated debts as the groundnut yields and prices fell, and farmers turned the rice project into subsistence farming; thus, the input credits could not be repaid. As the land productivity further diminishes, the food security worsens with increasing population, which was then growing at a rate of 19.3%, on average, per annum from 1921 to 1963[4], and consequently the demand for credit to purchase farm inputs, such as seeds, fertilizer and implements kept increasing. Also, the funds were not easily available to purchase the groundnut, in some instances produce were bought on credit, further complicating agricultural life for the farmers.

Two years before the colonial government handed the reign to the self-rule, they state "reports of experts received during the period have suggested that one of the main factors retarding the agricultural economy, in spite of the efforts of the Government's agricultural services with its long experience of farming in Gambia conditions, has been shortage of credit. Hopes of filling this gap lie largely in the co-operative movements; and these movements lead to the formation of Gambia Central Co-operative Banking and Marketing Union"[5]. Finances did not get better as agriculture continued to be dependent on government direct support, and the new independent Gambia was born on 18th February 1965 inheriting a fledging agricultural sector.

Agriculture in the first republic

The First Republic is the government that led the country to independence from the British and continued to reign until 1994 when it was overthrown through a military takeover. The British were in hurry to leave the Gambia as it did not represent any economic interest, but rather an economic draining. The Gambia Central Co-operative Banking and Marketing Union (later transformed into Gambia Produce Marketing Board) became a self-sustaining entity, purchasing the produce on cash and giving credits to the farmers to improve the production. By 1970, the first republic embarked on economic plans, calling for increased self-sufficiency, stabilized producer prices and expanded government agricultural extension services. A newly independent country yearning for other non-independent countries to be independent, the Gambia found Taiwan a sister in the struggle, and hence diplomatic relations were established. The latter responded to the call of the Gambia by sending technical missions to try the rice mechanization. In 1966, Taiwan developed 1200 ha of rice irrigation scheme water fed by diesel powered pumps. Fertilizers, seeds and fuel were provided free to farmers in the first year, tasking the farmers to be self-sustaining in the following years. The scheme faced technical difficulties, and was later taken over by personnel from the Mainland China in 1974 as the Gambia changed the diplomatic ties. The Chinese continued the work of the Taiwanese missions, and by 1980 less than half of the project area started by Taiwan was cultivated. The project resulted in high credit burdens on farmers, who had to shoulder the financings burdens of the project during dwindling groundnut yields and the 1970 to 1977 Sahelian droughts. The government cry for international help was heeded of by the World Bank, which in an attempt to combat food insecurity consolidated the Taiwanese irrigation scheme of 1200 ha, but maintained farm credit, further worsening the credit status of the farmers.

The situation could not be sustained, as farmers started abandoning the projects that would impound their produce sales for loan repayments. The rice project was abandoned and groundnut sales fell. The government had then no choice but to cancel and forgive all loan repayments in 1979 and 1980. But food insecurity did not go away, in fact, worsened by the droughts; and the government finances were strained. Again, food import substitution of rice was an option. The government supported by donors embarked on heavy mechanized rice irrigation called Jahally-Pacharr smallholder Rice Project in 1983, with a development cost of $46500[6] per hectare. It could only survive effectively for five years. Again, similar reasons for the failures were repeated, loan repayment difficulties (subsistence farmers were asked to pay capitalist loan repayments, which translated into asking a non-market to pay a market price), technical failures and fuel shortages. From the drought episodes of 1970 to 1977 and the failed ambitious rice projects combined with price controls to temper the food insecurity and stabilize the groundnut producer prices and subsidize farm inputs, the government continued to accumulate toxic assets on its budgets, which were found unsustainable. In 1985, it embarked on cleaning up its own finances, which was termed economic recovery programme. It removed subsidies, floated the prices and curtailed the public expenditures particularly in the areas of agricultural extension services. It removed taxes on groundnut exports, and encouraged private sector participation in the agriculture. It failed to understand that private sector seldom invests in industries characterized by diminishing returns and undifferentiated products. The lands have become too marginal due to traditional land tenure structures, where the lands were continuously split up to accommodate additional members as the population kept rising. The farmers were then farming on increasingly smaller and smaller plots of lands, which after some time had to be cultivated all year round. Thus, production increase demanded increasingly more farm improvements, which became costly to the farmers. Food insecurity continued to haunt the farmers, and hence more resources than before were allocated to food imports rather than capital accumulation, a phenomenon that in the long run would further impoverish the country.

In 1993, the government commissioned one Michael Roemer from the Agency for International Development to examine industrialization strategies that could bail out the country from its long protracted poverty and place it on a path of sustainable development and growth. He examined three industrialization options for the Gambia, and tried to look for a model country in the South East Asia or elsewhere. The options were resource based industrialization, which would involve exporting what the country has from land, water and tourism, then moving slowly to the second option of manufacturing goods that would substitute for the re-exported goods, and working towards positioning the Gambia as all open economy and haven for the investors or a managed openness. He concluded by stating that "the keys to rapid income growth and reduction in poverty are rapid and sustained growth of exports, especially of manufactures, and of productivity, together with high saving and investment ration". That is, agriculture cannot be, and has never been, a long term sustained source of income growth and poverty reduction. The first Republic did not have much time to digest Michael's proposal as it was toppled the following year in a military takeover.

Agriculture in the second republic

The second Republic expanded and revitalized the frontiers of agriculture that were earlier opened by the previous regimes. The frontiers were geared towards battling the long term enemies of the country, food insecurity and poverty. The solution, as can be deduced from the practice of the regime lies in "Back to Land", and "Operation Feed the Nation" through intensive mechanization to increase the production of cereals, rice and groundnut. In 1996, it imported 20 four wheel tractors, and before 2001 more than 70[7] tractors were imported and given to the farmers, and currently there are more than 500 tractors in the country for the agriculture.[8] Relations were re-established with Taiwan, which quickly dispatched its technical missions to the Gambia to revitalize the rice irrigation works, which used to be performed by the Taiwanese more than 21 years ago. Fertilizers, farm inputs and more tractors were imported and offered for sale to the subsistence farmers. The "Back to Land" clarion was somewhat responded to by the people as the cultivated land increased from 0.273 ha per person in 1994 to 0.323 ha per person in 2001, an increase of 18.39% in arable land cultivation. The yields increased from 1122 kg/ha in 1994 to 1230 kg/ha in 2001. The mechanization of agriculture still did not go far, as some farmers are of the view that tractors are not properly designed for the soil, which is too light, and tractors often over pulverize it resulting in erosion and loss of valuable topsoil. Kuye et al. (2006) find that as of 2002 only 1.7% of farm work was carried out by mechanical power; the majority of farm work, 73.4% was done with animals. Food security did not get better as food prices kept rising, the price of the stable diet, rice, has more than tripled in the span of 1994 to 2001, while the farm income measured by the groundnut price has kept crawling and sometimes declining. The country continues to import on average 66,000 tons[9] of rice annually. The 2003 Census has found that more than 50% of the Gambians live on less than a dollar a day.

The National Nutrition Agency has revealed that more than 50% of people in the affluent administrative regions, the Capital City Banjul and the Kanifing Municipality are suffering from mild to severe food insecurity[10]. The Government is weighing options again to increase rice production; a new design of farmer managed rice irrigation is underway to minimize pump irrigation schemes, which have often placed unbearable technical and other running costs on the farmers. To conclude this background review, we state that agriculture as a source of food security and foreign exchange still remains a distant dream for the Gambia after more than 100 years of struggle. The pre-independence failed to practice land enclosures that spurred agricultural industrialization in Europe; lands in the Gambia became too marginal with the increasing population as the lands were split up into numerous ownerships. Cash crop, groundnut, became the only hope for the farmers to modernize their living standards, because that was offering a small opportunity to participate in an exchange economy. Upland cereals were produced for subsistence, and rationally no one wanted to remain in the subsistence economy. Farmers have to then choose an optimal combination between exchange economy and subsistence economy, an equilibrium that was never reached due to increasing population, diminishing returns, droughts and socio-economic repercussions of the failing government interventions, such as rice irrigation schemes and agricultural extensions. The result is constant food insecurity, scary poverty and massive rural urban migration and overseas illegal migration.

The Greater Banjul Area, which has the best arable lands in the country, is being cut into smaller and smaller pieces of brick-wall fenced dwellings; or grabbed for speculation. The overseas illegal migration is costing more lives year after year; the costs are beyond quantity as the lives lost can be never restored. The situation does not call for more people back to land, which is in odd with the modern agricultural trend that has moved away from labor intensive technology to capital one. In the industrialized countries, less than 5% of population works on the land. But with best technology, they are able to feed themselves, accumulate capital, feed the other 95% of the population and ship the excess surplus

Table 1. Change of structural composition of GDP as income rises.

Sector	Low-income countries (%)	Middle-income countries (%)	High-income countries (%)
Agriculture	27	10	2
Industry	30	36	30
Services	43	36	68

Source: kidane et al. (2006).

to us in Africa. More people on land imply poverty specialization; they will on average produce the same undifferentiated things to be sold mostly to themselves and non-farmers who constitute a small portion of the population. This results in falling producer prices, and hence falling farm incomes. As farm incomes fall, no asset can be accumulated because the income will barely meet the food expenses. With an endowment of 550,000 arable lands, adding more labor will quicken the diminishing returns and clog the country deeper into subsistence economy. The country requires fewer labors with high technology on the land, where water, fertilizers and pesticide should be abundantly available almost at no cost to farmers. The excess farm labor should be exported to manufacturing and services industries that require minimum sophisticated education.

The thrust of this paper is to evaluate the effects of agricultural productivity and agricultural labor on the economic growth and economic well-being of the Gambia.

LITERATURE REVIEW

The importance of agricultural sector to the economic well-being, and especially for developing countries cannot be overemphasized. The developing countries, which are the low income countries, have higher agricultural GDP and higher agricultural employment than the high income countries. This can be seen from the structural composition of GDP in Table 1. As the income rises, people spend less on food, and agriculture consists mainly of food, the agricultural GDP tends to fall, and thus farm income left unsupported is destined to fall as the income rises. This becomes even fast occurring in the globalized economy where urban rich can further increase their income level by increasingly consuming cheap and subsidized food from the developed countries. The farmers, who often are also the rural people, will get trapped in the income poverty as they cannot move from the land to industries that rarely exist in the urban areas much more in the rural areas. This falling income will further depress rural trade and investment, because trade and investment flourish where purchasing power is. If the economy cannot move people from land to industries, then it should sustain them on the land to avoid and prevent rural urban migration, which leads to unsustainable pressure on the urban amenities and in turn to unsustainable economic prosperity. This principle of sustainable development and prosperity has been well adopted by the industrialized countries and recently industrialized countries. They introduced a complex array of agricultural support policies for price stabilization, input subsidies and direct income payments to prevent the decline of farmers' incomes.

In the periods of 1986 to 1988, and 1999 to 2001, the producer support equivalents (PSE) in OECD countries stood at 38 and 33%, respectively (Tokarick, 2005). The support was weighed heavily towards the price stabilization, which represented 82 and 70% for the period respectively. These were the same periods when World Bank and IMF were commanding African governments to remove agricultural supports and to adhere to textbook market principles. If it is a matter of textbook economics principles, Switzerland should just rely on other European countries with large scale agricultural economies for food, but no, Switzerland in 2001 had PSE of 69%, the highest in the world. To further prevent the decline of farmers' income, which as we mentioned earlier is natural effect of the rising income, the developed countries have resorted to direct income payments to farmers. In EU, farmers received prices that were 33% higher than the world price in 2001 (Tokarick, 2005). The United States of America has introduced a legislation to subsidize bio-fuel, which among other things, is a policy to directly support farm income; in 2005 farmers received a subsidy of $0.51 per gallon of ethanol blended and used as a motor fuel (Rubin et al., 2008). The European Union Council Regulation 1257/1999, in pursuit of sustainable rural development, has allowed the governments to massively fund the processing and marketing of agricultural products (Trefflich et al., 2007). The programme's objectives which are ideal for any developing country are to: 1) Preserve and provide new jobs in rural areas, 2) give rural people a prospective future and 3) stop and prevent rural urban migration.

The EU has learned that these objectives cannot be achieved without direct government intervention, and thus subsidy has been legislated. It pays to subsidize and support the agricultural sector; the case of the developed countries are clear evidence; the economies prospered with reduced food prices without the same time causing farmers to lose income and get trapped in poverty, because their incomes are supported through a complex sets of agricultural support policies that ultimately end up achieving the aforementioned objectives. This historical fact has missed the textbook economics, which has been dominated by the neo-classical mindset. To bring out some scientific evidence from the historical data, Harley and Knick (2002) applies CGE to 1841 British capitalist agriculture, and finds evidence that, despite its inefficient failure to equate marginal product of labor to various uses, has increased labor and peasant welfare. The agricultural support is to impact favorably on the regional economic growth after controlling for human capital. Subsidy increases the agricultural output and agricultural taxation lowers the per capita income in the economy (Lopez, 2000).

O' Ryani and Miller (2003) using a CGE analysis for Chile find that an increase in the agriculture increases the real income twice as much as compared to industry in the presence of price subsidy. Bednarikova and Doucha (2009) examine various agricultural support policies on certain regions in Czech Republic and find that all the policies have negative impacts except the direct income payment. Some studies, such as Xu (1993), Robinson et al. (1997) have looked specifically at the price support policy and they conclude that this particular policy has a negative impact on the economy.

These studies do help authorities shape certain policies, but have no evidence to make a strong case for or against the agricultural subsidy. In the current paper, we attempt to examine the agricultural impacts on the welfare of the economy. The study is motivated by the desire to provide some scientific evidence for the case of the Gambia, where agriculture has been the backbone of the economy but has recently been marginalized by macroeconomic policy makers.

The recent upsurge of interest in the sector by the government should be guided by some scientific evidence from the economy's data, and hence the thrust of this paper.

ESTIMATION TECHNIQUES AND DATA

Economic theory of production holds that output is a function of capital, labor and technological process. This paper argues that the per capita GDP in the Gambia is a function of capital, labor and agricultural per capita output. The agricultural output captures the economic activity level of 70% of the population and their purchasing power. Thus, the production function is specified in a log form as:

$$\ln Y_t = \alpha + \beta_1 \ln K_t + \beta_2 \ln AL_t + \beta_3 \ln A_t + \varepsilon_t \qquad (1)$$

Where, lnY = log per capita GDP current market prices in USA Dollars, lnK = log capital (capital formation) per labor in USA Dollars, lnAL = log agricultural labor per acre labor force (number of persons per acre), logA = log per capita agricultural output, which is crop yields in kilograms per acre and ε is the error term.

The data come from the World Bank database of World Development Indicators. There were some missing observations on the labor variables, which were then calculated by the author. More than 70% of the labor force is in the agriculture; and to bring out the relevance of this force and its use of land to cultivate, the paper introduces the agricultural labor per acre of lands used in crop production for the period of time. This is more meaningful than the gross labor force, and in this way, the model includes in its explanatory variables two agricultural factors, the agricultural output and the agricultural labor. The agricultural labor fluctuates between one person cultivating more than one acre to two persons sharing one acre. From 1960 to 1967, one person cultivated more than one acre; this changed to a one-to-one ratio from 1968 to 1979. The ratio stood at two persons to an acre from 1980 to 1991. The call of "back to land" was significant in the period from 1992 to 1998, as two persons cultivated less than an acre. It then stood at two persons to an acre from 1999 to 2009. The agricultural labor grew at an average rate of 3.3% per annum during the period from 1960 to 2009; whereas, the arable land conversion to the farming land grew at an average rate of 1.3% in the same period. Given the fact that the arable lands are fixed in amount, the natural law of diminishing returns will inevitably set in if the agricultural labor continues to grow at the rate of 3.3% per annum. More interestingly, these significant changes in the agricultural labor did not correspond with increased output as the crop yields per acre barely exceeded 1100 kg per acre for this period. For the estimation purposes, the paper uses the observations from 1966 to 2009 due to unavailability of data on capital formation and GDP before 1966.

Autoregressive procedure

Let Equation (1) be a long run relationship between the left side and the right side variables, and that Y^* is the equilibrium level of per capita GDP that the economic agents wish to attain given the capital stock, the labor force and the agricultural output. With this assumption, Equation 1 becomes:

$$\ln Y_t^* = \alpha + \beta_1 \ln K_t + \beta_2 \ln AL_t + \beta_3 \ln A_t + \varepsilon_t \qquad (2)$$

Equilibrium value of Y is unobservable, but at any given time, the agents adjust their per capita GDP towards its target level, and employing the partial adjustment process, this behavior of the agents can be described as:

$$\ln(Y_t - \ln Y_{t-1}) = \lambda(\ln Y_t^* - \ln Y_{t-1}), \qquad 0 \le \lambda \le 1 \qquad (3)$$

Equation (3) states that the current per capita GDP level moves partially from its previous level towards the desired equilibrium level, and the amount of adjustment is $\lambda(\ln C_t^* - \ln C_{t-1})$; where lambda λ indicates the speed of adjustment. By substituting for the value of the equilibrium Y from Equation (2) into Equation (3) and solving for the observed value of the current Y, an autoregressive model appears:

$$\ln Y_t = \lambda\alpha + (1-\lambda)\ln Y_{t-1} + \lambda\beta_1 \ln K_t + \lambda\beta_2 \ln AL_t + \lambda\beta_3 \ln A_t + v_t \qquad (4)$$

If there is an adjustment, the short run and long production elasticities can be derived by estimating Equation (4) as:

$$\ln Y_t = b_0 + b_1 \ln Y_{t-1} + b_2 \ln K_t + b_3 \ln AL_t + b_4 \ln A_t + v_t \qquad (5)$$

The coefficients of Equation (5) are the short run elasticities. The knowledge of these coefficients helps derive the long run elasticities in Equation (2).

Error correction method

Another way to look at the long relation in Equation (1) is to assume a gap between the left side variable and the right side variables and that it is corrected at every particular time to maintain the equilibrium balance between the two sides. This gap is the error term in Equation (1):

$$\ln Y_t - \alpha - \beta_1 \ln K_t - \beta_2 \ln AL_t - \beta_3 \ln A_t = \varepsilon_t \qquad (6)$$

The adjustment towards the equilibrium depends on the amount of error previously committed, or specifically:

$$(\ln Y_t - \ln Y_{t-1}) = \delta\varepsilon_{t-1} + e_t$$
$$\Delta \ln Y_t = \delta\varepsilon_{t-1} + e_t \qquad (7)$$

The coefficient of the lagged error term in Equation (7) is the amount of error corrected at every particular time towards the equilibrium. For this correction to take place, it should be significant and negative. Equation (7) is expanded to include not only the lagged error term of the long relation, but also the lagged adjustments of the dependent variable and the contemporaneous and lagged adjustments of the explanatory variables to capture all the potential adjustment processes in the relation towards the equilibrium. These adjustments disappear when the equilibrium is attained. Incorporating these potential adjustments processes in Equation (7), results in the error correction procedure as as follows:

$$\Delta \ln Y_t = b_0 + \sum_{i=1}^{q} b_{1i}\Delta \ln Y_{t-i} + \sum_{i=0}^{n} b_{2i}\Delta \ln K_t + \sum_{i=0}^{m} b_{3i}\Delta \ln AL_{t-i} + \sum_{i=0}^{p} b_{4i} \ln A_t + \delta\varepsilon_{t-1} + v_t \qquad (8)$$

When the adjustments processes are complete, the long equilibrium relation of Equation (1) emerges from Equation (8) as the coefficient of the lagged error term becomes negative one. However, the long run coefficients of Equation (1) are not easily derived from Equation (8), unlike the case of Equation (5). To derive the long coefficients, the value for the lagged error term from Equation (6) is to be substituted in Equation (8) which will consequently reduce the degrees of freedom. Given the limited observations, the paper stops at Equation (8) and analyzes the adjustment processes of the model. Two lags are introduced to adhere to the data limitation constraint.

Co-integration test

If the coefficient of the lagged error term in Equation (8) is significant and negative, and the variables in Equation (1) are integrated of order one, then a co-integration is established between the left side variable and the right side variables in Equation (1). Co-integration is a long co-movement between variables; the variables do not diverge, but co-move within a relation. This relation here can be identified as a production function of per capita GDP that co-moves permanently with capital stock, labor and agricultural output. This relation can be uni-directional or bi-directional depending on whether GDP Granger causes agricultural output, or agricultural output Granger causes GDP, or both Granger cause each other. This feedback process between the variables is tested using the Engle-Granger causality test.

Granger causality test

If agricultural output, A, Granger causes per capita GDP, Y, then the present value of per capita GDP can be predicted by using past values of the agricultural output, and vice versa. This test is performed as:

Feedback test from oil consumption and oil price to per capita GDP:

$$Y_t = \beta_0 + \sum_{i=1}^{q} \beta_{1i} \ln K_{t-i} + \sum_{i=1}^{m} \beta_{2i} \ln AL_{t-i} + \sum_{i=1}^{n} \beta_{3i} \ln A_t + u_t \qquad (9)$$

Feedback test from per capita GDP agricultural output:

$$A_t = a_0 + \sum_{i=1}^{q} a_{1i} \ln K_{t-i} + \sum_{i=1}^{m} a_{2i} \ln AL_{t-i} + \sum_{i=1}^{n} a_{3i} \ln Y_{t-i} + v_t \qquad (10)$$

If the sum coefficients of agricultural output in Equation (9) are significantly different from zero, then the changes in the per capita GDP can be predicted using the past values of the agricultural output. The feedback will run from per capita GDP to agricultural output, if the sum coefficients of per capita GDP in Equation (10) are significantly different from zero. The study uses 4 lags in

Table 2. Autoregressive model: Dependent variable is LY.

Sample (adjusted): 1967 2009				
Included observations: 43 after adjustments				
Variable	**Coefficient**	**Standard error**	**t-Statistic**	**Probability**
C	2.951571	0.843921	3.497449	0.0012
LY(-1)	0.517669	0.128973	4.013764	0.0003
LK	0.210255	0.055389	3.795966	0.0005
LAL	0.057358	0.102608	0.559001	0.5794
LA	-0.175863	0.120559	-1.458735	0.1529
R-squared	0.944237	Mean dependent variable		5.594529
Adjusted R-squared	0.938367	Significant difference dependent variable		0.400107
Standard error of regression	0.099330	Akaike info criterion		-1.671786
Sum squared resid	0.374928	Schwarz criterion		-1.466995
Log likelihood	40.94339	F-statistic		160.8634
Durbin-Watson statistic	1.620784	Probability (F-statistic)		0.000000

the test. This means past two year values are assumed to hold information about the current value of another variable if there are to be feedback processes.

EMPIRICAL ANALYSES

Autoregressive model

This model presents the short run and long run relationships between the per capita GDP and the explanatory variables of capital per worker, agricultural labor per acre and the crop yields per acre as shown in Table 2. The estimates show that no short run relations exist between per capita GDP and the agricultural labor per acre and the crop yields per acre. The capital per worker is found to have both short run and long run relation with the per capita GDP. In the short run and the long run, a unit increase of capital per worker corresponds on average, other things constant, with 0.48 unit increase and 0.99 unit increase respectively with the per capita GDP. A unit increase of agricultural labor per acre corresponds in the long run with 21 units increase of per capita GDP. The crop yields per acre have a wrong expected sign in this model. This could indicate the poor productivity contribution of the agricultural sector to the economy. This model might not fit the data as well as all the variables in the model are found to first difference stationary, and thus level OLS estimates could be spurious.

The contemporaneous relation between the per capita GDP and the explanatory variables is investigated in the error correction model.

Error correction method

In this method, the study examines the contemporaneous relations in the growth-agriculture nexus and searches, if any, for an equilibrium adjustment in the long run relation. Table 3 presents the error correction method estimates. Table 3 shows that the only significant contemporaneous relation in the model is between the per capita GDP and the capital per worker. The agricultural sector does not respond contemporaneously with the economic growth (changes in the per capita GDP). However, there is a significant long relation between the economic growth and the linear summation of the capital per worker, agricultural labor per acre and the crop yields per acre. Small deviations from the equilibrium level are adjusted fast towards the equilibrium as 0.466% of the deviations can be corrected annually. But large deviations, which could occur due to climatic shocks to the agriculture, would take several years to be corrected. For example, 10% deviation will take 21 years to adjust completely to the pre-shock level. The long run coefficients in this co-integrating relation are estimated using the vector error correction method.

Co-integration and vector error correction estimates

One lag is introduced in this model with the restrictions that the per capita GDP is an endogenous variable that could not feed back to the co-integrating relation, and having the long run co-integrating vectors normalized by the long run coefficient of the log per capita GDP, the following results in Table 4 are obtained (the details of the estimates are in Annex A). The restrictions have identified all the co-integrating vectors and the likelihood ratio test shows that these restrictions are valid and cannot be rejected at 10% significance level. Furthermore, the long run coefficients have the right expected economic signs. This can be seen by presenting the relation in the equation form as follows:

$$LY_t = -8.12 + 0.37LK_t + 0.3LAL_t + 1.7LA_t$$

$$t-ratios \quad (3.5968) \quad (1.063) \quad (3.6387)$$

Table 3. Error correction method.

Dependent variable: D(LY)				
Sample (adjusted): 1968 2009				
Included observations: 42 after adjustments				
Variable	**Coefficient**	**Standard error**	**t-Statistic**	**Probability**
C	0.005854	0.017350	0.337407	0.7379
D(LY(-1))	0.134681	0.187238	0.719305	0.4770
D(LK)	0.326793	0.079438	4.113832	0.0002
D(LK(-1))	0.006676	0.091702	0.072806	0.9424
D(LAL)	-0.006577	0.210426	-0.031256	0.9753
D(LAL(-1))	0.097794	0.210152	0.465351	0.6447
D(LA)	-0.070097	0.134972	-0.519348	0.6070
D(LA(-1))	-0.017543	0.131082	-0.133831	0.8943
ER(-1)	-0.466834	0.167407	-2.788616	0.0087
R-squared	0.432097	Mean dependent variable		0.030554
Adjusted R-squared	0.294424	Significant difference dependent variable		0.120470
Standard error of regression	0.101193	Akaike info criterion		-1.556159
Sum squared resid	0.337923	Schwarz criterion		-1.183801
Log likelihood	41.67934	F-statistic		3.138568
Durbin-Watson statistic	1.995807	Probability (F-statistic)		0.009431

Table 4. Vector error correction estimates.

Vector error correction estimates	
Sample (adjusted): 1968 2009	
Included observations: 42 after adjustments	
Standard errors in ( ) and t-statistics in [ ]	
Cointegration restrictions	
B(1, 1) = 1, A(1, 1) = 0	
Convergence achieved after 91 iterations.	
Restrictions identify all cointegrating vectors	
LR test for binding restrictions (rank = 1):	
Chi-square (1)	1.397011
Probability	0.237225
Cointegrating equation	CointEq1
LY(-1)	1.000000
LK(-1)	-0.369560 (0.10275) [-3.59687]
LAL(-1)	-0.301688 (0.28386) [-1.06282]
LA(-1)	-1.695782 (0.46604) [-3.63868]
C	8.119524

The capital per worker and the crop yields per acre are found to be significant and relevant factors for boosting economic growth. The presence of the crop yields per acre, other things constant, results in increasing returns to scale in the economy; whereas, the capital and the labor together, other things constant, will result in decreasing returns in the economy. This proves the profound importance of the agricultural productivity in the economy. As argued throughout this study, it is not more labor on the land but how much from land, which makes significant difference to the economic growth. The agricultural labor inputs are found irrelevant in the long run. Increased labor on land can set back the economic growth if yields are not correspondingly increased, because increased labor per land will make the law of diminishing returns quickly set in. The Granger causality tests in Annex B confirm these conclusions.

In these tests, there is a bi-directional relation between the per capita GDP and the crop yields per acre, and a uni-directional relation from the yields per acre to the capital per worker. This means that the yields per acre do not only boost the economic growth directly but also indirectly by enhancing capital formation in the economy as high yields often enable the 70% of the population to save and buy assets.

CONCLUSIONS AND POLICY RECOMMENDATIONS

Agriculture is a major engine for the economic well-being of the Gambia. The successive governments from the colonial period to the present have all recognized this importance of the agricultural sector. But most of their policies could bear any fruit, because they were working against natural laws. They in many instances encouraged slash and burn technology and putting increasingly more people onto the land instead of increasing the yields per acre. More people do remain still on the land, but they are now poorer than their forefathers. Their forefathers had one person to an acre producing on average of 1100 kg per acre. Today, there are two persons to an acre producing the same 1100 kg per acre. Using four econometric techniques to examine the growth-agriculture relation in the Gambia for the period from 1966 to 2009, it is found that the capital per worker is a significant and relevant factor input for the economic growth. The agricultural labor per acre is found to be irrelevant in both the short run and the long run. The agricultural productivity measured as crop yields per acre is the most important variable for the economic growth in the Gambia. It boosts both economic growth and the capital formation in the country. The authorities should refocus their agricultural policies from putting more people back to the land to increasing the crop yields per acre. Modern technologies suitable for small and fragile lands should be adopted. The fertilizers should be substantially subsidized to increase the yields, and labor inputs onto the lands should be reduced.

The processing, services and small manufacturing sectors should be developed and built into the agricultural policies in order to create redeployments for the agricultural labor surpluses.

ACKNOWLEDGEMENTS

The author acknowledges the comments and suggestions by Professor Abdoulaye Saine, Miami University, U.S.A, and Dr. Ebrima Ceesay. The endnotes and the language editing were done by them.

Endnotes

[1]Chr. Hennig, Rainer; "An Introduction to the Economic History of Rural Gambia", afrol News editor, http://www.afrol.com/archive/economic_history_gambia.htm

[2]Sessional Paper No. 133 of 1963, Report on the Census of Population of the Gambia Taken on 17th/18 April, 1963, the Government Printer, Bathurst (now Banjul), 1965.

[3]Kuye et al. (2006). "The Agricultural Health in the Gambia: Agricultural Practices and Development", Review Articles, Ann. Agric. Environ., vol. 13, p 1-12.

[4]The census of the Population of the Gambia 1963.

[5]Colonial report, The Gambia 1962/1963.

[6]Carney Judith (1998) "Women's land rights in Gambia Irrigated Rice Schemes: Constraints and Opportunities", Agriculture and Human Values, vol. 15, pp. 325-336, Kluwer academic Publishers 1998.

[7]http://www.statehouse.gm/agriculture/achievements.html

[8]http://observer.gm/africa/gambia/kerewan/article/2008/5/7/jammeh-warns-authorities-as-rice-tops-kerewan-meeting

[9]Republic of the Gambia: Appraisal Report, Farmer Managed Rice Irrigation Project, African Development Fund, Agricultural and Rural Development Department, Central and West, April 2005.

[10] Foroyaa News Paper, Serrekunda, the Gambia, May 8, 2009.

REFERENCES

Carney Judith A (1998). Women's land rights and opportunities", Agriculture and Human Values. Kluwer Academic Publishers 15:325-336,

Republic of the Gambia (2005). Appraisal Report, "Farmer Managed Rice Irrigated Project", African Development Fund, Agricultural and Rural development, Central and West, April 2005.

Kuye R, Kelley D, Shannon M (2006). Agricultural Health in the Gambia I: Agricultural Practices and Development", Review Articles. Ann. Agric. Environ. 13:1-1.

Roemer M (1993). "Strategies for industrialization: Lessons for the Gambia", Agency for International Development POL/CDIE/D1, Report Processing Form.

Xu D (1993). Price distortion in the transition process: a CGE analysis of China's case. Econ. Plann. 26:161-182.

Robinson S, San N, Nu, Suryana A (1997), Rice price policies in Indonesia: a computable general equilibrium analysis", Trade and Macroeconomic division, industrial Food Policy research Institute, 2033 K street, N.W, Washington, D.C.

Harley C Knick (2002). Computable general equilibrium models in economic history and an analysis of British capitalist Agriculture. Eur. Rev. Econ. History 6:165-191.

O' Ryani R, Miller S (2003) the role of agriculture in poverty alleviation, income distribution and economic development: a CGE analysis for Chile", Agricultural and development Economics division, food and Agriculture Organization of United nations, the role Roles of Agriculture International conference, Rome, Italy, October 20-22.

Bednarikova Z, Doucha T (2009). the Impacts of agricultural policy scenarios on development of remote rural areas – the case study of the Bruntal and Ostrava districts", Agric. Econ.Czech. 55(4):161-168.

Rubin, Ofir D, Carriquiry, Miguel H, Dermot J (2008). Implied objectices of U.S Bio-fuel subsidies", Center for Agricultural and Rural development, Iowa State University, working 08–WP 459.

Kidane W, Maetz M, Dardel P (2006). Food Security and agricultural development in Sub-Saharan Africa: building a case for more public support" food and Agricultural Organmization of Unietd Nations, Rome.

Lopez RA, Hathie I (2000). the structure of Government intervention in African Agriculture", the J. Develop. Stud. 37(1):57– 72.

Tokarick S (2005). who bears the cost of agricultural support in OECD countries. the World Economy 28:4.

Trefflich A, Uetrecht I, Efken J, Shaefer M, Steinbauer C, Wendt H (2007). Support scheme of food processing firms: a driving force for rural development?", IAMO Forum on 'sustainable rural development: what is the role of agric-food sector', June 27-29, Saale, Germany.

ANNEX

Annex A. Vector error correction estimates.

Included observations: 42 after adjustments				
Standard errors in ( ) and t-statistics in [ ]				
Cointegration restrictions				
B(1, 1) = 1, A(1, 1) = 0				
Convergence achieved after 91 iterations.				
Restrictions identify all cointegrating vectors				
LR test for binding restrictions (rank = 1)				
Chi-square(1)	1.397011			
Probability	0.237225			
Cointegrating equation	CointEq1			
LY(-1)	1.000000			
LK(-1)	-0.369560 (0.10275) [-3.59687]			
LAL(-1)	-0.301688 (0.28386) [-1.06282]			
LA(-1)	-1.695782 (0.46604) [-3.63868]			
C	8.119524			
Error correction	D(LY)	D(LK)	D(LAL)	D(LA)
CointEq1	0.000000 (0.00000) [NA]	0.425194 (0.13817) [3.07729]	-0.039842 (0.06582) [-0.60534]	0.225492 (0.09645) [2.33795]
D(LY(-1))	-0.160339 (0.21137) [-0.75857]	-0.289298 (0.33828) [-0.85521]	0.169003 (0.14119) [1.19698]	-0.002636 (0.20686) [-0.01274]
D(LK(-1))	0.154653 (0.10287) [1.50337]	0.199254 (0.16464) [1.21026]	-0.004553 (0.06872) [-0.06626]	-0.036074 (0.10068) [-0.35831]
D(LAL(-1))	0.034949 (0.24748) [0.14122]	-0.384372 (0.39607) [-0.97047]	0.111879 (0.16531) [0.67678]	-0.030919 (0.24220) [-0.12766]
D(LA(-1))	0.022322 (0.17928) [0.12451]	0.347941 (0.28692) [1.21269]	-0.028681 (0.11975) [-0.23950]	0.019401 (0.17546) [0.11057]
C	0.025988 (0.02072) [1.25433]	0.060957 (0.03316) [1.83838]	0.010920 (0.01384) [0.78904]	0.003942 (0.02028) [0.19441]

Annex A. Contd.

R-squared	0.094011	0.314329	0.058175	0.188813
Adj. R-squared	-0.031821	0.219097	-0.072635	0.076148
Sum of square resids	0.539096	1.380797	0.240544	0.516354
Standard error equation	0.122372	0.195845	0.081742	0.119763
F-statistic	0.747116	3.300661	0.444729	1.675885
Log likelihood	31.87073	12.11977	48.81758	32.77585
Akaike AIC	-1.231939	-0.291418	-2.038932	-1.275040
Schwarz SC	-0.983701	-0.043179	-1.790694	-1.026802
Mean dependent	0.030554	0.055572	0.018623	0.001043
Standard difference dependent	0.120470	0.221623	0.078926	0.124601
Determinant resid covariance (dof adj.)		3.92E-08		
Determinant resid covariance		2.12E-08		
Log likelihood		132.3483		
Akaike information criterion		-4.968967		
Schwarz criterion		-3.810520		

Annex B. Granger causality tests.

Lags: 4

Null hypothesis	Observation	F-Statistic	Probability
LK does not Granger cause LY	40	2.11750	0.10235
LY does not Granger cause LK		0.98764	0.42876
LAL does not Granger cause LY	40	0.48141	0.74914
LY does not Granger cause LAL		0.77203	0.55173
LA does not Granger cause LY	40	2.67605	0.05017
LY does not Granger cause LA		2.36908	0.07413
LAL does not Granger cause LK	40	1.76992	0.16010
LK does not Granger cause LAL		0.66905	0.61837
LA does not Granger cause LK	40	2.21461	0.09034
LK does not Granger cause LA		0.75811	0.56046
LA does not Granger cause LAL	40	0.25834	0.90233
LAL does not Granger cause LA		0.05992	0.99301

Effect of stabilizers on the quality of carbonated flavoured whey drink

Aysha Sameen, Muhammad Rizwan Tariq, Nuzhat Huma and Muhammad Issa Khan*

National Institute of Food Science and Technology, University Of Agriculture Faisalabad, Pakistan.

Carbonated Flavoured Whey Dink (CFWD) was prepared by blending sugar (8%), stabilizers (that is, carboxy methyl cellulose, carrageenan), colour and flavor (orange red) to liquid whey. The whey drink was subjected to carbonation. The whey drink was evaluated for their physico-chemical and sensory properties till 30 days of refrigerated (4°C) storage. The storage study revealed that fat, lactose, SNF, total solids, acidity and viscosity decreased while pH and total plate count increased with storage period. The significant changes were observed only for fat, lactose, solids-not-fat (SNF), total solids and viscosity. The sensory quality of CFWD containing carragenan was found to be highly acceptable.

Key words: Whey, stabilizer, carbonated drink.

INTRODUCTION

Whey was discovered accidentally about 3000 years back when enzyme chymosin in native to calf stomach coagulated the milk that was contained in a bag made of calf skin, resulting in formation of 'curd and whey' (FAO, 2006). Whey can be obtained from any type of milk. However, whey from cow's milk is quite popular in western countries, while in some regions of the world, goats, sheep and even camel's milk are being used in the manufacture of dairy products that results in the production of whey (Zadow, 1994).

Previously, whey was treated as an insignificant by-product of cheese making, used mainly in animal feed or surplus. With advances in technology and recent discoveries of functional and bioactive roles of whey proteins, whey and whey components are now viewed as precious ingredients. The recognition of whey as a source of unique physiological and functional attributes provides opportunity for the food industry to incorporate whey and whey components into a variety of foods (National Dairy Council, 2003).

De Wit and Moulin (2001) estimated that 700,000 tons of whey produced globally is used as ingredients in food.

The current world production of whey is about 125 million tones, in which about 64% is produced in European countries and 24% in North America. In the absence of systematic surveys/statistics, the predicted value of whey production in India is estimated at 4.84 million tones per annum (Raju et al., 2005).

Whey contains about 50% of the milk solids together with 100% of the lactose and 20% of the protein. The lactose makes up about 75% of the total whey solids (Siso, 1996). Whey comprises about 80 to 90% of the volume of milk from which it is obtained (Khamrui and Rajorhia, 1998).

Whey protein is a complete, high quality protein with a rich amino acid profile. It contains the full range of amino acids including essential amino acids (EAAs) and branched-chain amino acids (BCAAs – *viz.,* leucine, isoleucine, valine) which are important for tissue growth and repair. Leucine is a key BCAA in protein synthesis and plays a critical role in insulin and glucose metabolism. The BCAA comprises > 20% of total amino acids in whey. The EAAs and BCAAs in whey protein are not only present in higher concentrations than in other protein sources (that is, soy, corn, wheat), but they are more efficiently absorbed and utilized (Harper, 2000).

A growing acceptance of milk-based beverages has been reported in Brazil, where lactic beverages formulated with whey (made by mixing yoghurt and cheese whey) represented about one-third of the market

*Corresponding author. E-mail: drkhan@uaf.edu.pk.

Table 1. Particulars of mozzarella whey.

Parameter	Results
Fat (%)	0.40
Protein (%)	0.67
Lactose (%)	5.40
Ash (%)	1.0
MSNF (%)	7.30
Total solids (%)	7.66
Acidity (% LA)	0.50
pH	4.6

for yoghurt and other dairy-based lactic beverages (Penna et al., 2003). Whey based lactic beverages constitute an emerging segment of non-conventional dairy products that require sensory, physical and chemical characterization for quality control and product development. From consumer point of view, lactic beverages should be as viscous as milk. In this type of product, texture and mouthfeel are often matched to those found in traditional products (Jack et al., 1995). Whey drinks are formulated for kids, adults and enthusiast with the combination of yogurt, milk, fruit juices and soya solids (Kacvinsky, 2005).

The market demand for beverages is growing all over the world and Pakistan is no exception to it. Whey beverages have been recognized as a genuine thirst quencher, light, refreshing, healthful and nutritious (Prendergast, 1986).

Many attempts have been made on utilization of whey in the formulation of various dairy products but, still there is a scope to explore the possibilities for its utilization in beverage industries. Now-a-days, different dairy industries are looking for new product ideas and technologies to meet the consumer's requirement and to increase the profitability of the products. Product diversification is quite feasible using whey in formulated foods, especially beverages (Patel et al., 2007).

In Pakistan, 2500 tons of cheese is being produced per annum (PDDC, 2006). Hence, sufficient amount of whey is being produced. The development of technology has opened up new and cost effective ways of utilizing the whey constituents that are finding its place in newer applications, both in dairy and food industries.The project is planned with two main objectives: To develop a whey-based carbonated flavoured drink and to evaluate the storage changes.

MATERIALS AND METHODS

Liquid mozzarella whey was obtained from a local cheese industry. It was analyzed for fat, protein, lactose, total soluble solids, SNF, ash, acidity and pH. Fresh liquid whey was heated at 65°C for 30 min and divided into three equal portions. Additives like citric acid (0.07%), sugar (8%) and sodium benzoate (0.03%) were incorporated in whey. Carboxyl methyl cellulose (CMC), Carragenan and mixture of both (50:50) were used at 0.5% by weight of final whey beverage, after dissolving in hot water (specify temperature; proportion of stabilizer: water too) and added to the whey drink. The drink was made homogenous by subjecting the drink at 7000 rpm/min to reduce the size of the particles with the help of lab scale blender. Whey drink was filled into transparent 500 ml plastic bottles for carbonation. Carbonation was carried out at 50 kg/cm^2 (HUASHENG, DCGF14/12/5) and the whey drink cooled and finally stored in a refrigerator (4±1°C) for further analysis. The carbonated flavoured whey drink (CFWD) was evaluated for pH (pH meter 720, WTW 823362), acidity, lactose and total solids as per standard method (AOAC, 1990) while fat was analyzed by Gerber method and protein by Kjeldhal method (Kirk and Sawyer, 1991). Viscosity was measured by viscometer (Expotech USA model 35) The CFWD were evaluated organoleptically and rated for colour, odour, sediment, flavor, appearance and overall acceptability by a panel of five judges (Djuric et al., 2004). The data obtained was subjected to statistical analysis (Koffi et al., 2005).

RESULTS AND DISCUSSION

The particulars related to sweet whey is given in Table 1. The whey had 0.67% protein, 5.40% lactose and 7.66% total solids.

The influence of storage period on lactose and SNF content of carbonated flavoured whey drink (CFWD) was significant; the influence on acidity and viscosity of CFWD was highly significant. Rest of the constituents (*viz.,* protein, fat, ash, total solids) of CFWD remained unaffected by the storage period.

The highest pH value was recorded in sample S_2 (4.31) at 30th day of storage (Table 1). The pH of CFWD is in agreement with that reported by Girsh (2001) for whey drink. The acidity of the product was in decreasing order for S_1 (0.407), S_2 (0.433) and S_3 (0.437), when fresh (0 day) (Table 2). The protein content of CFWD of this present study is in accordance with that observed by Niketic and Marinkovic (1984) for cheese whey beverage. They found no change in the protein content of cheese whey beverage packaged in Tetrapak cartons during 6 weeks of storage at room temperature. The lactose content of CFWD did not vary within treatments but it tended to decrease during storage up to 30 days. The initial lactose content was 5.40% lactose at 0 day which decreased to 4.73% at 30th day of storage (Table 3). The decrease in lactose content during storage was due to its conversion into lactic acid (Goodnaught and Kleyn, 1976). The mean total solids content was 10.53% when fresh (0 day) which decreased to 9.15% at 30th day (Table 3). Ash represents the inorganic matter in the product (Ranhotra, 1985). The ash content remained almost constant throughout this study. Viscosity is directly related to thickness of whey drink. Polysaccharides contribute to the structural and textural properties of milk products (Kumar and Mishra, 2003). The initial mean value of viscosity was 3.88 poise which decreased to 3.57 poise at 30th day of storage (Table 3). Finally micro-structure showed the classical appearance of an acidic gel with the presence of polysaccharide appendages

Table 2. Influence of stabilizer on the physico-chemical characteristics of whey drink

Treatments	Fat (%)	Protein (%)	Lactose (%)	Ash (%)	SNF (%)	Total solids (%)	Acidity (% LA)	pH	Viscosity (poise)
S_1 (Carragenan)	0.24	0.67	4.98	1.38	9.52	9.87	0.407	4.23	3.82^{a}
S_2 (CMC)	0.24	0.67	4.98	1.38	9.52	9.87	0.433	4.27	3.72^{b}
S_3 (Carragenan + CMC)	0.24	0.67	4.98	1.38	9.52	9.87	0.435	4.20	3.62^{c}

^{abcd}, Different superscripts indicates significant difference between the means at $P \leq 0.05$ probability level.

Table 3. Effect of storage on physico-chemical characteristics of whey drink (S_1, S_2, S_3).

Storage period (days)	pH	Acidity (%)	Protein (%)	Fat (%)	Lactose (%)	Ash (%)	Total solids (%)	Viscosity (poise)	SNF (%)
0	4.18^{d}	0.463^{a}	0.67	0.30^{a}	5.40^{a}	1.39^{a}	10.53^{a}	3.88^{a}	9.54^{a}
10	4.23^{c}	0.435^{b}	0.67	0.25^{b}	5.04^{b}	1.38^{b}	10.13^{b}	3.77^{b}	9.52^{b}
20	4.25^{b}	0.419^{c}	0.67	0.21^{c}	4.74^{c}	1.38^{b}	9.68^{c}	3.65^{c}	9.51^{c}
30	4.27^{a}	0.383^{d}	0.67	0.20^{d}	4.73^{d}	1.38^{b}	9.15^{d}	3.57^{d}	9.51^{d}

Storage temperature = 4°C. ^{abcd}, Different superscripts indicates significant difference between the means at $P \leq 0.05$ probability level.

Table 4. Influence of stabilizer on sensory quality of whey drink.

Treatments	Colour (2)	Appearance (1)	Flavour (10)	Odour (2)	Sediment (5)	Total sensory score (20)
S_1 (Carragenan)	2.00	0.88^{a}	8.66^{a}	1.23^{a}	2.75	15.52^{a}
S_2 (CMC)	2.00	0.69^{b}	7.25^{b}	0.74^{b}	2.37	13.05^{b}
S_3 (Both)	2.00	0.69^{b}	5.50^{c}	0.75^{b}	2.37	11.31^{c}

^{abcd}, Different superscripts indicates significant difference between the means at $P \leq 0.05$ probability level.

Table 5. Effect of storage on sensory characteristics of whey drinks (S_1, S_2, S_3).

Storage period	Flavour	Odour	Colour	Sediment	Appearance	Total sensory quality
Day 0	7.66^{a}	1.16^{a}	2.00	3.00	1^{a}	14.82^{a}
Day 10	7.66^{a}	0.91^{b}	2.00	3.00	1^{a}	14.57^{a}
Day 20	6.88^{ab}	0.89^{b}	2.00	2.33	0.58^{b}	12.70^{b}
Day 30	6.33^{b}	0.67^{c}	2.00	1.66	0.42^{b}	11.08^{c}

Different superscripts a b c indicates significant difference between the means at the level of probability: $P \leq 0.05$

(Oliveira et al., 2002). The initial mean value of SNF was 9.54% which decreased to 9.51% at 30th day of storage (Table 3 and 4). The investigated values of SNF for CFWD are in line with the findings of Djuric et al. (2004) for whey beverage.

Sensory evaluation of carbonated flavoured whey drink

The sensory evaluation was conducted at ≤ 1 day and thereafter at an interval of 10 days up to 30 days (that is, at 10, 20 and 30th day). Organoleptic evaluation of CFWD showed that storage period had a significant effect on all of the sensory parameters evaluated. During storage the scores for flavour, odour, sediment, appearance and total quality varied from 7.66 to 6.33, 1.16 to 0.67, 3.00 to 1.66, 1.00 to 1.42 and 14.82 to 11.08 respectively (Table 5). At 0 day, maximum and minimum total sensory score (16.47) was associated with samples S_1 (16.47) and S_3 (13.00) respectively (Table 5). After 10, 20 and 30 days of storage, the judges observed a slight decline in the total sensory score in all of the CFWD samples. Hence it was found that CFWD sample S_1 obtained maximum scores regarding the organoleptic evaluation and remained the best. Incorporation of stabilizer exceeding 0.3% adversely affected the sensory properties of milk products. The sensory score of CFWD increased with level of stabilizer up to 0.35%; use of 0.5% resulted in undesirable flavor (Mehanna and Mehanna, 1989).

Conclusion

Carrageenan is the best choice as a stabilizer when compared to carboxymethyl cellulose during 30 days of refrigerated storage of carbonated flavoured whey drink. Carbonation increased the palatability of the whey drink and extended its shelf-life.

REFERENCES

AOAC (1990). Official methods of analysis, 15th ed. The Association of Official Analytical Chemists, Arlington, USA.

De Wit JN, Moulin J (2001). Whey protein isolates: manufacture, properties, and applications. Ind. Proteins 9:6-8.

Djuric M, Caric M, Milanovic S, Tekic M, Panic M (2004). Development of whey-based beverages. Europ. Food Res. Technol. 219:321–328.

FAO (2006). Food outlook no. 2, December 2006 (http://www.fao.org/docrep/009/j8126e/j8126e11.htm)

Girsh LS (2001). US Patent US 2001/0022986 A1.

Goodnaught, Kleyn (1976). Physical properties of yoghurt made from milk treated with proteolytic enzyme. J. Dairy Sci. 74:1503-1511.

Harper WJ (2000). Biological properties of whey components: A Review. The American Dairy Products Institute. Chicago, IL pp. 23-29.

Jack FR, Paterson A, Piggott JR (1995). Perceived texture: Direct and indirect methods for use in product development. Int. J. Food Sci. Tech. 30:1-12.

Kacvinsky AM (2005). Whey beverages: Stabilizer and process technology options. 4th Intl. Whey Conf., Sep. 11-14, 2005, Chicago pp. 45-57.

Khamrui K, Rajorhia GS (1998). Making profits from whey. Indian Dairyman 50:13-18.

Kirk RS, Sawer R (1991). Pearson's composition and analysis of foods. 9th ed. Longman Scientific and Technol, London, UK. pp. 77-98.

Koffi E, Shewfelt R, Wicker L (2005). Storage stability and sensory analysis of UHT processed whey-banana beverage. J. Food Quality 28:386-401.

Kumar P, Mishra HN (2003). Mango soy fortified set yoghurt: effect of stabilizer addition on physicochemical, sensory and textural properties. J. Food Chem. 87:501-507.

Mehanna NM, Mehanna AS (1989). On the use of stabilizer for improving some properties of Egypt. J. Dairy Sci. 17:289-296.

National Dairy Council (2003). Whey protein. www.nationaldairycouncil.org.

Niketic G, Marinkovic S (1984). Production of refreshing beverages from whey under aseptic conditions. Ml Jekarstvo 34:105-109.

Oliveira MN, Sodini I, Remeuf F, Tissier JP, Corrieu G (2002). Manufacture of fermented lactic beverages containing probiotic cultures. J. Food Sci. 67:2336-2341.

Pakistan Dairy Development Company (PDDC) (2006). The white revolution-Doodh darya. Pakistan Dairy Development Company, Township, Kot Lakhpat, Lahore.

Patel S, Prasanth S, Choudhary PL, Sahu C (2007). Techno-economic feasibility of whey based mango herbal (Ginger) beverage. Indian J. Dairy Sci. 60:149-155.

Penna ALB, Oliveira MN, Tamime AY (2003). Influence of carrageenan and total solids content on the rheological properties of lactic beverage made with yogurt and whey. J. Text. Stud. 34:95-113.

Prendergast K (1986). Whey drinks - technology, processing and marketing. Intl. J. Dairy Technol. 38:103-105.

Raju PN, Rao LH, Devi NL (2005). Whey protein and their uses in food industry. Indian Food Ind. 24:19-27.

Ranhotra GS (1985). Nutritional profile of corn and flour tortillas. Cereal Foods World. 30:703-704.

Siso MIG (1996). The biotechnological utilization of cheese whey: A review. Bioresource Tech. 57:1–11

Zadow JG (1994). Utilization of milk components. In: Modern Dairy Technology - Advances in Milk Processing, Vol. I., Robinson RK, 2nd ed. Chapman and Hall London, UK. pp. 313-373.

Quality of Black Teas in Indian market

Ch. Someswararao*, P. P. Srivastav and H. Das

Agricultural and Food Engineering Department, Indian Institute of Technology, West Bengal 721302, India.

Tea (*Camellia sinensis*) is the second most popular consumed drink in the world after water. Tea has numerous medicinal benefits mainly due to its antibacterial and antioxidant properties. In this study, the levels of Theaflavin (TF), Thearubigin (TR), high polymerized substances (HPS) and total polyphenols (TPP) of different tea brands found in Indian market were determined using UV/Vis spectrophotometric methods. Four brands of black teas were selected [Tata tea (CTC), Red label tea (CTC), Lipton Darjeeling tea (orthodox, Meri Chai and (CTC tea bag)]. Among the different market teas Lipton Darjeeling (orthodox) tea was found to have greater amount of 19.57% polyphenols and Darjeeling tea was found to have the highest aroma index of 15.9.

Key words: *Camellia sinensis*, Red label tea CTC tea, orthodox tea, polyphenols, aroma index.

INTRODUCTION

Tea (*Camellia sinensis*) is the most widely used ancient beverage in the world. Tea, consumed as a beverage worldwide, is prepared from systematically processed dried tea leaves by boiling in water (milk and/or sugar are added sometimes) (Sharma and Rao, 2010). India is the major producer, consumer and exporter of tea. India accounts for almost 31% of the global production of tea. Teas are usually classified as black tea, green tea, Oolong tea, yellow tea, white tea and dark compressed tea. Black tea is the most important one consumed across the world. Black tea quality depends mainly on the components and colour of the tea infusions and tea prices vary greatly, depending on the quality which has traditionally been assessed by a tea taster who has developed a language of his own to describe various quality attributes of a tea infusion (Liang et al., 2003). The quality and prices of tea are different according to different brands in the market. Borse et al. (2002) generated a chemical and bio chemical finger print to assess the intrinsic characteristics of black teas from different regions of India. Tata Tea, Brooke Bond Red Label Tea, Taj Mahal Tea, Lipton Tea are some of the leading tea brands in India. In the present study theaflavin (TF), thearubigin (TR), high polymerized substances (HPS) and total liquor color (TLC), total polyphenols, *L, a, b* color values and aroma index of some Indian market teas (CTC, Orthodox and Tea bag) were determined.

*Corresponding author. E-mail: chsomeswararao@gmail.com.

MATERIALS AND METHODS

Determination of soluble solid content

The soluble ingredients in black tea accounts for as much as 30% of the tea weight. Tea polyphenols constitute the major portion of the soluble ingredients. Tea catechins are the primary polyphenols in the tea, and accounts for 75 to 80% of soluble ingredients. The enzyme pectinase improved the extractable-solids-yield up to 11.5%, without much improvement in polyphenols recovery, while tannase pre-treatment showed a significant improvement in polyphenols recovery (14.3%) along with an 11.1% improvement in extractable-solids-yield (Chandini et al., 2011b). About 2.5 g of black tea was added to 150 ml of boiled water in a beaker. The beaker was then covered with a lid and allowed to brew for 5 min. The mixture was filtered through whatman No.4 filter paper. The liquor was then cooled to room temperature and weighed. An aliquot of liquor was removed, placed in a pre-weighed vessel and evaporated to constant weight in an oven at 105˚C for 24 h. From the weight of residue, the amount of soluble solids in the extract was obtained (Liang and Xu, 2001).

Determination of TF, TR, HPS and TLC of tea samples

The theaflavin (TF), thearubigin (TR), high polymerized substances (HPS) and total liquor color (TLC) of the tea samples were determined by the process shown in Figure 1. The tea samples were analyzed for various chemical constituents by standard procedure (Masoud et al., 2006).

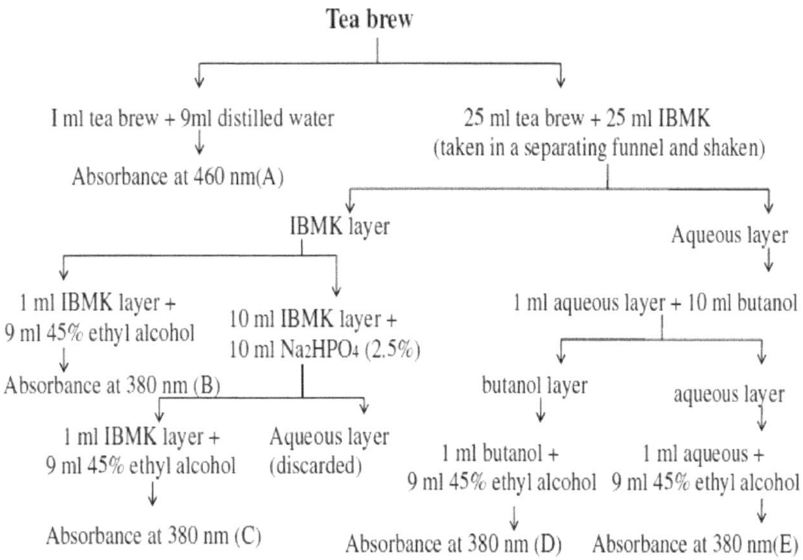

TF (%)	=	4.313 x C	TR (%)	=	13.643 x (B+D-C)
HPS (%)	=	13.643 x E	TLC	=	10 x A
TF	=	theaflavin	TR	=	thearubigin
HPS	=	high polymerized substances	TLC	=	total liquor color
IBMK	=	iso-butyl methyl ketone			
Na_2HPO_4	=	disodium hydrogen phosphate			

Figure 1. Method for determination of TF, TR, HPS and TLC.

Determination of total polyphenols of tea samples

For determination total polyphenols two grams of tea is added to 200 ml of water in a 250 ml conical flask and boiled in a water bath at 90°C for 10 min. The solution was filtered and then cooled to room temperature and finally diluted to 250 ml with distilled water (Yao et al., 2006). 1 ml of tea solution, 4 ml of distilled water and 5 ml of tartarate solution were taken in a volumetric flask and then diluted to 25 ml with phosphate buffer solution. The mixture absorbance was measured at 540 nm using spectrophotometer.

$$\text{Total polyphenols (\%)} = \frac{3.914\ E\ V_o\ 100}{1000\ V_1\ w}$$

Where, E = absorbance reading; V_o = total volume of the solution (250 ml); V_1 = the volume used for the measurement (1 ml), and W = dry weight of the tea sample.

The infusions were prepared by adding 2.5 g of tea granules to 150 ml of boiling water. Hunterlab calorimeter was used to measure the *L, a, b* values of 30 ml of each of the tea infusion. 'L' indicates lightness (L = 100 white, and L = 0 black), 'a' indicates redness when positive and greenness when negative and 'b' indicates yellowness when positive and blueness when negative. Flavour and aroma are important quality attributes of tea. Tea tasters usually assign scores to samples of tea under evaluation in a scale of 1 to 10 depending on the flavour, aroma and appearance of the sample. Electronic nose is a unique tool that is capable of sensing the volatile compounds of the tea sample and reliably predicts tea taster like scores with a high degree of accuracy. Neural network based soft computing techniques are used to tune near accurate co-relation smells print of multi-sensor array (aroma index) with that of tea tasters' scores. The aroma index of tea samples were determined by integrated electronic nose and vision (ENV) system developed by CDAT, Kolkata. The mean values of TF, TR, HPS, TLC and total polyphenols were analyzed for statistical significance by SPSS using one- way analysis of variance (ANOVA).

RESULTS AND DISCUSSION

The soluble components of tea are responsible for the taste, aroma and strength. Table 1 shows the average soluble solids in a cup of different tea infusions viz. Tata tea (CTC), Red label tea (CTC), Lipton Darjeeling tea (orthodox), and Meri chai (CTC tea bag). Meri chai (CTC tea bag) was found to have the highest average soluble solids of 0.231 g/g dry solids followed by Tata tea (CTC), Lipton Darjeeling (orthodox) and Red Label (CTC) with 0.22, 0.201 and 0.187 g soluble solids/g dry solids respectively (Table 1). In a cup (150 ml) of tea liquor Meri Chai (CTC tea bag) was found to have 0.515 g soluble solids, followed by Tata tea (CTC), Lipton Darjeeling (orthodox) and Red Label (CTC) with 0.486, 0.478 and 0.432 g soluble solids respectively (Table 1). Manson and Zhao (1994) reported similar results in extraction of tea soluble solids using ultrasound. Chandini et al. (2011a) also observed an increase in extractable-solids-yield in black tea extracts with increase in water-to-tea ratio and extraction time.

The enzyme polyphenol oxidase is responsible for oxidizing the catechins to the aflavins (TF) and thearubigins (TR), the tea pigments responsible for the colour and taste of black teas (Sharma et al., 2005). The tea samples were analyzed for various chemical

Table 1. Soluble solids in a cup of different tea infusions.

Tea sample	Average soluble solids (g/g dry solids)	Average soluble solids in a cup (150 ml) of tea liquor (g)
Tata tea	0.220^{bc}	0.486^{bc}
Red Label tea	0.187^{a}	0.432^{a}
Lipton Darjeeling tea	0.201^{ab}	0.478^{ab}
Meri Chai	0.231^{c}	0.515^{c}

All data are presented as the mean of three experiments; within columns, different superscripts indicate significant differences ($P < 0.05$).

Table 2. TF, TR, HPS and TLC of tea sample.

Tea sample	TF (%)	TR (%)	HPS (%)	TLC	TF:TR
Tata tea	0.335^{a}	9.121^{a}	14.453^{a}	4.765^{a}	0.036
Red label tea	0.263^{b}	7.296^{b}	14.899^{b}	3.893^{b}	0.036
Lipton Darjeeling tea	0.397^{c}	1.843^{c}	10.824^{c}	2.519^{c}	0.215
Meri Chai	0.283^{d}	6.451^{d}	22.169^{d}	5.733^{d}	0.044

All data are presented as the mean of three experiments; within columns, different superscripts indicate significant differences ($P < 0.05$).

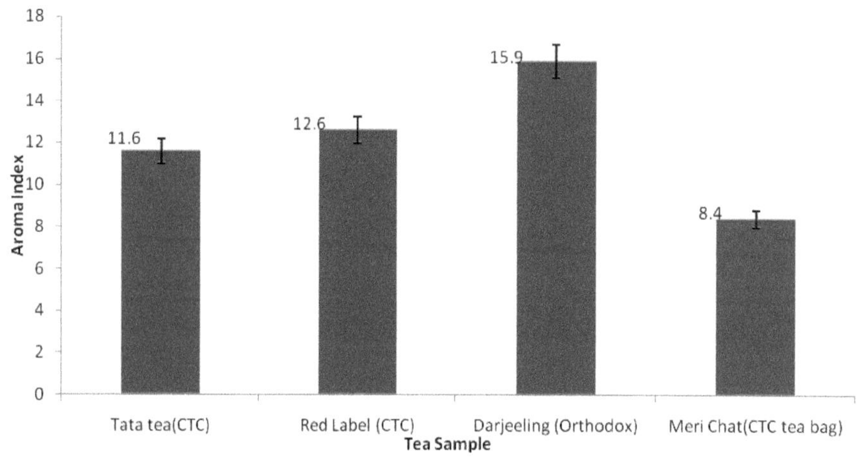

Figure 2. Aroma index of tea samples.

constituents by standard procedure (Masoud et al., 2006). The results are shown in Table 2. The TF:TR ratio should be within 0.08 to 0.1 for a good quality tea. The TF:TR ratio of Tata Tea, Red Label tea, Darjeeling Tea and Meri Chai are 0.036, 0.036, 0.215 and 0.044, respectively. Low TF content in black tea may be due to over fermentation and/or long periods of storage (Yao et al., 2006). Similar observations were made by Masoud et al. (2006) on different manufactured orthodox and CTC teas.

The primary polyphenols are oxidized during the fermentation process and the oxidized polyphenols in black tea are responsible for briskness, strength, colour, taste and pungency of the black tea infusion. Among the different market teas, Lipton Darjeeling (orthodox) tea was found to have a greater amount of polyphenols (19.57%), followed by Tata tea (CTC), Meri chai (CTC tea bag) and Red Label (CTC) with 15.56, 15.11 and 13.66% total polyphenols, respectively (Table 3). As each tea was processed from different tea cultivars that were grown in various agro-climatic conditions, the difference in polyphenol content of these various teas be reasonable. The *L, a, b* values of different tea infusions are shown in Table 4. Darjeeling tea was found to have the highest *L* and *b* value of 13.28 and 18.96 and lowest *a* value of 12.65 compared to other teas.

The major quality attributes of tea are flavor, aroma, color, and strength. Out of these, flavor and aroma are the most important attributes (Bhattacharyya et al., 2008). Darjeeling tea was found to have the highest aroma index of 15.9, followed by Red label (12.6), Tata tea (11.6) and Meri chai (8.4) (Figure 2).

Table 3. Total polyphenols of tea samples.

Tea sample	Total polyphenols (%)
Tata tea	15.56[a]
Red Label tea	13.66[b]
Lipton Darjeeling tea	19.57[c]
Meri Chai	15.11[d]

All data are presented as the mean of three experiments; within columns, different superscripts indicate significant differences ($P < 0.05$).

Table 4. *L, a, b* values of tea infusions.

Tea sample	*L*	*a*	*b*
Tata tea	16.84	21.73	10.92
Red Label tea	19.16	19.97	12.32
Lipton Darjeeling tea	31.28	12.65	18.96
Meri Chai	15.47	19.82	9.96

Conclusion

In conclusion Meri chai (CTC tea bag) was found to have the highest average soluble solids of 0.231 g soluble solids/g dry solids and Lipton Darjeeling (orthodox) tea was found to have the greatest amount of polyphenols (19.57%). Darjeeling tea was also found to have the highest aroma index of 15.9 among different market teas. Darjeeling tea is of superior quality both in chemical composition and liquoring characteristics compared to other market teas.

REFERENCES

Bhattacharyya N, Bandyopadhyay R, Bhuyan M, Tudu B, Ghosh D, Jana A (2008). Electronic nose for black tea classification and correlation of measurements with "Tea Taster" marks. *IEEE Trans.* Instrumentation Measure 57:1313-1321.

Borse BB. Rao LJ, Nagalakshmi S, Krishnamurthy N (2002). Fingerprint of black teas from India: identification of the regio-specific characteristics. Food Chem. 79:419-424.

Chandini SK, Rao LJ, Subramanian R (2011a). Influence of extraction conditions on polyphenols content and cream constituents in black tea extracts. J. Food Sci. Technol. 46(4):879-886.

Chandini SK, Rao LJ, Gowthaman MK, Haware DJ, Subramanian R (2011b). Enzymatic treatment to improve the quality of black tea extracts. Food Chem. 127:1039-1045.

Liang Y, Xu Y (2001). Efffect of pH on cream formation and solids extraction yield of black tea. Food Chem., 74:155-160.

Liang Y, Lu J, Zhang L, Wu S, Wu Y (2003). Estimation of black tea quality by analysis of chemical composition and colour difference of tea infusions. Food Chem. 80:283–290.

Manson TJ, Zhao Y (1994). Enhanced extraction of tea solids using ultrasound. Ultrasonics 32(5):375-377.

Masoud H, Bahram N, Forzani V (2006). Optimization of fermentation time for Iranian black tea production. Iranian J. Chem. Chem. Eng.25(1): 29-44.

Sharma V, Gulati A Ravindranath SD (2005). Extractability of tea catechins as a function of manufacture procedure and temperature of infusion. Food Chem. 93:141–148.

Sharma V, Rao LJM (2010). A thought on the biological activities of black tea. Critical Rev. Food Sci. Nutr. 49:5, 379-404.

Yao LH, Jiang, YM, Caffin N, Arcy , BD, Datta N, Liu X, Singanusong R, Xu Y (2006). Phenolic compounds in tea from Australian supermarkets. Food Chem. 96:614-620.

An assessment of support provided to smallholder farmers: A case study of the Vhembe District, South Africa

Phineas Khazamula Chauke, Thizwilondi Dennis Nekhavhambe and Tichaona Kudakwashe Pfumayaramba

The School of Agriculture, University of Venda, Thohoyandou, 0950, South Africa.

This paper assesses some support strategies provided to randomly selected smallholder irrigation farmers in Vhembe District, South Africa. Although irrigation farmers were selected as units of study, the focus is on all smallholder farmers. In particular the study investigates the extent to which smallholder farmers benefited from the new government policy of redressing injustices in South Africa by providing support to the previously disadvantaged in the country's rural areas. The study established the dominance of men and the elderly, less educated members, poor access to credit, non-membership to cooperatives, and substantial access to irregular extension services, and dependence on other farmers for market information. Inferential statistical analyses detected a negative impact of educational achievement and investment in production costs (inputs) to accessing credit. Amongst others the study recommended planned exit strategies for elderly farmers, support to new entrants (graduates), exposure to beneficial effects of cooperation and dependence on traditional market information sources, rather than ill-informed colleagues in the smallholder farming sector.

Key words: Support, smallholder farmers, extension, cooperative, credit, market information, production inputs.

INTRODUCTION

Failure of large-scale farming to address the plight of the rural poor and thus to bring about anticipated development shifted the policy agendas of many developing countries to supporting the smallholder farming sector, particularly in the 1980s to 1990s. One of the first African countries that made a swift shift from a focus on the former was Ghana, in response to observations that rather than benefitting the rural masses, large scale farming ended up accruing benefits to fewer rich commercial farmers and bureaucrats (Torah, 2001). At independence in 1994, the government of South Africa was faced with massive challenges of redressing inequalities within the country's rural landscape following many years of apartheid-based policies. In particular support to the white commercial farming sector occurred within an environment in which unsupported black dominated subsistence farming characterized the heavily congested communally-based homeland areas. Redressing inequalities entailed concerted efforts at the formulation of policies that would transfer effective support to the black small-scale farming sector, including those that are the focus of this study (that is, creating a conducive environment for credit availability, access to markets, promotion of cooperative enterprises, provision of extension services, access to inputs and farmer education) (Cooperatives Act, 2005; Department of Justice, 2011).

As a starting point, the new democratic government amalgamated all the former departments that were initiated to serve various agencies established under the apartheid agricultural sector regime (Department of Justice, online). This paper seeks to evaluate the extent of support provided to small-scale irrigation farmers within one of South Africa's rural district, that is, Vhembe District Municipality (VDM) in Limpopo Province. The study does not purport to investigate the levels of productivity (and thus income) nor the effectiveness of the irrigation system employed, but rather the level of support provided. The central problem to be unravelled by this study therefore centres on whether support services that are being provided in South Africa are in line with the new policy agenda that focuses on the previously disadvantaged. In the end both the targeted beneficiaries and government agencies will benefit from subsequent report back sessions to be cascaded within the district. To do justice to such an assessment however, it is imperative to provide an expose' of previous support services as attested by literature. The following section therefore provides a general review of the above mentioned support strategies, prior to articulation of the methodology, results, discussions and conclusions, and recommendations based on an empirical study of the VDM.

This paper sought to prioritise the contribution of credit to successful smallholder farming, particularly the extent to which it was impacted by other support strategies identified above. This view emanates from previous studies that have provided both success and challenging outcomes that relate to this variable. Some positive spinoffs that emanated from credit provision to smallholder farmers included participation in income generating activities, contribution to total family income and general improvement in the quality of life of rural people (Torah, 1994; Ferdoushi et al, 2011). As shown by the study conducted by Torah (1994) credit provision that is attached to funder's pre-conceived outcomes such as shifting crop production away from traditional systems could prove detrimental. In response to the failure of large scale commercial farming, the Bangladeshi government under the auspices of Agricultural Extension Services (AES) provided credit in kind to smallholders (seeds, equipments etc) with the proviso that they replace traditional staple food crops with their high value market oriented counterparts. In turn government assured farmers of procuring all their produce at acceptable prices. However, factors beyond farmers' control (bad weather, diseases etc) resulted in low and poor quality crop yields that were unsuitable for the market. As government agencies were reluctant to purchase the crop, higher repayment defaults were quite prevalent. Measures instituted by AES to improve repayment rates (group loans, loan delivery via traditional leaders, holding extension staff accountable for loans advanced to smallholders under their supervision) only succeeded in perpetuating the predicament as traditional leaders were also defaulters, whereas extension officers were reluctant to be associated with poor performing smallholders.

This paper is therefore premised on the economic theory of capital enhancement that supports intervention strategies leading to efficient allocation of resources, increased production and higher welfare (Schallock et al., 1994). Specific strategies for achieving the envisaged enhancement include access to credit, extension services, market and marketing information, membership to agricultural cooperatives (coops), investment in productive farm inputs and improved educational opportunities. A literature search on support and impact of capital enhancing strategies yielded a gap that this paper sought to close. As stated above, smallholder irrigation farmers in the Vhembe District of South Africa were targeted as units of investigation. The following paragraphs present a review of literature on associated support strategies.

Availing credit to smallholder farmers has both positive and negative spinoffs. Lack of credit facilities for smallholder farming in South Africa was sought to be one major contributing factor towards the collapse of many smaller enterprises and a major cause for higher transaction costs in both input and output markets (Randela et al., 2008; Drimie et al., 2009). Previous attempts at providing credit support to female farmers in particular focussed on bank guarantee and government credit schemes, banks for the poor and village based banks (Abee, 1996). While these had strengths and weaknesses, selection of one against another depended on management style and centres of decision making processes, adoption of centrally managed strategies over centralized and owner based management were crucial for to determination of the level of success that finally ensued. As noted by du Toit et al. (2010), the major challenge that is closely associated with credit provision to female farmers is entrapment into vicious circles of indebtedness as experienced recently in South Africa. As compared to their male counterparts, women were generally excluded from participation in the formal credit market, with commercial banks according preferential treatment to men (Mikalista, 2010).

Lack of support related to market access is a critical challenge that confronts smallholder farmers in South Africa (Mbatha et al., 2010). These farmers are faced with adverse produce marketing choices that are characterized by many producers and fewer buyers confining them to be price takers with little opportunity for flexibility as market players. The challenge is further exacerbated by lack of access to market information, a critical factor in the reduction of high transaction costs that are usually associated with agricultural markets (Randela et al., 2008). In response to this challenge some smallholders in South Africa negotiate with bigger supermarkets for preferential procurement of their produce under reasonable price arrangements. Such an arrangement allows smallholders to participate in regional trade as most South African supermarkets have invested extensively in neighbouring countries such as Botswana, Zambia and Namibia (Emongor and Kirsten, 2009).

Cooperatives have been recognized in most developing countries as important vehicles for the facilitation of both vertical coordination and horizontal integration of smallholder farmers and thus providing them with opportunities for venturing into the mainstream supply chain (Ortman and King, 2010, Lyne and Collins, 2008). In South Africa, several programmes such as those inclined towards the provision of direct cash grants, training, business advisory services, access to loans and technical support have been established (DTI, 2010; DWCPD, 2011). As advanced by Holloway et al. (1999) and Kherallah and Kirsten (2001) cooperatives assure members uninterrupted supplies of the right inputs at the right time, leading to a reduction in the transaction costs for both input and product markets. Cooperatives are also critical for improving the negotiation power of smaller farmers visa-a-versa large buyers or sellers. Bulking and bargaining services also enable cooperatives to increase access to market outlets and to avoid the hazards associated with perishability of crops and low rural produce demand. Cooperation also addresses the challenge that usually confronts smallholder farmers particularly in supplying produce that meet the required grades, quality standards and safety regulations (Kherallah and Minot, 2001; Mangisoni, 2006; Barnard et al., 2002).

A major challenge that confronts the South African cooperative sector is dependence on the above mentioned support strategies as major funding mechanisms. In most developing countries cooperatives are initiated to address specific economic needs of members rather than promotion of state dependent business entities (Ngwama et al., 2010; Ortman and King, 2010). Some challenges that have characterized many cooperatives relate to lack of adherence to cooperative principles (purpose, rights, obligations and management of cooperatives) (Chibanda et al., 2009), South African cooperatives are faced with the added challenge of poor bargaining power as most supermarkets preferred direct dealing with individuals rather than a collective of farmers (Emongor and Kirsten, 2009; Machethe, 1990).

Extension services have been adopted by many sub-Saharan governments as strategies for strengthening national agricultural support and increasing agricultural production (Evenson and Mwabu, 2001). It has however, been noted that agricultural transformation in many Sub-Saharan countries has largely been negatively impacted by the exclusion of women in the provision of extension support (Farinde and Taiwo, 2003). Poor extension services in South Africa have been identified as critical contributors to the marginalization and decrease in agricultural production activities of smallholder farmers, more especially in the country's poverty nodes (both rural and urban) (Drimie et al., 2009). Poor extension services were also observed in a Kenyan study (Mikalista, 2010). Any support strategy that enhances access to educational opportunities will indirectly impact on agricultural productivity. It can therefore be safely concluded that the level of education attained will contribute significantly to farm efficiency. Research work has attested to a positive relationship between income generation and the level of education held by smallholder farmers (Hellin et al., 2011; Evenson and Mwabu, 2001). This observation has also been confirmed in a South African study on modernizing agriculture, which has shown that the application of technological inputs called for certain skills such as reading instructions on packaging material (Kidane et al., 2005). In addition, as noted by Talimi and Masuku (2010), poorly educated and illiterate farmers were more likely to engage in illegal activities, such as employment of children in their farm production activities. In their study, Talimi and Masuku (2010) found that almost a third of households that employed child labour were illiterate as compared to a lower proportion of their literate counterparts. Illiteracy amongst smallholder farmers in South Africa is a long-term creation of the pre-1994 apartheid education policy that discriminated against blacks (Jogo and Hassan, 2010).

One major concern that impacts on support to smallholder farmers is the dominance of the elderly, especially women, in farming activities (Blake, 2011). In Nebraska for example, the average age of female farmers (59.7 years) was eight years higher than that of their male counterparts. This is a disturbing observation as individual job performance has been observed to decrease from around the age of 50 years, particularly for tasks that require problem-solving, learning and speed (Verhaegen and Salthouse, 1997). Elderly farmers are however, desirable in jobs requiring experience and verbal abilities, particularly in maintaining a high level of productivity. Arguments in support of the former observation centre around the slow pace at which elderly people grasp new technological concepts that differ from what they have already mastered as against in those that they had acquired many years of experience (Colonia-Willner, 1998). Minois and Boug (1997) also observed age related reductions in memory and learning capabilities among non-human species, ranging from fruit flies to primates. Some recent studies have however, negated the hypothesis of an inverse relationship between age and productivity (Nakana and Mkhabela, 2011; Barbierie and Mshenga, 2008). This paper however subscribes to the former views.

Support to farmers in the form of inputs is also critical for improved farmer participation in farming activities. James et al. (2009) have conveniently grouped agricultural inputs into four categories, namely, land, labour, capital and material inputs. The land input is further sub-divided into pastureland, non-irrigated crop lands and irrigated cropland while labour is usually sub-categorised into family and hired components. Employing the multi-factor productivity (MFP) approach, a measure of the ratio of the index of aggregate output quantity to the index of aggregate quantity of input. James et al. (2009) noted substantial growth in output against declining but high quality inputs. The study also found that labour productivity was positively influenced by higher levels of educational achievements. The positive role of input application for increased agricultural production has also been observed in a Malawian study that sought to improve the distribution of agricultural inputs to rural smallholder farmers (Farrow et al., 2011).

MATERIALS AND METHODS

Location of the study site

The study was conducted in the VDM, which is situated in the northern side of Limpopo Province, South Africa. The district shares international borders with Zimbabwe in the north, Mozambique in the east and Botswana on the north western side. It is composed of four local municipalities, namely; Makhado, Musina, Thulamela and Mutale. The VDM covers an area of 21 407 km^2 with a population of over 1.1 million inhabitants. It is predominantly rural with critical basic services backlogs, characterised by semi-arid conditions with rainfall ranging from 300 to 1000 mm per year. Farming under smallholder system is characterized by low levels of production and small farm sizes of approximately 1.5 ha with production being primarily for subsistence and little marketable surplus. Maize is the dominant cereal grain despite the dry and drought prone agro-ecology of much of the district (Statistics South Africa, 2001). Although the district is strategically located on the N1 corridor that connects South Africa to the rest of Africa, it mainly serves as throughway for trade traffic with very little direct spin-offs accruing to the local economy (apart from the limited shopping in Musina and Beit Bridge).

Study design and sampling procedure

A cross-sectional survey of smallholder farmers was conducted in the VDM in May 2011. The district had 30 irrigation schemes that were located within the three local municipalities, that is, Thulamela (14), Makhado (9) and Mutale (7).

Water sources were mainly storage dams and boreholes. These schemes accommodated 2100 farmers (900 in Thulamela, 700 in Makhado and 500 in Mutale respectively). A multistage clustered random sample of 370 farmers was carried out in the three local municipalities, that is, 162 from Thulamela, 126 from Makhado and 90 from Mutale respectively. In the first stage only smallholder irrigation farmers from the VDM were considered. Secondly smallholder irrigation farmers were selected on the basis of the local municipality in which they were located. The final stage involved selecting approximate proportional samples from three local municipalities (as Musina had no small-scale irrigation farming).

Data collection

Both qualitative and quantitative approaches were used to collect primary data using a pretested questionnaire between May and September 2010. The questionnaire was specifically designed to collect selected biographical data (gender and level of education), age composition (youth: 35 years and younger, adults: between 36 and 64 years, aged: 65 years and above), access to credit, extension services (access, extent of access and services provided), cooperatives (membership and benefits), markets (access to information, sources of information and marketing channels). The questionnaire was also designed to collect qualitative data, particularly as regards challenges faced by irrigators.

Data analyses

The IBM SPSS Version 19 computer program was used to capture and analyse data. Collected data were subjected to a descriptive analytical technique that focused on percentages of female and male led irrigation schemes accounted for within a particular variable and as a percentage of the total sampled farmers (gender biased). A logistic regression model, which considered credit availability as a dependent variable versus age, gender, level of education attained; access to extension services, market information, membership to cooperatives and input cost as predictor variables was also run on the basis of the probability equation as derived from the logistic regression model (Field, 2005):

$$P(Y_i) = 1/\ 1 + e^{-(b_0 + b_1 AGE + b_2 EDU + b_3 GEN + b_4 EXT + b_5 MINF + b_6 MCOP + b_7 INPCO + E_i)}$$

Where: $P(Y_i)$ = probability of credit being accessed by the ith irrigation scheme leader (1 = accessed credit, 0 = credit not accessed); e = base of natural logarithm; b_0 to b_6 = coefficients of t predictor variables; AGE = Age of irrigation scheme leader (categorical); EDU = Level of education attained by irrigation scheme leader (categorical); GEN = Gender of irrigation scheme leader. (1= female, 0 = male); EXT = access to extension services (1 = access, 0 = not accessed); MIF = access to market information (1 = access, 0 = not accessed); MCOP = membership to agricultural cooperatives. (1= member. 0 = non –member); PRODCOST = Production inputs cost (Rands), and E_i = Error term.

Diagnosis of the data was achieved through the log-likelihood and Wald statistics. The log-likelihood ratio is a measure of how well a particular model fits the data by comparing the observed and predicted values of the outcome. For example, for a given irrigation scheme leader, Y could either be 1 (credit has been provided) or 0 (credit was not provided). As noted by Field (2005), the log-likelihood is analogous to the residual sum of squares in multiple regression (an indicator of the extent of unexplained information after fitting the model, that is, larger values of the log-likelihood reflect poorly fitting statistical models). Log-likelihoods of different models are calculated and compared by looking at their differences, that is, comparing the state of a logistic regression model against a baseline, usually the logistic model when only the constant is included. The improvement of the model as additional predictors are added is computed using the following equation (Field, 2005):

$$X^2 = 2[LL(New) - LL(Baseline)]$$
$$(df = k_{new} - k_{baseline})$$

Multiplication of the value by 2 produces an X^2 distribution, and k = degrees of freedom.

Whereas in linear regression the estimated regression co-efficients (b_s) and their standard errors are used to compute the t-statistic, an

Table 1. Biographical characteristics and access to support services by irrigation scheme project leaders in the Vhembe District., May 2011.

Description		Gender				
		% Within main descriptor			% Within sub-descriptor [d]	
		Female	Male	Total	Female	Male
Level of education attained	No school	3.8	9.7	13.5	28.15	71.85
	Primary	11.4	30.8	42.2	27.01	72.99
	Secondary	10.8	25.7	36.5	29.59	70.41
	Tertiary	1.9	5.9	7.8	24.36	75.64
Age category	Youth (< 35 years)	3.8	7.7	11.4	33.33	67.54
	Adults (35 ≥ 65 years)	13.2	25.7	38.9	33.93	66.07
	Elderly (> 65 years)	10.8	38.9	49.7	21.73	78.27
Credit	Accessed	1,6	3.2	4.9	32.65	65.31
	Not accessed	26.2	68.9	95.1	27.55	72.45
Extension services	Accessed	20.5	57.8	78.4	26.15	73.72
	Not accessed	7.3	14.3	21.6	33.8	66.2
Agricultural cooperative	Member	2.2	5.4	7.6	28.95	71.05
	Non member	25.7	66.8	92.49	27.79	72.22
Market information	Accessed	27.8	71.6	99.5	27.94	71.96
	Not accessed	0	0.5	0.5	0	100
	Total	27.8	72.2	100		

Source: Study results; n =370, [d] = (Gender %/Total %) × 100.

Table 2. Logistic regression model results: Small-scale irrigators in the VDM.

Variable	B	S.E.	Wald	Sig.	Exp(B)
PRODCOST	-0.51	0.024**	4.645	0.031	0.950
EDUCL (1)	-2.992	1.049***	8.132	0.004	0.050
EDUCL(2)	-3.667	1.136***	10.427	0.001	0.026
AGE	0.011	0.014	0.635	0.425	1.012
MCOP	-0.223	0.830	0.073	0.788	0.800
Constant	5.174	1.310***	15.608	0.000	176.698

Source: Study results; n=370; ** = 5% level of significance; *** = 1% level of significance; Hosmer and Lemeshow Test: x^2 = 2.81; df = 8, p = 0.946; Cox & Snell R^2= 0.076; Nagelkerke R^2 = 0.234. Where: EDUCL (1) = Secondary education, EDUCL (2) = Tertiary Education, MCOP = membership to Agricultural Cooperative, PRODCOST = Production cost in thousands (R '000) of Rands.

analogous statistic for logistic regression is the Wald statistic, with its special distribution (the chi-square distribution). The Wald statistic reflects whether the b-coefficient of the predictor variables are significantly different from zero, that is, if significantly different, then the predictor is assumed to be making a significant contribution to predicting the outcome (Field, 2005).

RESULTS

Descriptive and inferential statistical results of the study are presented in Tables 1 and 2 respectively.

The first noticeable observation from Table 1 was the overwhelming dominance of males (72.2%) in the leadership of irrigation schemes in the VDM, in contravention of government policy towards empowering the vulnerable, including women. The dominance of males above the total average of 72.2% is also discernible under the column "% within sub-descriptor" and calculated as reflected in the footnote to Table 1. Accordingly, males were above their average in tertiary and primary education (75.64 and 72.99% respectively), participation by the elderly (78.27%) and access to extension (73.72%). As evidenced in the review of literature for this paper, credit is a quintessential for the success of farming enterprises. Table 1 also revealed that a very small proportion (4.9%) of small scale irrigation

Table 3. Observed vs. predicted frequencies for access to credit.

Observed		**Predicted**		
		Access to credit		**Correct (%)**
		Yes	**No**	
Access to credit	Yes	1	17	5.6
	No	0	352	100
Overall percentage				95.4

Source: Study results, 2011.

schemes had access to this resource, inferring that most rely on self-funding for survival. This result is not surprising and is thus expected, more especially due to high risks associated with defaults as attested by the reviewed literature above. However, one specific objective of this paper is to investigate inherent factors that could impact on credit access, given that few would be willing to venture into the associated risks. This will be achieved through the employment of inferential statistical analysis, with credit as a dependent variable.

Extension service support was relatively substantial (78.4%). When asked to reflect on the extent of access to this important resource, the majority of respondents pointed out that the process was quite irregular with almost a quarter (25.4%) of respondents reporting complete lack of contact with extension officers. The proportion that reported regular access was however, satisfactory at 36.2%. Table 1 also shows that most irrigation scheme leaders in the Vhembe District were either illiterate (13.5%) or semi-literate [attended primary schooling (42.2%)]. A substantial proportion (36.5%) had participated within the secondary school band (grades 8 to 12) with a small proportion that had reached tertiary level education (7.8%). Table 1 further illustrates that almost half (49.7%) of small-scale irrigation leaders were the elderly (above 65 years of age). In the SA context the elderly are expected to retire from active economic participation. The latter finding raises concerns regarding the sustainability of small-scale irrigation farming in the district. The situation is further exacerbated by the low participation of youth (11.4%) in leadership positions within small-scale irrigation schemes. Small-scale irrigators have made some inroads into the national fresh produce market (reported by 10.8% of respondents). An interesting finding regarding gender bias is however, that within the adult age category, female-controlled irrigation schemes were dominated by the adult group (36 to 65 years), in contrast to dominance by the elderly, as observed above for a combination of all irrigation schemes. A very small proportion (7.6%) of irrigation schemes had acquired membership to agricultural cooperatives with many reporting minimal benefits (97% of members).

The output for the logistic regression model that was fitted to the data is presented in Table 2. The model was run to assess the relationship between access to credit (dependent variable) and four predictor variables [that is, age (AGE), level of education (EDU), membership to agricultural cooperatives (MCOP) and production costs (PRODCOST)] following the elimination of some highly insignificant variables. It was noted in the descriptive analyses that the proportion of farmers that accessed credit was very small. However, the use of regression analyses allows researchers "to predict outcome" variables' from some predictor variables (Field, 2005: 144), without regard to pertaining situations.

As can be observed from Table 3, the model correctly predicted 95.40% of the cases. The Hosmer and Lemeshow test produced a chi squared value of 2.81 with a p-value of 0.95, indicating that the model's predicted estimates did not differ significantly from the observed data and thus an indication of an acceptable goodness of fit (Field, 2005). The high Wald statistics for all significant variables depicts that their b-values are significantly different from zero, thus making a significant contribution to credit access.

The categorical variable EDUCL had two categories EDUCL(1) (having secondary education) and EDUCL(2) (having tertiary education) being compared to a base category of no education/ primary education only. From the results in Table 2, it can be seen that production cost and having both secondary and tertiary education, negatively affected access to credit. This meant that for every R1000 increase in a farmer's production costs, the odds of accessing credit decreased from 1 to 0.95. The more educated a farmer was, the less likely for him or her to have accessed credit since having secondary or tertiary education both affected credit access negatively relative to no education at all or having only primary education. Being a member of an agricultural cooperative also had a negative impact on farmers' access to credit although this variable was not significant. Age, although not significant had a positive impact on the likelihood of accessing credit.

Apriori expectations were that access to credit would be positively influenced by all the variables. The negative impact of both higher educational attainment and membership production costs was unexpected. People who have been exposed to both higher educational opportunities and beneficial effects of investing in production costs (inputs) were expected to realize the critical role that credit provides in enhancing farm productivity. However, as noted in descriptive analyses of this study, most irrigation scheme leaders have low levels of education and are elderly, factors that impact negatively on access to new information and promotion of active economic participation in farming.

DISCUSSION

The finding that smallholder farmers in the VDM were male-dominated does not auger well with the South African government policy of women empowerment and capacity-building. The low proportion of women in leadership positions perpetuates inequality, poverty and unemployment in the country. As attested by several studies, many agricultural tasks that are performed in male-dominated projects were accomplished by women who made up to 57% of the labour force and produced 60% of domestic exports. The marginalization of women is more prominent amongst Indians, Latin Americans and African cultures (Rangnekar, 1998). African customary practices, for example, assigns property rights to males or male heirs, a practice that is widely applied in South Africa, as attested by the findings of this paper. According to the last census that was conducted in 2001, the proportion of females in South Africa was 60.15% (Statistics South Africa, 2001). The government's agenda is seemingly not achieving its intended objectives. The situation will continue unabated unless it is arrested through appropriate intervention strategies, as recommended in this paper.

The observation that the leadership of most irrigation schemes had acquired primary education, with a large proportion that was illiterate, translates into a number of bottlenecks for their future sustainability. In the first instance such leaders will be denied access to published information particularly as regards to produce market activities and price trends. National fresh produce markets and parastatals such as the National Marketing Council in South Africa produce vital market related data that farmers can take advantage of on a daily basis. Such data can only be accessed by farmers who are not only literate but also exposed to secondary and tertiary educational opportunities. Low educational achievement as attested through descriptive statistical analyses for this study could be associated with reliance on other farmers and dependence on the already glutted local market for disposal of produce. Consequently, due to large numbers of sellers and a relatively low customer base, farming enterprises that target such a market are bound to be price takers with high attrition rates.

Lack of access to credit as noted for this study has negative consequences for the sustainability of smallholder farmers in the VDM. Almost all (save for about 5%) relied on own funding for their farming activities. Although lack of collateral (a characteristic of most smallholders farming) could be attributed to this observation, the dominance of the elderly and a high combination of leaders with lower educational qualifications or illiteracy cannot be excluded. A study conducted in Pakistan noted that increasing access to credit was not only associated with increasing technical efficiency, but also that farmers who had access to the resource were more efficient (technical efficiency score of 0.8 and above) than their counterparts (technical efficiency score of 2.0) (Ayaz et al., 2010). Through the application of the Tobit model to collected data, that study identified farming experience (5% level of significance) and number of schooling years (5% level of significance) as additional contributors to technical efficiency of farmers.

The finding of the irregular access to extension services, as articulated above, could be associated with failure of small holder farmers to take advantage of profitable marketing channels and investment into productivity enhancing inputs. In the absence of appropriate guidance to accessing the formal market, extension services become crucial as marketing agents. In particular the revelation of the study that only 0.3% of the irrigation schemes obtained market information from extension officers could have contributed immensely to reliance on other farmers for this vital service to the farming community. Lack of access to formal markets for smallholder farmers in the VDM could persist unless drastic measures are taken to transform the extension service.

Agricultural cooperatives have been considered as important vehicles for the provision of affordable inputs, market access and technical production skills to farmers. In appropriate access to formal produce markets by small-scale irrigators in the VDM could be linked to the high level of non-membership to agricultural cooperatives. Effective membership is based on the extent to which potential benefactors are exposed to the principles and perceived advantages of cooperation. The revelation from literature regarding access to government funding as a precursor to the formation of cooperatives, needs urgent attention and complete reversal. For effective participation, members should not only be educated but also exposed to productivity-enhancing mechanisms such as extension services. As this study has clearly articulated, these are limiting factors in the Vhembe District. One critical principle of cooperation relates to internal rather than external drive in the establishment of cooperatives. The observations for this study regarding the few members that had affiliated to local primary cooperatives and the derivation of minimal benefits from such affiliation attest to external influence in the formation of most cooperatives in the VDM. As reflected in the literature review above, most cooperatives in South Africa were initiated in order to source government funding. As soon as such funds dry-up, the affected cooperative enterprises cease to exist.

CONCLUSIONS AND RECOMMENDATIONS

The main objective of this paper was to assess the support provided to smallholder irrigation farmers especially in the VDM, South Africa. Some support strategies that the study assessed included credit provision, extension services, access to markets and membership to agricultural cooperatives. A multistage random selection process yielded a sample of 370 small scale irrigation schemes scattered in three of the four local municipalities of the district. A pre-piloted questionnaire was administered to the leadership (either chairperson or his/her representative) of the irrigation schemes. The gender, education level and age of the respondents were also considered as important variables for assessing demographical impact on service provision. The collected data were captured and analysed through the IBM SPSS version 19 computer programme. A logistic regression model that considered credit availability as a dependent variable and seven other variables (age, level of education attained, gender, extension services, market information, membership to cooperatives, production costs) as predictor variables was fitted into the data. Through an elimination process, a number of variables were omitted from further analyses due to their insignificant predictive powers to the model.

The dominance of males in accessing most of the support services provided to small-scale irrigation farmers has been identified as a critical challenge for the VDM. This does not auger well in terms of food security as males have been observed to venture into farming mostly for self-enrichment, with little regard for family welfare. Strategies to increase women participation could include targeted interventions towards improvements in educational qualifications and preferential provision of grant funding. Such intervention strategies should also encompass training in financial management and technical production skills. Reducing the dominance of the elderly (65 years and above) will be a daunting task as many retirees look upon farming as an alternative to active participation in the formal job market. Exit strategies from the formal job market should therefore include workshops that would discourage new retirees from investing their life-long savings into new farming ventures as these will require physical strength and high level decision making skills. Retirees should instead be encouraged not only to look upon retirement as a period of rest but also as an opportunity to hand over existing farming ventures to their siblings. The benefactors (young farmers) should be exposed to training programmes and support mechanisms such as cheap finance, government grants and gradual exposure to private funding institutions. In the earlier years of their retirement, elderly farmers could be critical in monitoring and transmission of skills to younger farmers. As income from farming only accrues in the medium to long term, one strategy that could be used to attract youth into farming could be the provision of allowances - consistent with that prevailing in the formal job market. Perhaps the most critical challenge that prevents elderly farmers from abandoning their farming ventures could be linked to asset ownership and associated loss of esteem. To counteract the challenge, elderly farmers could be encouraged to hire out their farms to unemployed university and college graduates that are widespread in the VDM and other regions of South Africa. Targeted graduates should undergo relevant training in entrepreneurship, financial management and market development. Finance should initially be provided to cover both rental and living expenses for a predetermined period aligned to produce market readiness.

The high proportion of irrigation schemes lacking membership to agricultural cooperatives could be associated with lack of knowledge regarding benefits that can be derived from these business entities. Rather than encouraging farmers to form cooperatives in lieu of grant funding, public agencies should utilize their extension officials to expose farmers to the principles of cooperation as espoused by the International Cooperative Alliance (ICA) and the SA Cooperatives Act of 2005. In particular cooperatives should be established on the basis of perceived internal benefits rather than extrinsic rewards.

Reliance on other famers for market information should be mitigated through exposure to formal produce channels, especially fresh produce and export markets. The role of public agencies, particularly in exposing farmers to new technologies such as market based computer websites and cell phones, could be vital in opening up new markets for smallholder irrigation schemes in the Vhembe District. Funding received from the Land Bank of South Africa and the National Research Foundation is highly acknowledged.

REFERENCES

Abee A (1996). Beyond 'Banking for the Poor': Credit Mechanisms and Women's Empowerment. Gend. Dev. 4(3):48-53.

Ayaz S, Hussain Z, Sial MH (2010).Role of Credit on Production efficiency. World Acad. Sci. Eng. Tech. 66:1042-1047.

Barbierie C, Mshenga PM (2008).The role of the farm and owner characteristics on the performance of agri-tourism farms. Sociol. Rur. 48(2):166-18.

Barnard J, Cooper D, McIntosh A, Reynolds N, Vaughan A, Vawda S. (2002). Review and Policy Directed Evaluation of Community Production Centers: From Supply Driven to Demand-Led Approaches which Foster Community Ownership and Local Development. Report submitted to the Department of Public Works by McIntosh Xaba and Associates. December 2002.

Blake K (2011). Women in agriculture. Agricultural Economics Department, University of Nebraska – Lincoln. Posted at Digital Commons at University of Nebraska, Lincoln. [: http://digitalcommon. Unl.edu/agecon.cornhusker/190. Downloaded on 24 July 2011.

Chibanda M, Ortman GF, Lyne MC (2009). Institutional and government factors influencing the performance of selected smallholder agricultural cooperatives in KwaZulu Natal. *Agrek* 48(3):293-306

Colonia-Willner R (1999). Investing in Practical Intelligence: Ageing and Cognitive Efficiency among Executives. Int. J. Behav. Dev. 23(3):591-614

Cooperative Act No. 14 (2005). Government Gazette, Republic of South Africa. 482:27912, Government Printers, Pretoria.

Department of Justice (2011). Constitution of the Republic of South Africa, Act 108 of 1996, Government Printers, Pretoria.

Department of Justice (online). Agricultural Laws Extension Act 87 of 1996. http://www.Saflii.org/Za/Legis/num_Act/alca1996312/. Downloaded on 12 July 2012.

Department of Women, Children and Persons Living with Disabilities (DWCPD). (2011). Know Your Rights. www.dwcpd.gov,za. Downloaded on 16 June 2011.

DTI (Department of Trade and Industry) (2010). The DTI Cooperative Handbook.. http://led.co.za/node/4047. Downloaded on 19 July 2011.

Drimie S, Germishyse T, Rademeyer L, Schwabe C (2009).Agricultural production in Greater Sekhukhune: the future for food security in a poverty node of South Africa. *Agrek* 48(3):245-275

du Toit JP, Ortman G, Ramroop S (2010). Factors influencing the long-term competitiveness of commercial milk producers: evidence from panel data in East Griqualand, South Africa. Agrek 49 (1): 80-101.

Emongor R, Kirsten J (2009). The Impact of South African Supermarkets on Agricultural Development in SADC: A case Study in Zambia, Namibia and Botswana. *Agrek* 48(1):60-84.

Evenson RE, Mwabu G (2001). Effects of Agricultural Extension on Farm Yields in Kenya. Afr. Dev. Rev. 13(1):1-23

Farinde AJ, Taiwo KA (2003) Moving women from subsistence to commercialisation. Issues and Policy Imperatives. Paper Presented at the Workshop organised by FAO/IITA on "The Gender Impact of the Commercialisation of Smallholder agriculture Institute for International Tropical Agriculture, Ibadan, May 14 -16.

Farrow A, Resinamhodzi K, Zingere S, Delve RJ (2011). Spatially targeting the distribution of agricultural inputs stockists in Malawi. Agric. Syst. 104(9):694-702

Ferdoushi A, Chamhuri S, Nor AHI (2011). Improving Well-being of Rural Women Through Participation in Microcredit Programme: Evidence from Bangladesh. Aust. J. Bas. Appl. Sci.. 5(8): 857-862.

Field A (2005). *Discovering Statistics Using SPSS.* 2nd Edition, SAGE Publications, London.

Hellin J, Dixon J, Higman S, Keleman A (2011). High-Value Agricultural Products and Poverty Reduction: Smallholder Access to Maize Markets. J. Cr. Imp.. 25(4):371-391.DOI: 10.1080/15427528.2011.574224.

Holloway G, Nicholson C, Delgado C (1999). Agro-industrialization through Institutional Innovation: Transaction Costs, Co-operative and Milk Market Development in the Ethiopian Highlands. MSSD Discussion Paper No. 35.International Food Policy Research Institute, Washington DC, Washington.

James SJ, Alston JM, Pardey PG (2009). Structural Changes in US Agricultural Production and Productivity. Mag. Food. Farm Res. Issues 24(4):51-57

Jogo W, Hassan R (2010). Balancing the use of wetlands for economic wellbeing and ecological security. The case of Limpopo wetland in Southern Africa. Ecol. Ergon. 69(7):1569-1579.

Kherallah M, Kirsten J (2001). The New Institutional Economics: Applications for agricultural policy research in developing countries. Markets and Structural Studies Division, Discussion International Food Policy Research Institute, Washington DC. p. 41.

Kherallah M, Minot N (2001). Impact of Agricultural Market Reforms on Smallholder Farmers in Benin and Malawi. IFPRI Collaborative Research Project. International Food Policy Research Institute, Washington DC.

Kidane H, Alemu ZG, Kundhlande G (2005). Causes of Household Food Insecurity in Koredegaga Peasant Association, Oromiya Zone, Ethiopia. Agrekon 44(4):543-560

Lyne M, Collins R (2008). South Africa's new Co-operative Act: A missed opportunity for small farmers and land reform beneficiaries. Agrekon, 47 (2): 180-197

Machethe CL (1990). Factors contributing to poor performance of agricultural cooperatives in less developed areas. *Agrekon.* 29 (4): 305-309.

Mangisoni J (2006). Markets, Institutions and Agricultural Performance in Africa. ATPS special paper series the African Technology Policy Studies Network, Nairobi, Kenya. p. 27,

Mbatha NC, Antrobus GG, van Rooyen J (2010). Land Price Premiums in South Africa's land redistribution process: A case study of Northern KwaZulu Natal sugarcane farms. *Agrekon* 49(2):235-254

Minois N, Boug E (1997). Hypergravity and Aging in Drosophila Melonagster Conditioned Suppression and Habituation of the Proboscis Extension Response. Ageing, Exper. Clin. Res. 9(4):1-11

Mikalista S (2010). Gender-specific constraints affecting technology use and household food security in Western Province of Kenya. Afr.J. Food Agric. Nutr. Dev. 10 (4):2324-2340

Nakana E, Mkhabela T (2011). Performance determinants of wine farms in the Western Cape: An interval modelling approach. *Agrekon* 50(1):53-70.

Ngwama P, Lyne M, Ferrer S (2010). What will South Africa's new Cooperative Act do for small producers? An analysis of three studies in KwaZulu Natal. *Agrekon.* 49(1):39 – 55.

Ortman GF, King RP (2010). Research on Agri-food supply chains in Southern Africa involving small-scale farmers: Current status and future Possibilities. *Agrekon* 49(4):397-417

Randela R, Alemu ZG, Groenewald (2008). Factors enhancing market participation by small-scale cotton farmers. Agrekon, 47 (4):451-469.

Rangnekar S (1998). The Role of Women in Small -holder Rain-fed and Mixed Farming in India. Women in Agriculture and Modern Communication Technology: Proceedings of a Workshop, Dolberg, F Petersen, P.H: Copenhagen (Denmark): DSR 1999. ISBN 87-7432-515-9. p.181.

Schallock MD, Fredericks B, Dalke BA (1994). The house that traces built: A conceptual Model of Service Delivery Systems and Implications for Change. J. Special Educ. 28(2):203-223.

Torah S (1994). Agricultural Extension Services and Smallholder Farmers' Indebtedness in North-eastern Ghana, Journal of Asian and African Studies, 29 (1/2): 119-134.

Talimi T, Masuku MB (2010). Determinants of household participation in the child labour force: the case study of Iwo local government area in Osun State, Nigeria. Agrekon 49(3), 358-72.

Statistics South Africa (2001). Census Results, Government Printers, Pretoria.

Verhaegen P, Salthouse TA (1997). Meta-analysis of Age-Cognition Relations in Adulthood. Estimates of Linear and Non-linear Age Effects and Structural Models. Psychol.Bull.122 (:3):124-137.

An economics survey of barley crop, implications on optimized farm size and land consolidation: A case of Tehran Province of Iran

Syamak Mashayekhi[1] and Hamed Ghaderzadeh[2]

[1]Agricultural Economics, Natural Resources and Agricultural Research Center, P. O. Box: 15715 – 151, Taleghani Ave., Tehran, Iran.
[2]Department of Agricultural Economics, College of Agriculture, University of Kurdistan, Kurdistan Province, Iran.

The current study attempted to investigate the economic issues of Barely crops in Tehran Province of Iran. The data were collected in association with Tehran province Agricultural – Jahad Organization. Two-stage stratified random sampling method used to select samples and sample size is about 250 farms. The Trans log and polynomial function was used to estimate cost functions. The results showed that, the optimum size of barley farms in the case of small and medium, large and total farms groups were 17.08, 41.96 and 46.43 ha, respectively. It was greater than the average sample farm size in the case of small and medium farms group (2.96 ha) and total farms group (18.21 ha). The average total cost decreases with increase in farm size and reaches its minimum when area is 17.08, 41.96 and 46.43 ha in small, medium, large and pooled sample farmers groups, respectively. Beyond these levels of area, the average total cost increases at increasing rate. The estimated optimum size of farms showed the fact that the most profitable farm size for all farm size groups belongs to large category of farms.

Key words: Cost function, optimum size, Tehran province, barely crop, Trans log cost function.

INTRODUCTION

Agriculture contributed to more than 11.5% of the GDP at current prices in Iran in 2005 (Anonymous (2005). Barley as the fourth largest grain crop after wheat, rice, and corn in the world and second major crops in agricultural sector in Iran need more attention. Its area under cultivation occupied about 15.6% of total cultivated area under agricultural crops. It was extensively used as animal feed. Iran as one of the largest barley importers ranked fifth out of 127 Barley-importing countries in 2005 (FAO, 2008).

Tehran province is the smallest but most populous of the provinces in Iran. Agriculture is the main occupation in sub-urban plains and highlands. Land is the scarcest resource and improvement in crop productivity is the main source of growth.

The paper attempt to study the optimum size of barley crop and it may help the policy makers regarding the scope for consolidation of holdings in order to harness the economies of scale in barley production. This also may help to reduce the dependence on imports as well as the costs of production of livestock and poultry products. In attention to the importance of barley crop as major crop of Iran, this leads to system improvement of its economy. 1990, The matter of size-productivity relations is a major and continuing interest in Iranian agriculture. Afrakhteh (1997) in his study in Tea cultivation in north of Iran showed a positive and significant relationship between these two variables. Helfand and

Levine (2004) explored the determinants of technical efficiency, the relationship between farm size and efficiency in the center-west of Brazil in the agricultural year 1995 to 1996 using data envelopment analysis (DEA). The results indicated that the relationship between farm size and efficiency is non-linear with efficiency first falling and then the rising with size. Debates concerning the 'optimal farm structure' and 'optimal farm size' have a long history in agricultural economics (2004). In this study, the economics of farm size – productivity is analyzed using the long run average cost (LRAC) framework with the hypothesis that farm size expands when there is opportunity to gain efficiencies from size. The main object of the study is to find optimum size of farm in the study area.

MATERIALS AND METHODS

The data on Barley crop were collected in association with Tehran province Agricultural – Jahad Organization affiliated to Ministry of Agricultural – Jahad of Iran, during the agricultural year 2000-1. In this study, two-stage stratified random sampling method is followed. At first, 5% of villages with a population of below than 5000 from each of the eight townships, where Barley is largely cultivated (that is, Ray, Robatkarim, Firoozkooh, Savojbolag, Shahriyar, Varamin, Islamshahr and Karaj) were selected. In addition, this has to be marked that in Iran, each township includes some number of towns and a set of villages. Thus, total number of 55 sample barley growing villages covering all barley-cultivated areas in the Tehran province was chosen. At the second stage, the sampling units were selected randomly based on 20% of barley farm households of the sample villages. Thus, the total of 179 farms was the respondents for primary data. The farms were post-stratified into small, medium and large based on the criteria of small (below two hectares), medium (two to ten hectares) and large (above ten hectares). The data were collected in the year 2002 using a structured pre-tested questionnaire.
The cost function is defined as:

$$c(w,y) = \min_{x \geq o} \; [w \cdot x : x \in V(y)] \tag{1}$$

$$(\sum w_i x_i) \tag{2}$$

Where w = Vector of strictly positive input prices and $w \cdot x$ = Inner product.

The cost function thus is the minimum cost of producing a given output level during a given time period expressed as a function of input prices and output (Chambers, 1988). The cost function must be non-negative, non-decreasing, concave and linear homogeneous in input prices. Second, it should be capable of estimation with zero values of some outputs, which implies that it should allow for economies of scale and scope and sub-additively (Greer, 2003).

The most commonly employed functional forms in cost estimation are the Translog and the polynomial, both of which offer flexibility to avoid a priori restrictions on the elasticities of substitution among the variables. Studies by Carter and Dean (1961), Binswanger (1974), Paul (1987), Kuroda (1988), Mukhtar and Dawson (1990), Paul and Mehta (1991), Kuroda (1995), Salman (1999), Koo et al. (2001), Emadzadeh et al. (2002), demonstrate the application of Translog cost function in agriculture and farm production economics. Bacsi (1997) applied polynomial (quadratic) cost function to describe the production technology of agricultural producers. In this study polynomial cost function is estimated. Karaki (1998) studied the cost of production in cattle farms in Jordan using a cubic cost function and found that both borrowers and non-borrowers were far from the optimum size of farm. Kiani (1999) applied the polynomial cost function to determine the optimal input use (seed, fertilizer, labor, and so on) in irrigated wheat cultivation in 21 provinces of Iran. Elhendy and Alzoom (1961) examined the cost of Tilapia farming in the central region of Saudi Arabia from 23 intensive fish farms using the cubic cost function which provided the best fit. The optimum level of production per farm was 201 tones of tilapia per year.

A polynomial function is given by:

$$F(X) = a_n + a_{n-1}X^{n-1} + \ldots + a_1X + a_0 \tag{3}$$

Where n = Non negative integer, the degree of the polynomial function is the highest value of "n".

The average total cost function is formulated as:

$$ATC = f(X) \tag{4}$$

Where ATC = Average total Cost (Rupees / hectare)

X = Area (ha)

The linear, quadratic and cubic cost functions as in (5), (6) and (7) were estimated using OLS in order to analyze the cost per hectare - area relationship.

$$ATC = a_0 + a_1X \tag{5}$$

$$ATC = a_0 + a_1X + a_2X^2 \tag{6}$$

$$ATC = a_0 + a_1X + a_2X^2 + a_3X^3 \tag{7}$$

The linear and quadratic polynomial cost function are estimated using ordinary least squares (OLS) method and the method of fitting a linear and quadratic polynomial cost function is extended to polynomials of higher degrees. A particular degree of polynomial regression is selected as the best regression model based on error sum of squares, R^2, F-value and significance level of "F". The best model is one that has least amount of 'ESS', highest 'R^2 'and 'F' values and high significance level of 'F'.

As the variance of farm size in small size group was (0.432), the small and medium categories of farmers were pooled and a single cost function is estimated. The cost function is estimated for two size groups namely small and medium (below ten hectares) and large (above ten hectares) as: To select a particular degree of polynomial regression as the best regression model, theoretically, the subsequent two degrees of polynomial regression models should have "F" non-significant because the objective is to find the degree of polynomial cost function that provides an adequate fit. Consequently, the reduction in error sum of squares is tested by F - test as each successive term is added (Snedecor and Cochran, 1989).

For practical purposes, one can choose the model based on the quantity of error sum of squares (ESS), which is given by:

$$ESS = TSS - RSS \tag{8}$$

Where; TSS = Total sum of squares, RSS = Regression sum of

squares.

The researcher can decide based on the objective of research whether the ESS quantity is small or large. In other words, if reduction in error sum of squares (ESS) and increase in F-value or increase in coefficient of multiple determinations (R^2) in higher degree of polynomial function is found to be negligibly more than previous degree of polynomial function; because of simplicity of the model, it is better to select simpler model.

The optimum size of farm (X) is derived from the estimated cost function solving the first and second order conditions as in (9) and (10):

$$\frac{dATC}{dX} = 0 \tag{9}$$

$$\frac{d^2ATC}{dX^2} > 0 \tag{10}$$

In order to understand whether small and medium farms significantly differ from the large group, Chow's test (Gujarati, 2003) is applied.

The test procedure involves the residual sum of squares (RSS) of three cost functions (for small and medium size group, large size group and aggregate). The entire samples are broken into small and medium, and large sub samples. By using the RSS of these two sub samples and the entire samples, the Chow F-test is applied as follow:

$$F = \frac{\{RSS_A - (RSS_S + RSS_L)\} / K}{(RSS_S + RSS_L) / (n_1 + n_2 - 2K)} \tag{11}$$

Where RSS_A = Residual sum of squares for the aggregate samples, RSS_S = Residual sum of squares for small and medium samples, RSS_L = Residual sum of squares for large samples, K = Number of parameters involved in the cost function model, n_1 = Number of observations in small and medium size group, n_2 = Number of observations in large size group, The 'F' calculated is compared to 'F' table which is given by:

$$F(\alpha, K, n_1 + n_2 - 2K) \tag{12}$$

Where α = Significance level, K = Degrees of freedom for numerator, $n_1 + n_2 - 2K$ = degrees of freedom for denominator, If 'F' calculated is higher than 'F' table, then the null hypothesis of equity of the coefficients of cost function of two sub- samples is rejected. Therefore, two sub-samples are significantly different from each other, which is due to impact of size on cost of cultivation.

RESULTS AND DISCUSSION

The different degrees of polynomial cost function were estimated for each of these groups separately by using ordinary least squares (OLS) method and the results shown in Tables 1 to 4.

It should be noted that if reduction in error sum of squares (ESS) and increase in F-value or increase in coefficient of multiple determination (R^2) in higher degree of polynomial function is found to be negligibly more than previous degree of polynomial function, it is better to select more simple model because of simplicity of the model. The selection of best degree of polynomial cost function for different size groups is presented in Table 1. For all small and medium farmers group, large farmers group and entire sample farmers group cubic cost function provided the best fit.

The results of regression coefficients in respect of the cubic cost function for small and medium farmers group is presented in Table 2. The results showed that the coefficient of multiple determinations (R^2) was found to be 0.995 indicating that the variables included in the model had explained 99.5% of variations in the average total cost of barley cultivation per hectare. The adjusted coefficient of multiple determinations ($\bar{R}^2$) was found to be 0.995, which is equal to coefficient of multiple determinations indicating the goodness of fit. The observed "F" was found to be highly greater than the theoretical "F" value even at one percent level of significance which also reveals the goodness of fit.

The results of estimation of cubic polynomial cost function for large farmers group is presented in Table 3. This table shows very high values of R^2 (0.969) and adjusted R^2 (0.967) indicating goodness of fit. The coefficient of area (X) was negative and significantly different from zero, at one percent level of significance. The coefficient of quadratic term (X^2) was positive and significant at one percent level of significance. In the case of cubic term (X^3), the coefficient was found to be negative and significant at one percent level of significance. The F-value was found to be high and significant at one percent level of significance indicating goodness of fit.

The cubic polynomial cost function estimates for entire sample farmers is presented in Table 4. As already mentioned for the small and medium farmers group, and large farmers group, the estimation results of cubic polynomial cost function for entire sample farmers also showed the goodness of fit.

The Chow test was applied to test whether there is a significant change in the structural relationships of the regression coefficients of cost function of small and medium size with large size group. The F-ratio calculated in Chow test was found to be 866.193, which was highly greater than the theoretical "F" value ($F(\alpha, K, n_1 + n_2 - 2K) = F(0.01, 4, 171) = 3.32$) even at one percent level of significance, therefore the null hypothesis of the equity of the coefficients of two sub-samples is rejected. It indicates that small and medium farmers group and large farmers group are significantly different from each other which are due to impact of size on cost of cultivation.

In economic terminology, the "most profitable' amount can also be called the "optimum" amount (Doll and Orazem, 1978). The optimum size of farm, which is the most profitable amount of area, in the cases of small and medium, large and pooled groups was 17.08, 41.96 and 46.43 ha, respectively. It indicates that the average total cost decreases with increase in farm size and reaches its minimum when area is 17.08, 41.96 and 46.43 ha in small and medium, large and pooled sample farmers

Table 1. Selection of best degree of polynomial cost function.

Particulars (1)	No. of obs. (2)	ESS (3)	F-value (4)	R^2(%) (5)	$\bar{R}^2$(%) (6)
Small and medium farmers (≤ 10 ha)					
Linear	122	22,208,822	386.74*	76.3	76.1
Quadratic	122	2,482,635	2188.18*	97.4	97.3
Cubic	122	444,850	8253.02*	99.5	99.5
Quartic	122	512,922	7152.50*	99.5	99.4
Quintic	122	418,522	6525.23*	99.6	99.5
Large farmers (> 10 ha)					
Linear	57	14,513,860	152.86*	73.5	73.1
Quadratic	57	12,959,589	87.28*	76.4	75.5
Cubic	57	1,695,169	553.99*	96.9	96.7
Quartic	57	1,619,998	580.51*	97.0	96.9
Quintic	57	1,682,825	558.18*	96.9	96.8
Overall (0.15 - 200 ha)					
Linear	179	226,000,000	2.59	1.4	0.9
Quadratic	179	139,000,000	57.72*	39.6	38.9
Cubic	179	45,500,676	235.80*	80.2	79.8
Quartic	179	35,269,094	239.47*	84.6	84.3
Quintic	179	36,491,274	229.99*	84.1	83.7

*Significant at 1% level.

Table 2. Estimated cubic cost function for small and medium farmers groups.

Particulars (1)	Coefficient (2)	Standard error (3)	t-value (4)
Variable			
Intercept	8580*	16.344	524.981
X	- 851.147*	17.183	- 49.534
X^2	- 154.932*	4.456	- 34.771
X^3	7.021*	0.302	23.249
Model summary			
R^2	0.995*		
$\bar{R}^2$	0.995		
ESS	122		
N	2.96		
Average farm size (ha)	17.08		

*Significant at 1% level.

groups, respectively. No, this is not the result in a cubic function. Just look at the cubic function scatter and scatter diagram.

Conclusion

The optimum size of farm which gives the most profitable output or produces the output at the least cost is crucial for farmers in agriculture sector. The results showed that the optimum size of barley farms in the case of small and medium, large and total farms groups were 17.08, 41.96 and 46.43 ha, respectively. It was greater than the average sample farm size in the case of small and medium farms group (2.96 ha) and total farms group (18.21 ha).

The study of the optimum size of farm revealed that the average total cost decreases with increase in farm size

Table 3. Estimated cubic cost function for large farmers groups.

Particulars (1)	Coefficient (2)	Standard error (3)	t-value (4)
Variable			
Intercept	6288*	97.708	64.364
X	-87.541*	5.352	-16.356
X^2	1.295*	0.066	19.550
X^3	-0.004*	0.0002	-18.767
Model summary			
R^2	0.969		
$\bar{R}^2$	0.967		
ESS	1,695,169		
F –statistic	553.987*		
N	57		
Average farm size (ha)	50.84		

*Significant at 1% level.

Table 4. Estimated cubic cost function for entire farmers.

Particulars (1)	Coefficient (2)	Standard error (3)	t-value (4)
Variable			
Intercept	7327.113*	53.715	136.408
X	-153.643*	6.062	-25.346
X^2	2.119*	0.095	22.255
X^3	-0.0067*	0.0004	-18.918
Model summary			
R^2	0.802		
$\bar{R}^2$	0.798		
ESS	45,500,676		
F –statistic	235.801*		
N	179		
Average farm size (ha)	18.21		

*Significant at 1% level.

and reaches its minimum when area is 17.08, 41.96 and 46.43 ha in small, medium, large and pooled sample farmers groups, respectively. Beyond these levels of area, the average total cost increases at increasing rate. The estimated optimum size of farms showed the fact that the most profitable farm size for all farm size groups belongs to large category of farms.

Policy implication

Based on the results of the study, most of the barley farms were less than optimum size which directly influences the farm's profitability. Therefore, land reform could contribute to bring close together actual farm size with optimum farm size. The government could encourage farmers to come forward in the locally suitable land consolidation programs and agricultural cooperatives to operate around the optimum size of the farm which is the most profitable point.

REFERENCES

Afrakhteh H (1997). "The effective factors in quantity and quality of tea in Iran". Dev. Agric. Econ. 30:85-108.

Anonymous (2005).Iran Statistical Yearbook, . Statistical Center of Iran, Government of Iran, Tehran, Iran.

Bacsi Z (1997). "Modelling chaotic behaviour in agricultural prices using a discrete deterministic nonlinear price model". Agric. Syst. 55(3):445-459.

Binswanger HP (1974). "A cost function approach to the measurement of elasticities of factor demand and elasticities of substitution". Am. J. Agric. Econ. 56(2):377-386.

Carter OH, Dean GW (1961). "Cost-size relationship for cash crop farms in a highly commercialized agriculture". J. Agric. Econ. 43:264-277.
Chambers RG (1988). Applied production analysis a dual approach. First Edn., Cambridge University Press, Cambridge, U.K.
Doll JP, Orazem F (1978). Production Economics: Theory with Application. First Edn., John Wiley & Sons inc., New York, USA.
Elhendy AM, Alzoom AA (1961). "Economics of fish farming in Saudi Arabia:analysis of costs of tilapia production". Aquacult. Econ. Manage. 5(3-4):264-277.
Emadzadeh M, Akbari N, Ghasemi M (2002). "Estimation of translog cost function and floor price of Date production in Fars province of Iran". Dev. Agric. Econ. 40: 67-88.
FAO (2008). Statistical database, On-line data (http//: www.faostat.fao.org), Accessed 2 July 2008.
Gujarati DN (2003). Basic Econometrics. Fourth Edn., Boston, MA: McGraw-Hill.
Greer ML (2003). "Can rural electric cooperatives survive in a restructured US electric market? An empirical analysis". Energy Econ. 25(5):487-508.
Helfand SM, Levine ES (2004), "Farm size and determinants of productive efficiency in the Brazilian center-west". Agric. Econ. 31(2-3):241-249.
Karaki A (1998). "The impact of credit on small farm cost of production: The case of cattle farms in Jordan". Dirasat Agric. Sci. 25(2):7-41.
Kiani KH (1999). "Determination of optimal input use in irrigated wheat cultivation in Iran". Dev. Agric. Econ. 37:7-41.
Koo WW, Weining M, Sakurai T (2001). "Wheat demand in Japanese flour milling industry: a production theory approach". Agric. Econ. 24:167-178.
Kuroda Y (1988). "The output bias of technological change in postwar Japanese agriculture". Am. J. Agric. Econ. 70(3):663-673.
Kuroda Y (1995). "Labor productivity measurement in Japanese agriculture: 1956-1990". Agric. Econ. 12(1):167-178.
Mukhtar SM, Dawson PJ (1990). "Herd size and unit cost of production in the England and Wales dairy sector". J. Agric. Econ. 41:9-20.
Paul S (1987). "A cost function analysis of wheat production in India". Eur. Rev. Agric. Econ. 14:221-238.
Paul S, Mehta R (1991). "Technology, factor demand and substitution in Indian agriculture 1960-1983". Indian J. Agric. Econ. 46(4):530-542.
Salman AZ (1999). "Econometric analysis of Barley production in low rainfall areas in Jordan". J. Dirasat-Agric. Sci. 50(4):584-605.
Snedecor GW, Cochran WG (1989). Statistical Methods. Eighth Edn., Lowa state University Press, Lowa, U.S.

Performance evaluation of a manually operated cotton picker

Adebija J. A. and Jackson B. A.

National Centre for Agricultural Mechanization, KM 20, Ilorin-Lokoja Highway, PMB 1525, Ilorin, Kwara State, Nigeria.

Cotton harvesting/picking may be accomplish manually or mechanically using cotton picker. A manually operated cotton picker was developed and on-farm performance evaluation was carried out. The cotton picker was compared with the existing manually operated cotton picker, and the traditional method of harvesting cotton on the bases of time input, labour requirement, ground and plant harvest losses. Results indicate that labour requirement for the developed cotton picker; the existing cotton picker and manual cotton picker were 166.17, 173.8 and 93.3 man-hour-ha^{-1} respectively. With their respective ground harvest losses of 15, 12 and 5% at the branch moisture content of 39.5% wet base and cotton moisture content of 2.8% wet bases. The study also indicates that time input, labour requirement, ground plant harvest losses were statistically significant at 5% level.

Key words: Cotton picker, picking, efficiency, picking losses, manual.

INTRODUCTION

Importance of cotton in Nigeria

Cotton has been grown in Nigeria for many centuries to meet the need of the local spinning and weaving industries (Mustafa et al., 1968). Among the several different types of cotton introduced into Nigeria for research purposes was "Allen long staple" an American upland type (*Gossypium hirsutum*), a onetime popular variety in parts of the United States of America for its high quality. It came to Nigeria from Uganda in 1912 (Mustafa et al., 1968). This cotton proved better and adapted very well to the local Nigerian farming condition such that subsequent improvements were easily made from it. This lead to what is known as "Nigerian Allen" and later the "Samaru Allen" from which satisfactory levels of yield resistance to diseases and pest as well as high lint's quality were obtained. It was from this seed that improved varieties such as Samaru 26C, Samaru 26J, Samaru 68, Samaru 70, Samaru 72 and Samaru 77 were obtained (Jones, 1968). Later in the sixties many cultivars and strains/varieties were introduced into Nigeria from different parts of the world to aid the cotton research programme. Some of these cultivars possessed very distinctive characteristics such as high lint quality, high yield, resistance to pest and diseases and early maturity. These were used to develop the current commercial varieties of cotton (Jones, 1968).

Generally, the entire cotton growing regions are divided into three zones based partially on climatological and other related conditions. These divisions are:

1. Northern cotton zone comprises of Kano, Kaduna, Katsina, Sokoto, Kebbi and Jigawa States.
2. Easter cotton zone comprises of Borno, Bauchi, Adamawa, Yobe and Taraba States
3. Southern cotton zone comprises of Kwara, Niger, Kogi, Oyo, Osun and Ondo (Gbadegesin and Uyovbisere, 1994)

Figure 1. Top view of the manual cotton picker.

With all these improvements in the quality of cotton and the increase in the output of cotton, there was an establishment of mills for spinning and weaving of cloths. The total output of cotton was so high that there was need for export of the product to earn foreign exchange (Mustafa et al., 1968). Apart from meeting the need of weaving industry, exportations was enhanced and other use for the cotton seed was also developed which is, the extraction of oil from the seed. The cotton oil seed is very rich in mineral content. The oil in cotton seed is 35% protein (Hui, 1992). The cotton seed left after extraction of oil is the meal which is completely edible to ruminants and monogastric animals. As such it is used as protein additive to animal feeds (Monogastric animals). It is used to provide raw materials for our cloth industries and the lint is used to spin thread which is used for sawing. It provides employment in cotton ginning industry. Cotton is the most important raw material for textile industries and it is used extensively in medicine and surgery for dressing wounds. The oil is also used as carriers for agriculture sprayers and the by-product from the refining process are used in soap manufacturing. It is also used as a source of fatty-acids (Hui, 1992).

Economic importance of cotton

About 20 million tons of cotton are produced each year in about 90 countries. China, United States, India, Pakistan, Uzbekistan and West Africa account for over 75% of global production (Internet document). Cotton represents nearly half the fibre used to make clothes and other textiles worldwide, with much of the rest coming from synthetic products.

U.S. textile mills presently consume approximately 7.6 million bales of cotton a year. Eventually, about 57% of it is converted into apparel, more than a third into home furnishings and the remainder into industrial products.

Objectives

1. To carry out on farm evaluation of a developed manually operated cotton picker.
2. To compare the performance of the machine with an earlier developed manually operated picker and the traditional method of cotton picking.
3. To assess its adaptability

Justification

Manual harvesting of cotton is drudgery prone and time consuming. Pate et al. (1963), Mechanical harvesting of cotton is limited to large scale farmers in Nigeria because they are the only ones who can afford these machines. This make the production of this vital commodity limited to very few hands. Even the few large scale cotton producers who have these machines spend a great deal of their cost in maintaining them. This is equally not making it economical for them to be in the production of cotton Prentice (1972).

Description of the cotton picker

The cotton picker consists of picking finger which are fixed to the chain. They are fastened to the chain by brazing on a sheet metal which is brazed to the chain. The fingers are made of three pieces of wire of length 5 cm each. A throttle wire is used because it has a springing effect which is needed in the operation of the picking fingers. The chain is driven by two shafts (OD = 55 mm each) the shaft at the picking end of the machine passes through a bearing (ID = 25 mm) fitted with a casing. The crank handle is connected to the shaft at the power end. Human power is transmitted to the machine for cotton picking operation by the sprocket and chain arrangement, the spikes are fitted in front of the fingers at the rear end of the machine to disengage the picked cotton which then fall by gravity through the passage provided on the casing Smith and Wilkes (1976) . A hanged bag to casing serves as the collection point. Three ground wheels (one in front and two at the rear) are provided to carry the assembled components of the machine. The pictorial view of the machine cotton picker is shown in Figures 1 and 2.

Figure 2. Pictural view of the cotton picker.

Table 1. Table of subjects data.

Subjects	A	B	C
Sex	Male	Male	Female
Age	26	18	19
Occupation	Secondary school leaver	Student	Student
Height (m)	1.8	1.6	1.8
Health condition	Good	Good	Good
Body weight (kg)	57	54	58

Figure 3. Cotton picker been used on the field.

RESEARCH METHODOLOGY

The harvesting efficiency, timeliness, and energy/power utilization by the operator/user of the cotton picker was evaluated on the field (Figure 3). The experimental field of Faculty of Agriculture in A.B.U. Zaria was used for the test. Three subjects were used for the test and their personal data was taken. The table of the data of the body condition of the subjects is shown in Table 1.

The time taken to harvest all the cotton stands contained in each ridge was measured using stop watch for both traditional hand method and mechanized method using both the existing and the improved version Bates (1963).

During the test the necessary time of operation of each of the subjects using the options were taken and their body conditions such as pulse rate before and after were taken Saidu (1990).

Evaluation criteria for the manually-operated cotton picker

The layout of the test field was such that planting of cotton was done in inter-row width of 120 cm and the inter-row distance between plants was 90 cm as recommended by Prentice (1972) to increase the efficient operation of the machine. The plant height between 45 and 60 cm fall within the working range of the

Table 2. Harvesting test result showing losses and labor requirements.

Data no.	Branch moisture content (% w.n)	Cotton moisture content (% w.n)	Field area covered (m^2)	Time Input (min)			Labour requirement (man-hour-ha^{-1})			Ground harvest loss (%)			Plant harvest loss (%)		
				Mech^1	Mech^2	Man'^{1}	Mech^1	Mech^2	Man'1	Mech^1	Mech^2	Man'1	Mech^1	Mech^2	Man'1
1	57.8	2.8	50	92.50	105.0	67.50	308.3	350.0	225.0	17	13	10	19	18	5
2	37.2	2.7	50	125.0	128.0	68.00	416.6	426.6	226.7	15	12	5	17	15	3.5
3	58.4	2.9	50	165.0	170.0	92.00	550.0	566.6	306.7	10	8	5	18	16	10
1	39.5	2.83	50	50.00	56.15	38.00	166.7	173.8	93.3	15	12	5	14	12	12
2	41.6	1.86	50	85.00	93.00	71.00	283.3	310.0	236.7	15	8	7	13	10	7
3	36.4	2.35	50	115.0	118.0	98.30	383.3	393.3	326.3	17	14	9	11	10	5
1	33.4	1.73	50	100.0	120.4	64.00	333.3	401.3	180.0	10	9	10	12	10	8
2	31.5	2.04	50	62.5	66.38	30.00	208.3	221.6	100.0	11	10	9	16	15	7
3	34.6	2.3	50	70.00	74.45	48.00	233.3	238.1	116.7	10	10	10	15	13	6

M^1 = Test result from the developed cotton picker; M^2 = Test result from Asota (1990) machine (existing machine).

machine in order to reduce field losses during harvesting. Two system of cotton harvesting, (i) manual system (manual picking of cotton with hands) and (ii) mechanical system (mechanical picking by the developed manually-operated cotton picker) were considered. Parameters such as ease of operation, machine efficiency, moisture content, field losses, energy expenditure and labour requirement were determined. The performance of the manually-operated cotton picker in terms of these parameters was evaluated and compared with manual system to evaluate the improvement in the performance of the developed machine and its feasibility for adaption (Table 2).

RESULTS AND DISCUSSION

The performance evaluation was carried out when majority of the bolls (about 80% of the total crop) were opened to reduce field losses and increase the quality of the cotton harvested. According to cotton improvement conference (1963), cotton should be 75 to 80% matured before picking/harvesting. In the field where cotton stands are unharvested for 4 to 6 weeks, losses in fiber and seed quality are increased. The machine was made to operate at an average forward speed of 1m/s, higher speed could result to dirty cotton and increase in cotton losses Basil and Geoge (1955).

Analysis of field results

The result from the field tests have shown that the developed manually operated cotton picked had lower average labour requirement of 320.34 man-hour-ha^{-1} as against then average labour requirement of 342.37 man-hour-ha^{-1} recorded for the existing machine. However the lowest average labour requirement of 201.27 man-hour-ha^{-1} when the cotton moisture content was 2.8% (w.b.) was obtained for the manual picking of the cotton. The differences in their means were statistically significant at 5%. Similar trends in the results were obtained for other moisture level considered. However, there was no comparative advantage of the developed cotton picker over the existing machine in terms of ground and plant harvest losses (Tables 3 to 5).

Conclusions

1. The developed manually operated cotton picker had an efficiency of 60%.
2. Lower labour requirements were recorded for the developed manually operated cotton picker over the existing cotton picker.
3. There was a time input of 105 min when

Table 3. Paired sample t-test value of significant between machines 1 and 2 in terms of time input, labour requirement, ground harvest loss and plant harvest loss.

Performance characteristics	t-test (cal.)	t-test [Tab.(0.05)]	t-test[tab.(0.01)]	Remark
Time input	-3.86	2.31	3.36	**
Labour requirement	-3.19	2.31	3.36	*
Ground harvest loss	3.90	2.31	3.36	**
Plant harvest loss	3.35	2.31	3.36	*

Machine 1 = after modifications; Machine 2 = before modifications; * = statistically significant at 5% level; ** = statistically significant at 1% level.

Table 4. Paired sample t-test value of significant difference between machine 1 and hand picking in terms of time input, requirement, ground harvest loss plant harvest loss.

Performance characteristics	t-test (cal.)	t-test [Tab.(0.05)]	t-test[tab.(0.01)]	Remark
Time input	4.64	2.31	3.36	**
Labour requirement	5.46	`2.31	3.36	**
Ground harvest loss	4.16	2.31	3.36	**
Plant harvest loss	6.60	2.31	3.36	**

Machine 1 = after modifications;** = statistically significant at 1% level.

Table 5. Paired sample t-test value of significant difference between a manual cotton picker and hand picking in terms of time input, labour requirement, ground harvest loss and plant harvest loss.

Performance characteristics	t-test Cal.	t-test, Tab.(0.05)	t-test, tab.(0.01)	Remark
Time input	5.63	`2.31	3.36	**
Labour requirement	6.08	`2.31	3.36	**
Ground harvest loss	3.25	`2.31	3.36	*
Plant harvest loss	3.71	2.31	3.36	**

Machine 2 = before modifications; * = statistically significant at 5% level; ** = statistically significant at 1% level.

using the existing cotton picker in an area of 50 m^2. This time was reduced to 93 min when using the developed manually operated cotton picker on the same area of field.

RECOMMENDATION

The manually operated cotton picker can be improved for greater efficiency, increase in field capacity and speed of picking using a fuel (petrol or diesel) powered engine. Improvement can be made in the picking width by increasing the picking fingers in such a way that two rows can be picked at a time. With this improvement a specific planting space and ridge spacing will have to be used on the field. All these will reduce total time needed to work on a field and the man-ha-$hour^{-1}$ value of the machine. Speed of picking will also be enhanced if a fuel (petrol or diesel) powered engine replaces manual operation.

REFERENCES

Asota CN (1990). Cropping Scheme report, Institute for Agricultural Research, Ahmadu Bello University, Zaria, Nigeria.

Basil GC, Geoge JH (1955). Cotton growing problems Ms Grow Hill Book Company, Inc. New York. pp. 68-74.

Bates WN (1963). Mechanization of tropic crops, Temple Press Limited, London. pp. 38-60.

Jones JS (1968). Cotton Production in Nigeria and the possibilities of a cotton estate to produce longer staple cotton. Report to the committee. pp. 3-5.

Pate JB, Barker DH, Feaster CV (1963). Cotton improvement conference, Dallas, Texas. Pp. 105-120.

Prentice AN (1972). Cotton with special reference to Africa. Longman Group Ltd., London. pp. 156-163.

Saidu SM (1990). Field evaluation and modification of a manually operated cotton picker. Unpublished B. Eng. Project work, Department of Agricultural Engineering, Ahmed Bello University, Zaria, Nigeria. pp. 2-8.

Smith HP, Wilkes LH (1976). Farm machinery and equipment. M Mc Grow hill Book Company, New York pp. 64-108.

Mustafa S, Obionu JA, Wali A, Datti A, Faskari NaA,Yerima AK(1985).Report of the committee on cotton production in the Northern States. pp. 10-12.

Permissions

All chapters in this book were first published in AJAR, by Academic Journals; hereby published with permission under the Creative Commons Attribution License or equivalent. Every chapter published in this book has been scrutinized by our experts. Their significance has been extensively debated. The topics covered herein carry significant findings which will fuel the growth of the discipline. They may even be implemented as practical applications or may be referred to as a beginning point for another development.

The contributors of this book come from diverse backgrounds, making this book a truly international effort. This book will bring forth new frontiers with its revolutionizing research information and detailed analysis of the nascent developments around the world.

We would like to thank all the contributing authors for lending their expertise to make the book truly unique. They have played a crucial role in the development of this book. Without their invaluable contributions this book wouldn't have been possible. They have made vital efforts to compile up to date information on the varied aspects of this subject to make this book a valuable addition to the collection of many professionals and students.

This book was conceptualized with the vision of imparting up-to-date information and advanced data in this field. To ensure the same, a matchless editorial board was set up. Every individual on the board went through rigorous rounds of assessment to prove their worth. After which they invested a large part of their time researching and compiling the most relevant data for our readers.

The editorial board has been involved in producing this book since its inception. They have spent rigorous hours researching and exploring the diverse topics which have resulted in the successful publishing of this book. They have passed on their knowledge of decades through this book. To expedite this challenging task, the publisher supported the team at every step. A small team of assistant editors was also appointed to further simplify the editing procedure and attain best results for the readers.

Apart from the editorial board, the designing team has also invested a significant amount of their time in understanding the subject and creating the most relevant covers. They scrutinized every image to scout for the most suitable representation of the subject and create an appropriate cover for the book.

The publishing team has been an ardent support to the editorial, designing and production team. Their endless efforts to recruit the best for this project, has resulted in the accomplishment of this book. They are a veteran in the field of academics and their pool of knowledge is as vast as their experience in printing. Their expertise and guidance has proved useful at every step. Their uncompromising quality standards have made this book an exceptional effort. Their encouragement from time to time has been an inspiration for everyone.

The publisher and the editorial board hope that this book will prove to be a valuable piece of knowledge for researchers, students, practitioners and scholars across the globe.

List of Contributors

Shahin Jahangirzadeh Khiavi
Department of Horticulture Science, University of Tehran, Karja 31587, Iran

Zabihollah Zamani
Department of Horticulture Science, University of Tehran, Karja 31587, Iran

Mohsen Mardi
Agricultural Biotechnology Research Institute (ABRII), Karaj, Iran

Mohamadreza Fatahi Moghdam
Department of Horticulture Science, University of Tehran, Karja 31587, Iran

Joseena Mariam Iype
School of Bio Sciences and Technology, VIT University, Vellore-632014, TamilNadu, India

Rashmi Mishra
School of Bio Sciences and Technology, VIT University, Vellore-632014, TamilNadu, India

S. Karthikeyan
School of Bio Sciences and Technology, VIT University, Vellore-632014, TamilNadu, India

S. Babu
School of Bio Sciences and Technology, VIT University, Vellore-632014, TamilNadu, India

K. M. Gothandam
School of Bio Sciences and Technology, VIT University, Vellore-632014, TamilNadu, India

Kisu Simwaka
School of Economics and Finance, University of KwaZulu-Natal, J Block, 3rd Floor, Westville Campus, Durban, South Africa

Stuart Ferrer
South African Cane Growers Association, KwaZulu-Natal, South Africa

Geoff Harris
School of Economics and Finance, University of KwaZulu-Natal, J Block, 3rd Floor, Westville Campus, Durban, South Africa

Godwin O. Chukwu
National Root Crops Research Institute, Umudike, P.M.B. 7006, Umuahia, Abia State, Nigeria

Tessy U. Madu
National Root Crops Research Institute, Umudike, P.M.B. 7006, Umuahia, Abia State, Nigeria

Emilia C. Chinaka
National Root Crops Research Institute, Umudike, P.M.B. 7006, Umuahia, Abia State, Nigeria

Nkoli H. Anyaegbunam
National Root Crops Research Institute, Umudike, P.M.B. 7006, Umuahia, Abia State, Nigeria

Yuh-Tyng Chen
Department of Information Management, Tainan University of Technology, Yung-Kang District, Tainan 710, Taiwan

Lin-Fan Chen
Department of Applied English, Tainan University of Technology, Yung-Kang District, Tainan 710, Taiwan

P. K. Chauke
Department of Agricultural Economics, University of Venda, Thohoyandou, 0950, South Africa

M. L. Motlhatlhana
Department of Agricultural Economics, University of Venda, Thohoyandou, 0950, South Africa

T. K. Pfumayaramba
Department of Agricultural Economics, University of Venda, Thohoyandou, 0950, South Africa

F. D. K. Anim
Department of Agricultural Economics, University of Venda, Thohoyandou, 0950, South Africa

ZhaoSen Xie
College of Horticulture Science and Technology, Hebei Normal University of Technology and Science, Qinhuangdao 066600, Hebei, China

Xijun Guo
Liaocheng Municipal Forestry Bureau, Liaocheng, 252000, Shandong, China

Hongmei Cao
College of Horticulture Science and Technology, Hebei Normal University of Technology and Science, Qinhuangdao 066600, Hebei, China

W. Abera
African Centre for Crop Improvement, University of KwaZulu-Natal, Private Bag X01, Scottsville 3209, Pietermaritzburg, South Africa

S. Hussein
African Centre for Crop Improvement, University of KwaZulu-Natal, Private Bag X01, Scottsville 3209, Pietermaritzburg, South Africa

J. Derera
African Centre for Crop Improvement, University of KwaZulu-Natal, Private Bag X01, Scottsville 3209, Pietermaritzburg, South Africa

M. Worku
CIMMYT Kenya ICRAF House, P. O. Box 1041, Nairobi Kenya

M. D. Laing
African Centre for Crop Improvement, University of KwaZulu-Natal, Private Bag X01, Scottsville 3209, Pietermaritzburg, South Africa

Munyati Vincent
Department of Agricultural Economics, Education and Extension, Bindura University of Science Education, P. Bag 1020, Bindura, Zimbabwe

Mugabe Douglas
Department of Agricultural Economics and Development, Midlands State University P. Bag 9055, Gweru, Zimbabwe

Chipunza Nyasha
Department of Agricultural Economics and Development, Midlands State University P. Bag 9055, Gweru, Zimbabwe

Mafuse Never
Department of Agricultural Economics, Education and Extension, Bindura University of Science Education, P. Bag 1020, Bindura, Zimbabwe

Chagwiza Godfrey
Department of Agricultural Economics, Education and Extension, Bindura University of Science Education, P. Bag 1020, Bindura, Zimbabwe

Musara Joseph
Department of Agricultural Economics, Education and Extension, Bindura University of Science Education, P. Bag 1020, Bindura, Zimbabwe

V. B. Taru
Department of Agricultural Economics, Federal University of Technology, P. M. B. 2076 Yola, Adamawa State, Nigeria

A. A. Ndaghu
Department of Agricultural Technology, The Federal Polytechnic, P. M. B. 35 Mubi Adamawa State, Nigeria

C. I. Ofoka
Department of Agricultural Extension, University of Nigeria Nsukka, Nigeria

J. M. Chah
Department of Agricultural Extension, University of Nigeria Nsukka, Nigeria

M. C. Madukwe
Department of Agricultural Extension, University of Nigeria Nsukka, Nigeria

D. Kalisa
Ministry of Agriculture, Box 9192, Dar es Salaam, Tanzania

A. Majule
Institute of Resource Assessment, University of Dar es salaam, Box 35097 Dar es Salaam, Tanzania

J. G. Lyimo
Institute of Resource Assessment, University of Dar es salaam, Box 35097 Dar es Salaam, Tanzania

S. P. Deshmukh
Department of Agronomy, Navsari Agricultural University, Navsari, India

J. G. Patel
Department of Agronomy, Navsari Agricultural University, Navsari, India

A. M. Patel
Department of Agronomy, Navsari Agricultural University, Navsari, India

S. A. Mir
KVK Pombay, Kulgam, Kashmir- 192233, India

I. A. Mir
ARIS, Cell, SKUAST Kashmir, India

S. M. K. Quadri
Post Graduate Department of Computer Science, University of Kashmir, Hazratbal Campus-190006, India

N. Ahmad
Central Institute of Temperate Horticulture, (ICAR), Old Airport, Srinagar, India

S. Angchuk
KVK Kadakh (Leh), India

Jose de Assuncao Sambo Tomas
Department of Agricultural Economics, College of Management, Zhejiang University, Hangzhou, China

Ezekiel N. Ndunda
Centre for Environmental Economics and Policy in Africa (CEEPA), Department of Agricultural Economics, Extension and Rural Development, Faculty of Natural and Agricultural Sciences, University of Pretoria, Pretoria 0002, South Africa

Eric D. Mungatana
Centre for Environmental Economics and Policy in Africa (CEEPA), Department of Agricultural Economics, Extension and Rural Development, Faculty of Natural and Agricultural Sciences, University of Pretoria, Pretoria 0002, South Africa

Adeosun Olubunmi
Department of Fisheries Technology, Oyo State College of Agriculture, Igbo-ora, Oyo State, Nigeria

Arif Semerci
Department of Agricultural Economics, Faculty of Agriculture, Mustafa Kemal University, Tayfur Sökmen Kampüsü, Antakya, Hatay, Turkey

Carlo Ponzio
Dipartimento di Scienze Agrarie, Alimentari ed Ambientali. Università Politecnica delle Marche, Via Brecce Bianche, 60131 Ancona, Italy

Ramesh Gangatharan
Dipartimento di Scienze Agrarie, Alimentari ed Ambientali. Università Politecnica delle Marche, Via Brecce Bianche, 60131 Ancona, Italy

Davide Neri
Dipartimento di Scienze Agrarie, Alimentari ed Ambientali. Università Politecnica delle Marche, Via Brecce Bianche, 60131 Ancona, Italy

Taye S. Mogaji
Department of Mechanical Engineering, Federal University of Technology Akure, Ondo State, Nigeria

Awolala D. Olufemi
Department of Agricultural and Resource Economics, Federal University of Technology Akure, Ondo State, Nigeria

Olorunisola P. Fapetu
Department of Mechanical Engineering, Federal University of Technology Akure, Ondo State, Nigeria

A. O. Ajayi
Department of Agricultural Extension and Rural Development, Obafemi Awolowo University, Ile-Ife, Osun State, Nigeria

O. S. Alabi
Department of Agricultural Extension and Rural Development, Obafemi Awolowo University, Ile-Ife, Osun State, Nigeria

T. O. Akinsola
Department of Agricultural Extension and Rural Development, Obafemi Awolowo University, Ile-Ife, Osun State, Nigeria

Robert M. Ogeto
Kenya School of agriculture, P. O. Box 1909-10100, Nyeri, Kenya

Erick Cheruiyot
Department of Crops, Horticulture and Soil Sciences, Egerton University, P. O. Box 536, Njoro, Kenya

Patience Mshenga
Department of Agricultural Economics and Agribusiness Management, Egerton University, P. O. Box 536, Njoro, Kenya

Charles N. Onyari
Embu University College, P. O. Box 6-60100, Embu, Kenya

Magicson Mthembu
Faculty of Natural and Agricultural Sciences (Centre for Sustainable Agriculture), University of the Free State, South Africa

Cliff S. Dlamini
WWF South Africa, C/O Table Mountain Fund, Centre for Biodiversity Conservation, Kirstenbosch National Botanical Gardens, P. O. Box 23273, Claremont, 7735, South Africa

Banele Nkambule
World Vision Swaziland, P. O. Box 2870 Mbabane Swaziland

J. A. Yidana
Department of Horticulture, Faculty of Agriculture, University for Development Studies (UDS), Tamale, Ghana

M. Osei-Kwarteng
Department of Horticulture, Faculty of Agriculture, University for Development Studies (UDS), Tamale, Ghana

Y. Amadu
Department of Horticulture, Faculty of Agriculture, University for Development Studies (UDS), Tamale, Ghana

David R. Coventry
School of Agriculture Food and Wine, University of Adelaide, Waite Campus, SA, Australia

Randhir Singh Poswal
Directorate of Wheat Research, Agarsain Marg, Karnal-132001, Haryana, India

A. Yadav
CSS Haryana Agricultural University, Hisar, Haryana, India

Ramesh Kumar Gupta
Directorate of Wheat Research, Agarsain Marg, Karnal-132001, Haryana, India

B. S. Duggal
State Department of Agriculture, Haryana, Panchkula, India

R. S. Chhokar
Directorate of Wheat Research, Agarsain Marg, Karnal-132001, Haryana, India

V. Kumar
CSS Haryana Agricultural University, KVK, Bawal, Haryana, India

A. S. Riar
Directorate of Wheat Research, Agarsain Marg, Karnal-132001, Haryana, India

Subash Chandra Gill
Directorate of Wheat Research, Agarsain Marg, Karnal-132001, Haryana, India

Anuj Kumar
Directorate of Wheat Research, Agarsain Marg, Karnal-132001, Haryana, India

Ramesh Kumar Sharma
Directorate of Wheat Research, Agarsain Marg, Karnal-132001, Haryana, India

Ramesh Chand
Directorate of Wheat Research, Agarsain Marg, Karnal-132001, Haryana, India

R Sendhil
Directorate of Wheat Research, Agarsain Marg, Karnal-132001, Haryana, India

S. G. L. Kleemann
School of Agriculture Food and Wine, University of Adelaide, Waite Campus, SA, Australia

J. A. Cummins
Rural Solutions SA, Australia

Mahmoud Bakhshinejad
Department of Economics, Islamic Azad University, Shahrekord Branch, Iran

Esmail Hosseini
Department of Banking, Monetary and Banking Research Institute (MBRI), Central Bank of Iran

Eliud Birachi
International Center for Tropical Agriculture-Africa (CIAT), Box 1269 Kigali, Rwanda

Rachel Zozo
Consortium for Improvement of Agricultural-based Livelihoods in Central Africa (CIALCA), CIAT-TSBF, Bukavu, Democratic Republic of Congo

Bernard Vanlauwe
International Center for Tropical Agriculture-Africa (CIAT-TSBF), Nairobi, Kenya

Jonas Chianu
African Development Bank

W. Chiuri
International Center for Tropical Agriculture-Africa (CIAT), Box 1269 Kigali, Rwanda

N. C. Ehirim
Department of Agricultural Economics, Federal University of Technology Owerri, Imo State, Nigeria

N. N. O. Oguoma
Department of Agricultural Economics, Federal University of Technology Owerri, Imo State, Nigeria

Rajiv Sharma
CIMMYT-Afghanistan, Kabul, Afghanistan

B. C. Yadav
Indian Agricultural Research Institute Regional Station, Pusa, Samastipur (Bihar)-848 125, India

Ravish Choudhary
Indian Agricultural Research Institute Regional Station, Pusa, Samastipur (Bihar)-848 125, India

P. L. Saran
Indian Agricultural Research Institute Regional Station, Pusa, Samastipur (Bihar)-848 125, India

Bukhari Sillah
Department of Economics, College of Business Administration, King Saud University, Saudi Arabia

Aysha Sameen
National Institute of Food Science and Technology, University Of Agriculture Faisalabad, Pakistan

Muhammad Rizwan Tariq
National Institute of Food Science and Technology, University Of Agriculture Faisalabad, Pakistan

Nuzhat Huma
National Institute of Food Science and Technology, University Of Agriculture Faisalabad, Pakistan

Muhammad Issa Khan
National Institute of Food Science and Technology, University Of Agriculture Faisalabad, Pakistan

Ch. Someswararao
Agricultural and Food Engineering Department, Indian Institute of Technology, West Bengal 721302, India

P. P. Srivastav
Agricultural and Food Engineering Department, Indian Institute of Technology, West Bengal 721302, India

H. Das
Agricultural and Food Engineering Department, Indian Institute of Technology, West Bengal 721302, India

Phineas Khazamula Chauke
The School of Agriculture, University of Venda, Thohoyandou, 0950, South Africa

Thizwilondi Dennis Nekhavhambe
The School of Agriculture, University of Venda, Thohoyandou, 0950, South Africa

Tichaona Kudakwashe Pfumayaramba
The School of Agriculture, University of Venda, Thohoyandou, 0950, South Africa

Syamak Mashayekhi
Agricultural Economics, Natural Resources and Agricultural Research Center, P. O. Box: 15715 - 151, Taleghani Ave., Tehran, Iran

Hamed Ghaderzadeh
Department of Agricultural Economics, College of Agriculture, University of Kurdistan, Kurdistan Province, Iran

J. A. Adebija
National Centre for Agricultural Mechanization, KM 20, Ilorin-Lokoja Highway, PMB 1525, Ilorin, Kwara State, Nigeria

B. A. Jackson
National Centre for Agricultural Mechanization, KM 20, Ilorin-Lokoja Highway, PMB 1525, Ilorin, Kwara State, Nigeria